Developing and Applying Optoelectronics in Machine Vision

Oleg Sergiyenko
Autonomous University of Baja California, Mexico

Julio C. Rodriguez-Quiñonez
Autonomous University of Baja California, Mexico

A volume in the Advances in Computational Intelligence and Robotics (ACIR) Book Series

www.igi-global.com

Published in the United States of America by
IGI Global
Information Science Reference (an imprint of IGI Global)
701 E. Chocolate Avenue
Hershey PA 17033
Tel: 717-533-8845
Fax: 717-533-8661
E-mail: cust@igi-global.com
Web site: http://www.igi-global.com

Library of Congress Cataloging-in-Publication Data

Names: Sergiyenko, Oleg, 1969- editor. | Rodriguez- Quinonez, Julio C., 1985- editor.
Title: Developing and applying optoelectronics in machine vision / Oleg Sergiyenko and Julio C. Rodriguez- Quinonez, editors.
Description: Hershey, PA : Information Science Reference, [2017] | Includes bibliographical references and index.
Identifiers: LCCN 2016017801| ISBN 9781522506324 (hardcover) | ISBN 9781522506331 (ebook)
Subjects: LCSH: Computer vision--Equipment and supplies. | Optical pattern recognition--Equipment and supplies. | Optoelectronic devices. | Image converters.
Classification: LCC TA1634 .D483 2017 | DDC 621.39/93--dc23
LC record available at https://lccn.loc.gov/2016017801

This book is published in the IGI Global book series Advances in Computational Intelligence and Robotics (ACIR) (ISSN: 2327-0411; eISSN: 2327-042X)

British Cataloguing in Publication Data
A Cataloguing in Publication record for this book is available from the British Library.

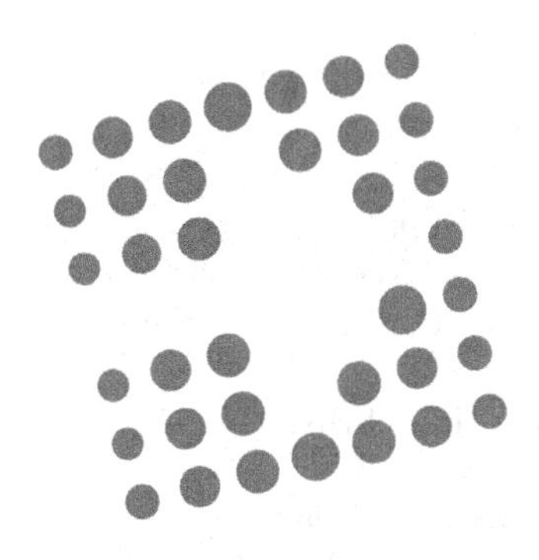

Advances in Computational Intelligence and Robotics (ACIR) Book Series

ISSN: 2327-0411
EISSN: 2327-042X

Coverage

While intelligence is traditionally a term applied to humans and human cognition, technology has progressed in such a way to allow for the development of intelligent systems able to simulate many human traits. With this new era of simulated and artificial intelligence, much research is needed in order to continue to advance the field and also to evaluate the ethical and societal concerns of the existence of artificial life and machine learning.

The **Advances in Computational Intelligence and Robotics (ACIR) Book Series** encourages scholarly discourse on all topics pertaining to evolutionary computing, artificial life, computational intelligence, machine learning, and robotics. ACIR presents the latest research being conducted on diverse topics in intelligence technologies with the goal of advancing knowledge and applications in this rapidly evolving field.

Mission

- Computational Intelligence
- Evolutionary Computing
- Algorithmic Learning
- Robotics
- Artificial Intelligence
- Artificial Life
- Synthetic Emotions
- Computational Logic
- Natural Language Processing
- Adaptive and Complex Systems

IGI Global is currently accepting manuscripts for publication within this series. To submit a proposal for a volume in this series, please contact our Acquisition Editors at Acquisitions@igi-global.com or visit: http://www.igi-global.com/publish/.

The Advances in Computational Intelligence and Robotics (ACIR) Book Series (ISSN 2327-0411) is published by IGI Global, 701 E. Chocolate Avenue, Hershey, PA 17033-1240, USA, www.igi-global.com. This series is composed of titles available for purchase individually; each title is edited to be contextually exclusive from any other title within the series. For pricing and ordering information please visit http://www.igi-global.com/book-series/advances-computational-intelligence-robotics/73674. Postmaster: Send all address changes to above address.

Titles in this Series

For a list of additional titles in this series, please visit: www.igi-global.com

Integrating Cognitive Architectures into Virtual Character Design
Jeremy Owen Turner (Simon Fraser University, Canada) Michael Nixon (Simon Fraser University, Canada) Ulysses Bernardet (Simon Fraser University, Canada) and Steve DiPaola (Simon Fraser University, Canada)
Information Science Reference • copyright 2016 • 346pp • H/C (ISBN: 9781522504542) • US $185.00 (our price)

Handbook of Research on Natural Computing for Optimization Problems
Jyotsna Kumar Mandal (University of Kalyani, India) Somnath Mukhopadhyay (Calcutta Business School, India) and Tandra Pal (National Institute of Technology Durgapur, India)
Information Science Reference • copyright 2016 • 1015pp • H/C (ISBN: 9781522500582) • US $465.00 (our price)

Applied Artificial Higher Order Neural Networks for Control and Recognition
Ming Zhang (Christopher Newport University, USA)
Information Science Reference • copyright 2016 • 511pp • H/C (ISBN: 9781522500636) • US $215.00 (our price)

Handbook of Research on Generalized and Hybrid Set Structures and Applications for Soft Computing
Sunil Jacob John (National Institute of Technology Calicut, India)
Information Science Reference • copyright 2016 • 607pp • H/C (ISBN: 9781466697980) • US $375.00 (our price)

Handbook of Research on Modern Optimization Algorithms and Applications in Engineering and Economics
Pandian Vasant (Universiti Teknologi Petronas, Malaysia) Gerhard-Wilhelm Weber (Middle East Technical University, Turkey) and Vo Ngoc Dieu (Ho Chi Minh City University of Technology, Vietnam)
Engineering Science Reference • copyright 2016 • 960pp • H/C (ISBN: 9781466696440) • US $325.00 (our price)

Problem Solving and Uncertainty Modeling through Optimization and Soft Computing Applications
Pratiksha Saxena (Gautam Buddha University, India) Dipti Singh (Gautam Buddha University, India) and Millie Pant (Indian Institute of Technology - Roorkee, India)
Information Science Reference • copyright 2016 • 403pp • H/C (ISBN: 9781466698857) • US $225.00 (our price)

Editorial Advisory Board

List of Reviewers

Table of Contents

Detailed Table of Contents

Machine vision is supported and enhanced by optoelectronic devices, the output from a machine vision system is information about the content of the optoelectronic signal, it is the process whereby a machine, usually a digital computer and/or electronic hardware automatically processes an optoelectronic signal and reports what it means. Machine vision methods to provide spatial coordinates measurement has developed in a wide range of technologies for multiples fields of applications such as robot navigation, medical scanning, and structural monitoring. Each technology with specified properties that could be categorized as advantage and disadvantage according its utility to the application purpose. This chapter presents the application of optoelectronic devices fusion as the base for those systems with non-lineal behavior supported by artificial intelligence techniques, which require the use of information from various sensors for pattern recognition to produce an enhanced output.

In the image sensing optical receiver, such as CMOS Image Sensors (CIS), in general, comprise of photodiode and analog-mixed -signal circuits to amplify small photocurrent into digital signals. The CMOS image sensors are now the technology of choice for most imaging applications, such as digital video cameras, scanner and numerous other. Even though their sensitivity does not reach the one of the best actual CCD's (whose fill factor is about 100%), they are now commonly used because of their multiple functionalities and their easy fabrication. CMOS analog-mixed -signal circuits play a very important role for CMOS image sensors. This chapter will present the recent works on the CMOS analog-mixed -signal circuit for image sensing application, particularly for device related to machine vision application. A review of designing aspect of CIS are done. A proposed light integrating CIS was developed and measured.

There is a crescent need to produce Printed Circuit Boards (PCB) in a customized and efficient way, therefore, there is an effort from the scientific and industrial community to improve image processing techniques for PCB inspection. The methods proposed at this chapter aim the formation of a system to inspect SMD (Surface Mounted Devices) components in a SSP (Small Series Production), ensuring a satisfactory production quality. This way, a 3-step inspection system is proposed, formed by image preprocessing, feature extraction and evaluation components, based on characteristics related to shape, positioning and histogram of the component. The inspection machine used in this project is inserted in a cooperation among machines context, in order to provide a fully autonomous factory, coordinated by a multi-agent system. Experimental obtained results show that the proposed inspection system is suitable for the case, reaching a success rate above 89% when using actual components.

The presented book chapter provides an overview and detailed description about actual used laser scanner systems. It explains and compares the mainly used coordinate measurement methods, like Time-of-flight, Phasing and Triangulation and Imaging. A Technical Vision System, developed by the engineering institute at the Autonomous University Baja California (UABC) is presented. The mostly used mechanical principles to position a laser beam in a field of view are described and which mechanical actuators are applied. The reflected laser beam gets measured by light sensors or image sensors, which are explained and some principle measuring circuits are provided. The received measuring data gets post-processed by different algorithm or principles, which close the chapter.

Face detection, tracking and recognition is still actual field of human centered technologies used for developing more natural communication between computing artefacts and users. Analyzing modern trends and advances in this field, two approaches for face sensing and recognition have been proposed. The first color/ shape-based approach uses sets of fuzzy saturated color regions providing face detection by Fourier descriptors and recognition by SVM. The second approach provides fast face detection by adaptive boosting algorithm, and recognition based on SIFT key point extraction into eye-nose-mouth regions has been improved using Bayesian approach. Designed systems have been tested in order to evaluate capability of the proposed approaches to detect, trace and interpret faces of known individuals registered into facial standard databases providing correct recognition rate in range of 94.5-99.0% with recall up to 46%. The conducted tests ensure that both approaches have satisfactory performance achieving less than 3 seconds for human face detection and recognition in live video streams.

Preface

Optoelectronics is the study of technologies and physical phenomena's, which incur from the combination of optics and semiconductor electronics. Optoelectronics combines the benefits of electronic data processing with the advantages of fast, electromagnetic and electrostatic noiseless transmission of light. Thereby electrical signals are converted and processed into optical signals and vice versa, wherein the produced light can either propagate in free space or in solid light-transmissive mediums.

HISTORY

The history of optoelectronics is closely connected to the historical development of light sensors and optical communication. Before the electrical measurement technology was used for light measurement, light was quantified by comparing visually the intensity of two light sources. Thereby the contrast sensitivity of the human eye was taken in advantage. This method had the great disadvantage that the results depended on the subjective impression of the measuring person. Especially when measuring light with different colors and at low intensities, the comparisons were very critical. In the 19th century the Ohm's Law and the temperature dependency of the electrical resistance of metals was discovered. Temperatures were measured using these resistor-type thermometers and by development of more sensitive measurement circuits, even smaller temperature differences could be determined. The first *Bolometer* was developed around 1880 by Samuel Pierpont Langley (1834-1906) and possessed thin blackened strip of platinum, through which small temperature differences could be measured, caused by optical radiation incidence. The bolometer was further developed in Germany at the "Physikalisch-Technische Reichsanstalt" among others by the physicist Otto Lummer (1860-1925) and used to examine the radiation of a black body for the first time. Even today, bolometers are used in particular for measuring infrared radiation. However, the resistance temperature dependence is not the only effect that can be used to measure light. In

1821 Thomas Johann Seebeck discovered (1770-1831) the so-called thermoelectricity, which explains the appearance of an electrical voltage, when two metals are brought together in two points using different temperatures. This effect can also be used for radiation measurements and based on *Thermocouples* the *Thermopile* was developed by Leopoldo Nobili (1784-1835) in 1830. Thermopiles are used today for heat flow or broadband radiation measurement. Another effect for measurement of light is given by the pyroelectric effect, which was 1824 defined by the Scot David Brewster (1781-1868) and named pyroelectricity. This effect describes the property of certain piezoelectric crystals to respond to a temperature change in time with a charge separation, which produces a measurable electrical voltage. Due to lack of technologies, that pyroelectric effect could take place for more intensive use by means of infrared sensors until the middle of the 20th century. There are other light sensors which base on the thermal effect of radiation. Marcel J. E. Golay (1902-1989) in 1947 developed the *Golay Cell*, in which a gas absorbs radiation, thus warms and expands itself. The expansion of gas is measured using optoelectronics and depends on the incident light intensity. Current detectors are mostly based on the *Photoelectric Effect*, which describes the interaction of photons with material. Thereby an electron is released from a bond (atomic bond, valence band or conduction bond of a solid) absorbing a photon, wherein the energy of the photon must be at least as large as the electron binding energy. It is usually distinguished between the inner and the outer photoelectric effect. The *InnerPhotoelectric Effect* describes the conductance increase in semiconductors by formation of free charge carrier pairs in form of electrons and holes using the energy of an absorbed photon. The electron is thereby increased from the valence band into the conduction band, which represents a higher energy level. Therefore, the energy of the single photon must be at least equivalent to the band gap of the irradiated semiconductor. Since the size of the bandgap depends on the material and the energy of a photon by the Planck constant is directly dependent on its frequency, there is a maximum light wavelength up to which the inner photoelectric effect occurs. The inner photoelectric effect is the physical principle of many modern light sensors, like photoresistors, photodiodes, phototransistors and CCD sensors. Based on this effect, Werner Siemens (1816-1892) built in 1875 the first photoresistor; with he could determine light magnitudes quantitative using a photometer. One year later in 1876 the English William Grylls Adams (1819-1892) found a related effect, which is the base of all modern solar cells. In 1930 Walter Schottky (1886-1976) was able to demonstrate at the German physicist conference the first selenium solar cell that could provide enough electrical energy to power a small electric motor. Beside selenium, the photovoltaic effect was also researched using germanium and silicon. The American engineer Russell Ohl (1898-1987) studied 1940 pure silicon and observed the photovoltaic effect, which was much stronger than in selenium. He discovered the p-n junction of semiconductors during resistance measurements of silicon wafers, whose resistance drastically altered when exposed to light. Thus the technique of the photodiode was born and represent since the state of the art. The *OuterPhotoelectricEffect* describes the process of an electron release from a semiconductor or metal surface using light irradiation. This effect was first discovered in 1839 by the Frenchman Edmond Becquerel (1820-1891) when working with galvanic cells, which provided an electrical voltage when irradiated with light. Heinrich Hertz (1857-1894) inspected in 1886 the stimulating effect of ultraviolet light on the formation of electrical sparks. The research of Hertz was continued by his student Wilhelm Hallwachs (1858-1922) in 1887. The Russian Alexander Grigorjewitsch Stoletow (1839-1896) explained the generation of an electric current under irradiation of UV light and use of the outer photoelectric effect. This electric current represents a measure of the incident UV light intensity. The first *Photocell* was developed in 1893 by Julius Elster (1854-1920) and Hans Friedrich Geitel (1855-1923). It consists of an evacuated glass envelope containing a photocathode and an anode. By incident light, electrons are released from the photocathode and discharged from the opposite anode, resulting in a measurable photocurrent. The photocell has been improved since then, at which the released electrons are accelerated using electric and magnetic fields and by the recovered energy so-called secondary electrons generated when hitting more electrons. Due to this amplifier effect the photomultiplier was developed in the 1930 years, with also single photons could be detected. Long time these vacuum tubes have been the most reliable light sensors, but were later replaced by light sensors based on semiconductor technology, using the inner photoelectric effect. In the 1950 years the first machine vision applications concepts were developed. James J. Gibson (1904-1979) introduced the concept of optical flow and based on his research the mathematical models for optical model were developed. In 1960 Larry Roberts wrote his PhD thesis, where he discussed about the possibility of extract 3D geometrical information from 2D perspective and which actually represents the base of stereo vision. In 1970's MIT opened a Machine vision course and researchers starts to work on "low-level" tasks as edge detection and segmentation. In 1980's the development of smart cameras or intelligent cameras generated an improvement in machine vision. OCR (optical character recognition) and face recognition popularized the practical applications of machine vision systems. In 1990's the applications of machine vision systems became common and a massive growth of machine vision industries leads to new innovations. In this decade neural networks started to work in conjunction with machine vision systems. In 1995 the first machine vision interface was standardized, the IIDC 1.04 (Fire Wire or IEEE1394) was created. IIDC was not created in the machine vision industry but it was adopted by the industry. In 2000's several machine vision interfaces were born, Camera link 1.0 in 2000, GigE Vision 1.0 in 2006, Camera Link 1.2 in 2007, IIDC 1.32 in 2008 and GigE Vision 1.2 in 2009. Finally in 2011 GigE Vision 2.0 and USB3 Vision 1.0 in 2012.

IMPORTANCE OF OPTOELECTRONIC DEVICES AND MACHINE VISION

The importance of optoelectronic devices cannot be neglected in today's life and business and the numerous applications of optoelectronic components can no longer be covered by a course, which is why these applications must be broken down into different areas to study. From the resulting different fields of applications for optoelectronic devices, the present book will inspect two special areas: the development of optoelectronic devices and the application of these optoelectronic devices in machine vision. The development of optoelectronics can be carried out at different levels, where the device level and the system level describe two most important one. The device level is represented by optoelectronic components, which can be classified by *transmitters* and *receptors*. Transmitters are optoelectronic actuators, which produce light from electricity that is laser and light - emitting diodes. Thereby the emission spectrum of the transmitters can be located in both visible and invisible (UV or IR) spectral range. Receptors are optoelectronic detectors, which convert light into electrical signals, like photoresistors, photodiodes, phototransistors, photomultipliers, CCD sensors, CMOS sensors and more. These receptors can be further divided into light sensors and image sensors. An optoelectronic system, containing a transmitter and a receptor is called a photoelectric sensor, among which belong the optocouplers for example. Besides transmitters and receptors exist other components, which are needed for light-transmission, light-amplification or light–modulation. The transmission of optical signals can be carried out through free space or in using waveguides and optical circuits. Optical modulators are devices, which modules light with a defined characteristic. This may be for example a time or spatial variation in amplitude or phase. Light receptors and optical modulators get described in detail in book chapter 4. On the device level, optoelectronics have made a major advance in the last two decades, especially in measurement precision and measurement time by use of new theoretical concepts. These advances allow a higher resolution of measured objects with shorter measurement time, which in turn facilitate the application of real-time methods in image processing. A mobile robot, for example, can thus detect its surroundings more accurate and correct the trajectory of his current path in real time.

Besides pure data transmission, optoelectronics is also used by machines, to receive and process visual information's. Machine vision is based on the abilities of the human visual system though, which allows machines to identify and classify clearly patterns or objects from their near surroundings. Especially machine vision systems are currently used in industrial processes, in automation technology and quality management. Further fields of applications can be found for example in traffic technology and in security technology. Few current research projects are dealing with the understanding of the real meaning and content of an image. Instead are developed algorithms for object recognition and description. Features of these objects are then measured, objects are classified and based on these results, decisions are taken or processes are controlled. Since image understanding is very much related with design and application of computational methods, it can be classified as a branch of computer science, which has strong relationships with photogrammetry, signal processing and artificial intelligence. The used methods of machine vision originate from mathematics, in particular geometry, linear algebra, statistics and functional analysis. Typical tasks of machine vision can be defined by object recognition and measurement of object geometric structures and object movements. These tasks are based on algorithms of image processing, for example segmentation and on algorithms of pattern recognition, for example for classifying objects. Machine vision is used mainly in industrial and natural environments. Industrial environments usually possess standardized and consistent conditions, such as camera position, lighting, speed of a conveyer, background color, or the location of the measured object. Thereby these techniques of machine vision nowadays can be successfully used in such environments. The situation is completely different in natural environments, where the conditions change continuously and may not even be predictable. Such environments provide far more difficult requirements on used machine vision algorithms and the conditions often cannot be manipulated. Therefore the development of robust and error-free running programs in natural environments is far more difficult, than in industrial environments.

IMPACT OF THE BOOK

The topic of this book *OptoelectronicsandMachine Vision* takes place in current research and application of devices that detect and measure their environment by optoelectronic sensors. Optoelectronic sensors can be found in a wide range of modern life, be it motion sensors, solar cells, facial recognition security systems, light sensors in digital cameras and in latest smartphones, just to name a few. Also, there is a wide range of applications for light transmitters, such as laser light for measurement of objects, optical fiber for transmission of light signals and in general lighting lamps with a special frequency band for excitation of chemical processes. The impact of the book can be defined by endorsement in studies and development of complex machine vision systems using optoelectronic devices. Through methodical presentation and description of individual machine vision systems, the engineer, scientist or practitioner shall be supported with fundamental knowledge and applications examples, which differ in complexity.

Nowadays machine vision can find in most of manufactures industries. Machine vision is used by electronic industry to perform inspections of integrated circuits, inspections that can't be performed by the human eye without the aid of microscopy technology. Throughout the manufacturing industry, machine vision is used to provide information about automatic machine movements, bonding and wire bonding and visual inspections, only to mention a few applications. In microelectronics industry, machine vision is used to provide position feedback of components and loose wires. The food industry uses machine vision systems for sorting, measure size, and analyze the content of packages. In television industry, machine vision is used in white balance applications, PCB tracking, etc. In automobile industry machine vision systems are used to look the flushness of sheet metals, control the paint quality, visual feedback in arm robot tracking. Practically every nowadays industry has been adopted machine vision system in some way or another.

The book provides a basic introduction to physical principles of optoelectronic sensors and their electrical wiring, as well as examples of computer algorithms for object recognition and analysis in machine vision systems.

Thus exists a necessary connection between optoelectronics and machine vision, which represents the main constituent of present book. The reader thereby is introduced to the development and implementation of different optoelectronic systems in various machine vision applications. The book shall serve as a reference work for students and researchers of applied physics and optics, as well as a reference book for engineers and technicians in today's industry of machine vision.

ORGANIZATION OF THE BOOK

The book contains nine chapters, which are briefly described in the following:

Chapter 1 gives a wide introduction and presentation of optoelectronic sensor fusion in machine vision. It describes the theoretical approach to fuse the measurement data of photodiodes and CCD devices, to obtain a lower detection error in angle measurement. The chapter focuses on sensor fusion techniques, provides a wide introduction to optoelectronic devices and deals with theoretically background of sensor and data fusion.

Chapter 2 reviews mostly used CMOS Image sensors in optoelectronic systems. It also explains some of the most important parameters such as noise, dynamic range, sensitivity and resolution in relation to the analog-mixed- signal circuit design. The chapter gives an overview about technology options and types of CMOS processes actually used in market and update some of well-known circuit techniques and architectures.

Chapter 3 describes in a general and complete matter the detection of PCB errors. It analyses the necessary steps of image acquisition, recognition and processing, to detect missing, rotated, shifted or wrong electrical parts on PCB boards. The chapter focuses on the aspect of image processing and analyzes different methods to detect failure in component placement during PCB production.

Chapter 4 gives an introduction and summary of actually laser scanners used as optical devices. A definition of laser scanner is given, some applications areas are described and laser scanners categorized by design principles, measurement values and measurement methods. The chapter describes in detail four commonly used methods for spatial coordinate's measurement, also provides a special example of a Technical Vision System prototype and specifies the mainly used mechanical principles for laser beam positioning, with the mechanical actuators for moving scanning mirrors. Furthermore the chapter describes the class of light and image sensors used for laser beam measuring and it gives some technical implementations about how to use light sensors. The post-processing of the measured physical signals gets described in the last section of this chapter.

Chapter 5 presents recent developments in the technical field of face sensing and particular pays attention to analysis of emergent pattern recognition approaches. The introduced novel approach of face recognition is able to learn and recognize new faces during less than three seconds, which is confirmed by several experiments using more than hundreds of images within indoor environments as well as images of 28 different individuals from Yale B and CAS-PEAL standard face databases.

Chapter 6 introduces a wide theoretically frame of statistical signal properties in optoelectronic systems and discusses a derivation of the signaltonoise relation when measuring optical radiation. The chapter develops a mathematical model of signals in output plane of optoelectronic systems and shows the possibility of signal detection characteristics improvement.

Chapter 7 summarizes systems and techniques of machine vision optical scanners for landslide monitoring. Thereby different systems are reviewed, which use fixed triangulation sensors with CCD or PSD sensors, Polygonal Scanners, Pyramidal and Prismatic Facets, Holographic scanners, Galvanometer and Resonant scanners or scanners with off-axis parabolic mirrors. This chapter is focused on the use of active laser scanners systems for landslide monitoring and compares the use of incoherent and coherent light in scanners using triangulation.

Chapter 8 provides a general introduction to 3D imaging systems for agricultural applications using Structure from Motion (SMF) and Shape from Focus (SFF) measuring methods. The chapter presents innovations on 3D tools and methods used in agriculture for different applications such as 3D characterization of crop and 3D use for root phenotyping using rhizotron systems.

Chapter 9 offers an introduction to microorganism detection using optoelectronic devices. The microorganism detection has the aim to elaborate tests in soils and determine the adequate microorganisms that can break down heavy metals.

Introduction

Optoelectronics is the combinational study of the physics of light and electronic devices that utilize light (Cardinale, 2003). The importance of optoelectronic devices cannot be understated; optoelectronic devices are merged in everyday lives and must be studied on deep. From the lecturer's perspective, the application of optoelectronic devices covers a wide area of opportunities that can be difficult to cover in only one course. As result, the area of optoelectronic devices must be studied by particular fields or applications. Due that particular need, this book is intended to study two particular fields: The development of optoelectronic devices, and the application of these optoelectronic devices in machine vision.

The development of optoelectronics can be understood in several ways, it can be analyzed at device level or system level. At device level exists representative components as photodiodes, phototransistors, optoisolators, photoresistors, LED, CCD, etc… where in last decades the trend is picoseconds devices. Picosecond optoelectronic devices are electronic components that can record in real time the physical and electronic events that take place on picosecond time scales (Lee, 2012). Several picosecond sensors systems have been developed as: picosecond pulse generation, picosecond photoconductors even high power picosecond lasers, and all this advances are taken place in the current machine vision systems. On other hand, at system level, optoelectronic devices have been taken major breakthroughs in the last 20 years with the development of the further technologies and application of theoretical concepts developed since early 70s that can now be applied due the current technology. The next section describes some of the most representative optoelectronic systems applied in machine vision, a review of machine vision systems is provided, trending topics in both areas are discussed and finally the scope of the book is presented.

3D OPTOELECTRONIC SCANNING TECHNOLOGIES

3D Optoelectronic Scanners plays an important role in machine vision applications; they are complete optoelectronic systems that analyze real world objects or scenes to collect digital information about the shape, and possibly appearance as color.

Different technologies can be used to develop 3D Optoelectronic Scanners, each technology with its own advantages and limitations so, it is important to mention the most widely used systems.

Laser Triangulation

Laser triangulation scanner is a kind of active optoelectronic system. The principle of triangulation cares on the projection of a light pattern, i.e., a line or a dot is projected by the laser over an object and captured by the digital camera (or other type of photodetector). The distance from object to system can be calculated by trigonometry, as long as you know a priori distance between the scanning system and the camera/photodetector (Franĉa, et al., 2005), (Básaca, et al., 2010).

The laser triangulation concept can be further categorized into three categories: 1) static triangulations (Petrov, et al., 1998), (Marshall, 2004); 2) dynamic triangulations (Sergiyenko, et al., 2009), (Sergiyenko[a], et al., 2009) (Sergiyenko, 2010), (Sergiyenko, et al., 2011); 3) laser line projection systems with complex sensory part suggesting multiple polyhedrons analysis at the same time in parallel processing (Bin, 2016).

Laser static triangulation principle (Marshall, 2004) in general can be based on two schemes represented in Figure 1 (a and b). The first one uses a fixed angle of emission and variable distance; the second one, on the contrary, fixed triangulation base and variable scanning angle.

Figure 1. Two principles of laser triangulation (a – "with fixed angle of emission"; b - "with fixed triangulation distance", consists of a laser dot or line generator and a 1D or 2D sensor located at a triangulation distance from the light source

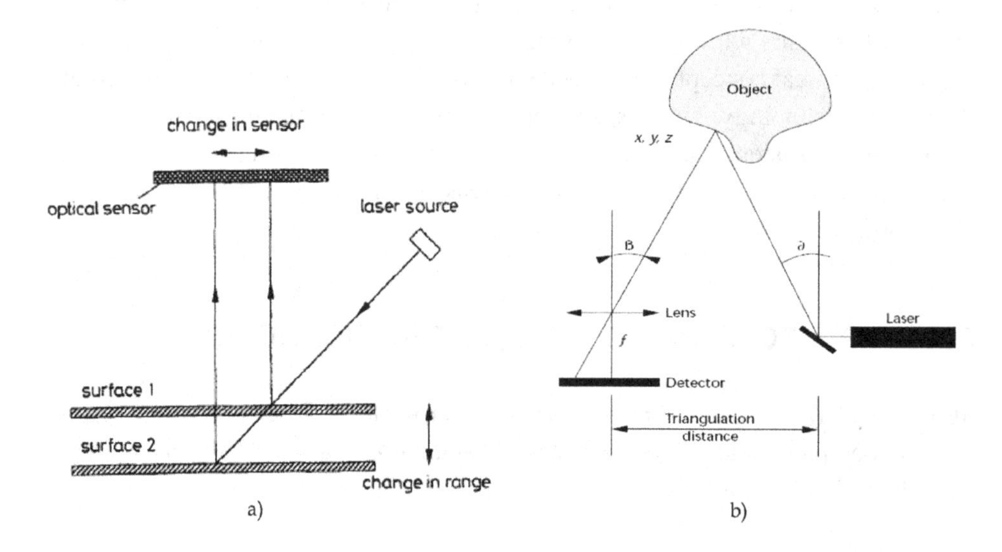

Figure 2. Spatial angle measurement system

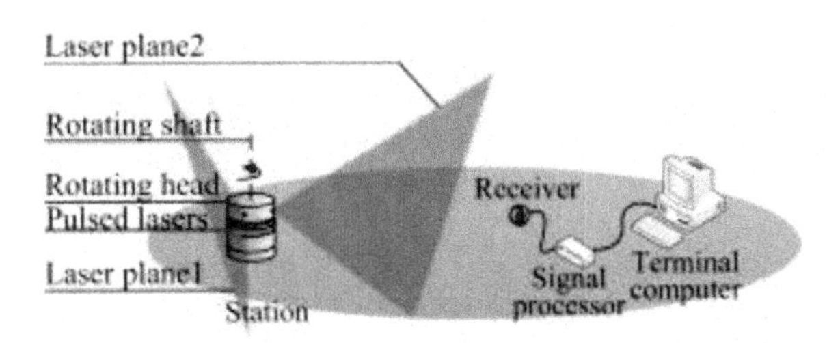

The laser line projection method in fact is based on the same principle, however to save operating time it use the complex laser projecting head with several lasers, as the rule in pulsed mode, making n triangulations in parallel simultaneously. The general principle of such system (Bin, 2016) can be viewed on Figure 2.

Stereo Vision

In Stereo Vision systems two or more cameras simultaneously capture the same scene. Every stereo vision system has their own function methodology, but in principle the stereo approach uses the following sequence of processing steps: Image acquisition, Camera Modeling, Feature Extractions, Correspondence Analysis and Triangulation. No further equipment and no special projections are required (Sansoni, Trebeschi, & Docchio, 2009). The advantages of the stereo approach are the simplicity and low cost; the major problem of stereo vision system is the correspondence problem. A variant of stereo vision system implies the use of multiple cameras or a single moving camera to obtain multiple images, in this case a constraint is that the scene doesn't contain moving parts (Sansoni, Trebeschi, & Docchio, 2009).

The general principle of such system (Chaumette, 2006) can be viewed on Figure 3.Any stereovision technical system is approach to a multicamera system. In the elementary case under consideration it is two cameras system. If a stereovision system is used, and a 3-D point is visible in both left and right images (see Figure 3), it is possible to use as visual features s (vector ***s*** contains the desired values of the scene/features):

$$\mathbf{s} = \mathbf{x}s = (\mathbf{x}l, \mathbf{x}r) = (xl,\ y\ l,\ xr,\ yr),$$

i.e., to represent the point by just stacking in **s** the x and y coordinates of the observed point in the left and right images.

For a 3-D point with coordinates $\mathbf{X} = (X, Y, Z)$ in the camera frame, which projects in two images as a 2-D point with coordinates $\mathbf{x} = (x, y)$, we have (Chaumette, 2006):

Figure 3. A stereovision system

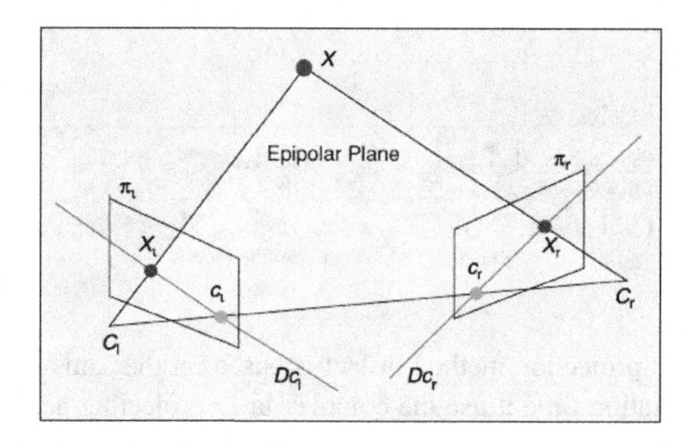

$$\begin{cases} x = X/Z = (u - c_u)/f\alpha \\ y = Y/Z = (v - c_v)/f, \end{cases} \tag{1}$$

where image measurements matrix **m** = (u, v) gives the coordinates of the image point expressed in pixel units, and **a** = $(c\ u, c\ v, f, \alpha)$ is the set of camera intrinsic parameters: $c\ u$ and $c\ v$ are the coordinates of the principal point, f is the focal length, and α is the ratio of the pixel dimensions. In this case, we take **s** = **x** = (x, y), the image plane coordinates of the point.

Taking the time derivative of the projection equations (1), we obtain the result which can be written in general form

$$\mathbf{X} = \mathbf{Lx}\ \mathbf{V}c, \tag{2}$$

where **V***c* is a spatial velocity of the camera be denoted by **v***c* = $(vc, \omega c)$, (with vc the instantaneous linear velocity of the origin of the camera frame and ωc the instantaneous angular velocity of the camera frame) and the interaction matrix **Lx** (we consider here the case of controlling the motion of a camera with six degrees of freedom) related to **x** is

$$L_x = \begin{bmatrix} \frac{-1}{Z} & 0 & \frac{x}{Z} & xy & -\left(1+x^2\right) & y \\ 0 & \frac{-1}{Z} & \frac{y}{Z} & 1+y^2 & -xy & -x \end{bmatrix} \tag{3}$$

Lx is an interaction matrix related to s, or feature Jacobian. In the matrix **Lx**, the value *Z* is the depth of the point relative to the camera frame. Therefore, any control scheme that uses this form of the interaction matrix must estimate or approximate the value of Z.

Structured Light

The structured light scanners project coded light patterns on the scene, such patterns can be: horizontal and vertical grids. However, more complex patterns as sine waves have been proposed (Scharstein & Szeliski, 2003). When locating such coded points in the image grabbed by a camera, the correspondence problem is solved with no requirement for geometrical constraint and the 3D structure of the scene is recovered by triangulation (Salvi, Pages, & Batlle 2004), (Gu, et al., 2013). Typical structured light 3D measurement system is made up of CCD cameras used for image capturing and a DLP projector used for digital fringe pattern image projection (Liu, et al., 2015).

Time of Flight

It is an active sensor that measures the travel time of infrared light, and consequently it does not interfere with the scene in the visual spectrum; its core components are a CMOS chip and an infrared light source (Cui, 2010). Optical range cameras use an emitted near-infrared light who is reflected back from the scene and capture by the camera. Depending on the camera model, the illumination can be modulated by sinusoidal or pseudonoise signal. The received signal is demodulated an the phase shift provides the time between emission and detection, indicating how far the light has traveled and thus allowing to indirectly measure the depth of the scene (Reynolds, et al., 2011).

CT Scan

Computed tomography uses a computer to combine many X-ray images taken from different angles, to produce a cross sectional tomographic. The CT scan refers to computer tomography and the most common used technology is X-ray, although, other types exists as positron emission tomography (PET) and single-photon emission computed tomography (SPECT). The uses of the mentioned technologies allow us to see inside of some objects without performing invasive procedures. The most common area of application is in medicine, in the detection of tumors, infarctions, hemorrhage, calcifications, bone trauma, etc..., (Hsieh, 2009).

MACHINE VISION

Machine vision is the technology and methods used to provide imaging-based automatic inspection and analysis for such applications as process control, and robot guidance in industry (Stanney & Hale, 2014), (Zou & Zhao, 2015). Machine vision is not a new topic, however, in the last 25 years the topic has moved on at accelerating rate, so, some topics that were considered tangential to machine vision fifteen or ten years ago, now need to be included on the literature. An example of this topic is the role of artificial neural networks in machine vision. Artificial Neural Networks can be useful in circumstances where conventional methods do not provide desired outputs and the problem is strongly correlated to conditional probabilities; an example is the corner, edges/boundaries detection. To understand a complete machine vision system is important to know its main elements at hardware level (Image acquisition), and software level (Image processing) and the link between them.

Image Acquisition

The image acquisition process involves three steps; (1) energy reflected from the object of interest, (2) optical system which focuses the energy and (3) a sensor which measures the quantity of energy (Moeslund, 2012).

Energy

The source of the energy can be natural (as the sun light) or artificial. The purpose of lighting system is to provide some source of measurable energy. It is needed to consider a reliable source of light, as with the human eye, the vision system is affected by the quality and intensity of light (Zou & Zhao, 2015). Light sources may include incandescent, lasers, fluorescent, X-Ray, infrared and ultraviolet sources. Is important to bear in mind that the human eye is sensitive to the visible spectrum of light 400-700nm (nn = nanometer) and the visible electromagnetic spectrum of most of the cameras is different. The Lighting or Illumination is one of the most important parts of machine vision systems, as rule-of-thumb you can consider that illumination is 66% of the entire system design.

Optical System and Image Sensor

The light reflected by the object of interest must be captured by a photodetector component; usually this component is a CCD or a camera; however in other cases the component can be a simple optical sensor. An optical system includes a diverse variety of components, a typical optical system in machine vision applications is composed of the sensor, lenses and filters; however, these components can vary according to the requirements. A first main component on an optical system is the Lens. A lens is an element which focuses the incoming light onto the sensor, a simple lens consists in a simple piece of material (typicaly glass) and compound lenses are made up of several simple lenses (Macnaughton, 2005), (Zou & Zhao, 2015). A second component to consider in the optical system is the Filter. Optical filters are devices most often used to selectively transmit certain wavelengths, (Hecht, 2015). There are different types of filters as Monochromatic filters who only allow a thin range of wavelengths, Dichroic filters allow to pass a selected wavelength and reflects the others, An absorptive filter absorbs some wavelengths and allow to pass others. Filters and Lenses work in conjunction to conditioning the image received by the photodetector.

The typical image sensor used in machine vision is the charge-coupled device CCD, however Active Pixel Sensors as complementary metal-oxide-semiconductor (CMOS) have acquired great popularity in the field of image sensors. CCD has higher light sensitivity, translates into better images in low-light conditions (Nilsson, 2008). CCD and CMOS sensors are often seen as competitors, however, each sensor has its own strengths and weaknesses. CCD sensors generally offer better light sensitivity than CMOS sensors; however CMOS sensors are less expensive than CCD. This can change in the following years since quality of modern CMOS sensors is rapidly increasing (Nilsson, 2008).

Image Processing

The image processing is the computer manipulation of image by mathematical operations or algorithms to enhance its quality or extract particular features. Typically image processing is implemented at software but with the development of current technology more algorithms of image processing are implemented at hardware. Image processing can be generally divided into three levels of processing: low level, intermediate level, and high-level. Low level processing performs image acquisition by transferring the electronic signal of the photodetector into a numeric form. Low level also realizes the pre-processing task, pre-processing of the image includes correction of distortion, noise removal, grey level correction and correction of blurring. Intermediate level processing performs image segmentation operations as: thresholding, edge-based segmentation and region-based segmentation. The intermediate level also includes the image representation, and description. High level processing performs recognition and interpretation, this can be done by classifiers or with multilayer neural networks (Brosnan, & Sun, 2004), (Zou & Zhao, 2015). The image processing application includes Classification, Feature extraction, Pattern recognition, Projection and Multi-Scale signal analysis.

TRENDS OF OPTOELECTRONIC SYSTEMS AND MACHINE VISION

The trends in optoelectronics and machine vision can be divided into two areas, commercially and industrially. In both cases, as the computing capabilities increase, the new systems allow to perform more complex operations.

In the commercial field the use of smartphone allows the average person to acquire a whole machine vision system. The current advances in visual odometry in conjunction with the inertial measurement unit (IMU) of a smartphones enable to create novel systems. Both visual odometry and the inertial navigation system (obtained from the IMU) can determine the movements and positions of a robot or a pedestrian according to the world coordinate system (Tomažič & Škrjanc, 2015). With camera-enabled mobile phones, the industry of application developers can create a new range of imaging applications, applications as automatic face detection for cataloging photos, user recognition for social media, room measurement for remodeling, camera based gesture control and augmented reality navigation are only a few trending applications for smartphones that support its technology in machine vision (Durini, 2014). It is noteworthy that vision systems that are based on statistical tools for classification as neural networks are already using these technologies in their algorithms.

Otherwise, in the industrial area, machine vision systems have acquired great acceptability in inspection and test process. The software trend in this area is to incorporate learning based methods as neural networks, Neuro-fuzzy systems and support vectors machine to machine vision systems to increase their classification capabilities (e. g. the classification of conforming products, and automatic optical inspection) (Bosque & Echanobe, 2014), (Shanmugamani, Sadique, & Ramamoorthy, 2015), (Zhang, et al., 2014). In the research area, the scientists are working with Biologically-Inspired new visual classifications (Tang & Qiao, 2014), (Kerr, et al., 2015), and new approaches as dynamic triangulation and fusion of different scanning technologies have been implemented (Rodríguez, 2011), (Rodríguez-Quiñonez, 2014), (Budzan & Kasprzyk, 2016).

SCOPE OF THE BOOK

The book is aimed to acquaint the reader with technical issues and commonly used in the area of machine vision; with chapters focused on visual inspection and recognition as: “Automated Visual Inspection System for Printed Circuit Boards for Small Series Production: A Multiagent Approach Context” in this chapter the reader will be introduced to the automatic optical inspection (AOI) approach, analyzing different methods to detect failure in component placement during a PCB production. The scope of this chapter is not only to provide a visual analysis, but also include artificial intelligence systems such as genetic-programming or classifiers in the analysis. Another technical chapter is: "Machine Vision application on science and industry: Real-Time Face Sensing and Recognition in Machine Vision: Trends and New Advances". This chapter includes recent developments in the area of face sensing and recognition and particular attention is given to analysis of emergent pattern recognition approaches. Experimental results with more than hundreds images within indoor environments as well as images of 28 different individuals from Yale B and CAS-PEAL standard face databases have been tested confirming that the proposed approaches are able to learn new faces and recognize them during less than three seconds.

This book has wide covered areas of machine vision, considering not only passive systems but also active systems as Laser scanners, providing a review of different scanning technologies and realizes a comparison between them. Different coordinate measuring methods as Time-of-flight, Phasing and triangulation are compared; these methods are described in terms of content and mathematically. The chapter "Laser scanners" provides an analysis of dynamic triangulation, and numerate several components of a full developed technical vision system. Additionally, the chapter includes a post-processing of measured values section, where the authors explain different approaches to improve the obtained 3D measurements. The chapter "Machine Vision optical scanners for Landslide monitoring" provides a summary of the systems and techniques for this common application of machine vision. The reviewed systems covers scanners with fixed triangulation sensors using CCD or PSD, Polygonal Scanners, Pyramidal and Prismatic Facets, Holographic scanners, Galvanometer and Resonant scanners, Scanners with off-axis parabolic mirrors. This chapter is focused on the use of active laser scanners systems for Landslide monitoring, and compares the use of incoherent vs coherent light in scanners for scanners that use the triangulation method with landslide monitoring applications.

Also, here are reviewed common CMOS Image sensors used in optoelectronics systems, this book explains some of the most important parameters such as noise, dynamic range, sensitivity and resolution in relation to the analog-mixed- signal circuit design, technology option/type of CMOS process, some of design methodologies used in the designing the CMOS Image sensor and update some of well-known circuit technique/architecture.

Finally, the book pays some attention to very important aspect of machine vision and optical scanning: the methods applicable for rectification of optical signals with the aim to improve the quality of surrounding perception. For example, the chapter "Applying Optoelectronic Devices Fusion in Machine Vision: Spatial Coordinate Measurement" introduce us into the problem of such techniques applications, on the example of optical ray axis search, both hardware and applying mathematical formalism.

This is a complete book about Developing and Applying Optoelectronics in Machine Vision; it provides an accessible, well-organized overview of machine vision applications and properties that emphasizes basic principles. Coverage combines an optional review from key concepts, up to the progress gradually through more advanced topics. This book includes the recent developments on the field, emphasizes fundamental concepts and analytical techniques, rather than a comprehensive coverage of different devices, so readers can apply them to all current, and even future, devices. This book discusses important properties for different types of applications, such as analog or digital links, the theoretical methods of image processing in optoelectronic systems; optoelectronic devices integration for spatial coordinate measurement; machine vision applications on science, industry, agriculture and analysis of optoelectronic devices.

This book generally combines the optical, electrical and software behavior of a whole machine vision system, including interwoven properties as interconnections to external components and timing considerations. It provides the key concepts and analytical techniques that readers can apply to current and future devices. It also emphasizes the importance of time-dependent interactions between electrical and optical signals.

This is an ideal reference for graduate students and researchers in electrical engineering and applied physics departments, as well as practitioners in the machine vision industry.

Julio C. Rodríguez-Quiñonez
Autonomous University of Baja California, Mexico

Oleg Sergiyenko
Autonomous University of Baja California, Mexico

REFERENCES

Básaca, L. C., Rodríguez, J., Sergiyenko, O. Y., Tyrsa, V. V., Hernández, W., Hipólito, J. I. N., & Starostenko, O. (2010, July). 3D laser scanning vision system for autonomous robot navigation. In *Industrial Electronics (ISIE), 2010 IEEE International Symposium on* (pp. 1773-1778). IEEE.

Bin, X., Xiaoxia, Y., & Jigui, Z. (2016). Architectural stability analysis of the rotary-laser scanning technique. *Optics and Lasers in Engineering*, *78*, 26–34. doi:10.1016/j.optlaseng.2015.09.005

Bosque, G., del Campo, I., & Echanobe, J. (2014). Fuzzy systems, neural networks and neuro-fuzzy systems: A vision on their hardware implementation and platforms over two decades. *Engineering Applications of Artificial Intelligence*, *32*, 283–331. doi:10.1016/j.engappai.2014.02.008

Brosnan, T., & Sun, D. W. (2004). Improving quality inspection of food products by computer vision—A review. *Journal of Food Engineering*, *61*(1), 3–16. doi:10.1016/S0260-8774(03)00183-3

Budzan, S., & Kasprzyk, J. (2016). Fusion of 3D laser scanner and depth images for obstacle recognition in mobile applications. *Optics and Lasers in Engineering*, *77*, 230–240. doi:10.1016/j.optlaseng.2015.09.003

Cardinale, G. (2003). *Optoelectronics: Introductory theory and experiments*. Cengage Learning.

Chaumette, F., & Hutchinson, S. (2006). Visual servo control PART I: Basic approaches. *IEEE Robotics & Automation Magazine*, *13*(4), 82–90. doi:10.1109/MRA.2006.250573

Cui, Y., Schuon, S., Chan, D., Thrun, S., & Theobalt, C. (2010, June). 3D shape scanning with a time-of-flight camera. In *Computer Vision and Pattern Recognition (CVPR), 2010 IEEE Conference on* (pp. 1173-1180). IEEE.

Durini, D. (Ed.). (2014). *High performance silicon imaging: Fundamentals and applications of CMOS and CCD sensors*. Elsevier.

Franĉa, J. G. D., Gazziro, M. A., Ide, A. N., & Saito, J. H. (2005, September). A 3D scanning system based on laser triangulation and variable field of view. In *Image Processing, 2005. ICIP 2005. IEEE International Conference on* (Vol. 1, pp. I-425). IEEE.

Gu, J., Nayar, S. K., Grinspun, E., Belhumeur, P. N., & Ramamoorthi, R. (2013). Compressive structured light for recovering inhomogeneous participating media. *Pattern Analysis and Machine Intelligence. IEEE Transactions on*, *35*(3), 1–1.

Hecht, J. (2015). *Understanding fiber optics*. Jeff Hecht.

Hsieh, J. (2009, November). *Computed tomography: Principles, design, artifacts, and recent advances*. Bellingham, WA: SPIE.

Kerr, D., McGinnity, T. M., Coleman, S., & Clogenson, M. (2015). A biologically inspired spiking model of visual processing for image feature detection. *Neurocomputing*, *158*, 268–280. doi:10.1016/j.neucom.2015.01.011

Lee, C. H. (Ed.). (2012). *Picosecond optoelectronic devices*. Elsevier.

Liu, X., Sheng, H., Zhang, Y., & Xiong, Z. (2015, May). A structured light 3D measurement system based on heterogeneous parallel computation model. In *Cluster, Cloud and Grid Computing (CCGrid), 2015 15th IEEE/ACM International Symposium on* (pp. 1027-1036). IEEE.

Macnaughton, J. (2005). *Low vision assessment*. Elsevier Health Sciences.

Moeslund, T. B. (2012). *Introduction to video and image processing: Building real systems and applications*. Springer Science & Business Media. doi:10.1007/978-1-4471-2503-7

Nilsson, F. (2008). *Intelligent network video: Understanding modern video surveillance systems*. CRC Press. doi:10.1201/9781420061574

Petrov, M., Talapov, A., Robertson, T., Lebedev, A., Zhilyaev, A., & Polonskiy, L. (1998). Optical 3D digitizers: Bringing life to the virtual world. *IEEE Computer Graphics and Applications*, *18*(3), 28–37. doi:10.1109/38.674969

Reynolds, M., Doboš, J., Peel, L., Weyrich, T., & Brostow, G. J. (2011, June). Capturing time-of-flight data with confidence. In *Computer Vision and Pattern Recognition (CVPR), 2011 IEEE Conference on* (pp. 945-952). IEEE. doi:10.1109/CVPR.2011.5995550

Rodríguez, J. C., Sergiyenko, O. Y., Tyrsa, V. V., Basaca, L. C., & Hipólito, J. I. N. (2011, March). Continuous monitoring of rehabilitation in patients with scoliosis using automatic laser scanning. In Health Care Exchanges (PAHCE), 2011 Pan American (pp. 410-414). IEEE. doi:10.1109/PAHCE.2011.5871941

Rodríguez-Quiñonez, J., Sergiyenko, O., Hernandez-Balbuena, D., Rivas-Lopez, M., Flores-Fuentes, W., & Basaca-Preciado, L. (2014). Improve 3D laser scanner measurements accuracy using a FFBP neural network with Widrow-Hoff weight/bias learning function. *Opto-Electronics Review*, *22*(4), 224–235. doi:10.2478/s11772-014-0203-1

Salvi, J., Pages, J., & Batlle, J. (2004). Pattern codification strategies in structured light systems. *Pattern Recognition*, *37*(4), 827–849. doi:10.1016/j.patcog.2003.10.002

Sansoni, G., Trebeschi, M., & Docchio, F. (2009). State-of-the-art and applications of 3D imaging sensors in industry, cultural heritage, medicine, and criminal investigation. *Sensors (Basel, Switzerland)*, *9*(1), 568–601. doi:10.3390/s90100568 PMID:22389618

Scharstein, D., & Szeliski, R. (2003, June). High-accuracy stereo depth maps using structured light. In *Computer Vision and Pattern Recognition, 2003. Proceedings. 2003 IEEE Computer Society Conference on* (Vol. 1, pp. I-195). IEEE. doi:10.1109/CVPR.2003.1211354

Sergiyenko, O., Hernandez, W., Tyrsa, V., Devia Cruz, L., Starostenko, O., & Pena-Cabrera, M. (2009). Remote sensor for spatial measurements by using optical scanning. *Sensors (Basel, Switzerland)*, *9*(7), 5477–5492. doi:10.3390/s90705477 PMID:22346709

Sergiyenko, O., Tyrsa, V., Basaca Preciado, L., Rodriguez-Quinonez, J., Hernandez, W., Nieto-Hipolito, J., & Starostenko, O. (2011). Electromechanical 3D optoelectronic scanners: resolution constraints and possible ways of its improvement. In Optoelectronic devices and properties. In-Tech.

Sergiyenko, O. Y. (2010). Optoelectronic system for mobile robot navigation. *Optoelectronics, Instrumentation and Data Processing*, *46*(5), 414–428. doi:10.3103/S8756699011050037

Sergiyenko, O. Y., Tyrsa, V. V., Devia, L. F., Hernandez, W., Starostenko, O., & Rivas-Lopez, M. (2009). Dynamic laser scanning method for mobile robot navigation. In *Proceedings of ICCAS-SICE, ICROS-SICE International Joint Conference*. Fukuoka, Japan: Academic Press.

Shanmugamani, R., Sadique, M., & Ramamoorthy, B. (2015). Detection and classification of surface defects of gun barrels using computer vision and machine learning. *Measurement*, *60*, 222–230. doi:10.1016/j.measurement.2014.10.009

K. Stanney, & K. S. Hale (Eds.). (2014, July). *Advances in cognitive engineering and neuroergonomics.AHFE Conference*.

Tang, T., & Qiao, H. (2014). Exploring biologically inspired shallow model for visual classification. *Signal Processing*, *105*, 1–11. doi:10.1016/j.sigpro.2014.04.014

Tomažič, S., & Škrjanc, I. (2015). Fusion of visual odometry and inertial navigation system on a smartphone. *Computers in Industry*, *74*, 119–134. doi:10.1016/j.compind.2015.05.003

Zhang, B., Huang, W., Li, J., Zhao, C., Fan, S., Wu, J., & Liu, C. (2014). Principles, developments and applications of computer vision for external quality inspection of fruits and vegetables: A review. *Food Research International*, *62*, 326–343. doi:10.1016/j.foodres.2014.03.012

Zou, X., & Zhao, J. (2015). Machine vision online measurements. In *Nondestructive measurement in food and agro-products* (pp. 11–56). Springer Netherlands.

Chapter 1
Applying Optoelectronic Devices Fusion in Machine Vision:
Spatial Coordinate Measurement

Wendy Flores-Fuentes
Autonomous University of Baja California, Mexico

Julio C. Rodríguez-Quiñonez
Autonomous University of Baja California, Mexico

Moises Rivas-Lopez
Autonomous University of Baja California, Mexico

Javier Rivera-Castillo
Autonomous University of Baja California, Mexico

Daniel Hernandez-Balbuena
Autonomous University of Baja California, Mexico

Lars Lindner
Autonomous University of Baja California, Mexico

Oleg Sergiyenko
Autonomous University of Baja California, Mexico

Luis C. Basaca-Preciado
Center of Excellence in Innovation & Design – CETYS University, Mexico

ABSTRACT

Machine vision is supported and enhanced by optoelectronic devices, the output from a machine vision system is information about the content of the optoelectronic signal, it is the process whereby a machine, usually a digital computer and/or electronic hardware automatically processes an optoelectronic signal and reports what it means. Machine vision methods to provide spatial coordinates measurement has developed in a wide range of technologies for multiples fields of applications such

DOI: 10.4018/978-1-5225-0632-4.ch001

as robot navigation, medical scanning, and structural monitoring. Each technology with specified properties that could be categorized as advantage and disadvantage according its utility to the application purpose. This chapter presents the application of optoelectronic devices fusion as the base for those systems with non-lineal behavior supported by artificial intelligence techniques, which require the use of information from various sensors for pattern recognition to produce an enhanced output.

INTRODUCTION

The present chapter surged in the research continuity of a 3D Vision System for mobile robot navigation application, a 3D medical laser scanner, and a structural health monitoring system. With the objective of increasing the accuracy of the systems, digital and analog processing signals methodologies have been developed in order to find the energetic center of the optoelectronic signal handled by these systems. Into the task of systems overall robustness, its measurement data has been submitted to statistical analysis, finding a non-linear behavior of the systems, leading to the need of artificial intelligence applications such as neuronal network (NN) and support vector machine regression (SVMR), in a modern approach to the prediction of the non-linear measurement error of the systems to compensate it. In the process of obtaining enough information from a measurement system to extract from it a model to predict its measurement error. It has been done a search of attributes to build the training dataset and test dataset. Ha been found that the pattern recognition can be enhanced by the sensor fusion and redundancy theory. This theory refers to the synergistic use of information from various sensors to achieve the task required by the system. Input data (attributes) are combined, fused and grouped for proper quality and integrity of the decisions to be taken by the intelligent algorithm. Besides, the benefits can be extracted from the redundant data, the reduction of uncertainty and the increasing of precision reliability. By these reasons, the photodiodes and charge coupled devices (CCD) are fusion in the task of robust systems building for machine vision by Spatial Coordinates Measurement (Weckenmann, 2009; Elfes, 1992; Zhang, 2008; Shih, 2015). The specific properties of both, their advantages and limitations have been considered, since, the photodiode is the sensor who gives place to the laser-scanning and the CCD is the sensors who gives place to the close-range photogrammetry. The energetic center of the laser optoelectronic signals from the photodiode and the energetic center of the image signal from the CCD sensor are detected to combine these sensors outputs, and to exploit their natural synergy new experimental results are presented to demonstrate the increase of systems accuracy.

BACKGROUND

Optoelectronics is the study of any devices that produce an electrically-induced optical output or an optically-induced electrical output and the techniques for controlling such devices (Marston, 1999), it includes generation, transmission, routing, and detection of optoelectronic signals in a widespread of applications (Dagenais, 1995). Wherever light is used to transmit information, tiny semiconductor devices are needed to transfer electrical current into optical signals and vice versa. Examples include light-emitting diodes, photodetectors and laser diodes (Piprek, 2003).

Most optoelectronics devices applications have focused on single sensors and relatively simple processes to extract specific information from the sensor, however the use of multiple sensors by an optoelectronic device fusion technology deliver more advanced information and enable to develop intelligent and sophisticated optoelectronic systems, in special for machine vision applications (Yallup, 2014). More than one optoelectronic sensor may be needed to fully monitor the observation space at all times. Methods of combining multiple sensor data are in developing due to the availability and computational power of communications devices that support algorithms needed to reduce the raw sensor data from multiple sensors to convert it to the information needed by the system user (Klein, 2003).

Optoelectronic systems with only one sensor are not recommended for those optoelectronics signals with non-lineal behavior, the observations made by one sensor could be uncertain and occasionally incorrect, a single sensor could not detect all the critical characteristics of the optoelectronic signal as could be done with multiple sensors, which could provide different information each one, in special if they are from different technologies, they can provide different information regarding the target under observation. Besides, if a unique sensor is used, and it fails, the optoelectronic system is shutdown, while this sensor it is not working. Several sensors deliver redundant information to increase reliability in case of errors or failure of a sensor. The redundant information can reduce uncertainty and increases the accuracy with which the characteristics are perceived by the optoelectronic system. Further information from various sensors is obtained to characterize the environment, perceived in a way that would be impossible using only the information for each sensor separately (Castanedo, 2013).

Machine vision is supported and enhanced by optoelectronic devices, the output from a machine vision system is information about the content of the optoelectronic signal, it is the process whereby a machine, usually a digital computer and/or electronic hardware automatically processes an optoelectronic signal and reports what it means. That is, it recognizes the content, it not only locates the content, but to inspect it as well. Machine vision includes two components, measurement of features and pattern classification based on those features. Defining the pattern classification as the process of making decision about a measurement, with a set of measurements knowledge about the possible classes could be obtained, leading to make a decision (Snyder, 2010).

The machine vision technology has a widespread of applications, as is the automatic target recognition, 3D vision, vehicle vision and the industrial inspection, that make use of it for replace or complement manual inspections and measurements, to automate the production, in order to increase production speed and yield, as well as to improve product quality.

In special, machine vision methods to provide spatial coordinates measurement has developed in a wide range of technologies for multiples fields of applications such as robot navigation, medical scanning, and structural health monitoring. Each technology has specific properties that could be categorized as advantage and disadvantage according its utility to the application purpose.

The computer vision has guide the machine vision to the tendency of duplicate the abilities of human vision by electronically perceiving and understanding an image with a high a dimensional data, increasing the data storage requirement, and the time processing due to the complexity of algorithms to extract important patterns and trends and understand what data says (learning data). For some applications it is not necessary to acquire the whole vision of a scene as makes the human vision to extract the features of interest, it is possible to only extract the significant characteristics from the scene by the use of a positioning laser (PL) (Rodriguez-Quiñonez, 2014), a rotatory mirror scanner and dynamic triangulation and/or by the extraction of important features from images captured through the use of camera with CCD (Charge Coupled Device) (Hou, 2011).

Researches previously published has found that the 3D modelling of objects and complex scenes constitutes a field of multi-disciplinary challenges and difficulties, in the going-over accuracy, reliability, quality, portability, low cost and automatization of the whole procedure. By these reasons although there are a wide variety of sensors, the optical scanning and the digital cameras play the main role in this application, because even though these two types of sensors can work in a separate fashion, it is when are fusion together when the best results are attained. The sensor fusion, in particular concerning these both technologies appears as a promising possibility to improve the data acquisition and the geometric and radiometric processing of these data (Gonzalez-Aguilera, 2010).

Optoelectronic Device Fusion in Machine Vision is expected to play an increasingly important role in the near future, by its desired characteristics, such as: optimization, deduction, robustness, flexibility, inference, accuracy, reliability, speed and cost-effectiveness.

By this motivation it is presented a review of modern optoelectronic devices, including light-emitting diodes, edge-emitting lasers, vertical-cavity lasers, electro absorption modulators, a novel combination of amplifier and photodetector (Piprek, 2003), and CCD, to know the diverse optoelectronic devices, and which are the most suitable according to the needs in numerous machine vision applications.

The available sensor fusion techniques are classified into three nonexclusive categories:

1. Data association,
2. State estimation, and
3. Decision fusion.

Also architectures and algorithms are introduced.

Machine vision operation are described, an introduction to the methods for imaging, processing, analysis, communication and actions are developed.

The current optoelectronic scanning technology and image processing for 3D reconstruction techniques are fusion, and a briefly introduction to the optimal hardware architecture is reviewed.

OPTOELECTRONIC SENSORS FUSION IN MACHINE VISION

Spatial Coordinate Measurement

Machine vision methods to provide spatial coordinates measurement has developed in a wide range of technologies, the computer vision has guide the machine vision to the tendency of duplicate the abilities of human vision by electronically perceiving and understanding an image with a high a dimensional data, increasing the data storage requirement, and the time processing due to the complexity of algorithms to extract important patterns and trends and understand what data says (learning data). For some applications it is not necessary to acquire the whole vision of a scene as makes the human vision to extract the features of interest (Sonka, 2014), it is possible to only extract the significant characteristics from the scene by the use of a positioning laser (PL), a rotatory mirror scanner and dynamic triangulation, for detailed system overall see (Basaca-Preciado, 2014; Rodriguez-Quiñonez, Sergiyenko, Preciado, et al, 2014; Flores-Fuentes, 2014a; Flores-Fuentes, 2014b).

Optoelectronic Devices

Optoelectronic devices take advantage of sophisticated interactions between electrons and light, it combines electronics and photonics. Photons are the smallest energy packets of light waves and their interaction with electrons, energy exchange between them is the key physical mechanism in optoelectronic devices. Optoelectronics brings together optics and electronics within a single device, a single material, semiconductors designed to allow for the transformation of light into current and vice versa.

1. **Photodetector Devices:** The purpose of any photodetector is to convert light (photons) into electric current (electrons). Photodetectors are among the most common optoelectronic devices; they are used to automatic open doors, detect the signal from infrared remote controls, and record pictures in modern cameras. Photodetectors are fabricated from various semiconductor materials, since the band gap needs to be smaller than the energy of the photons detected. Photon absorption generates electron–hole pairs which are subsequently separated by the applied electrical field. Depending on the desired use, photodetectors are designed in many different ways, and can be classified in photodiodes, infrared-detectors and charge coupled devices (CCD), as shown in Figure 1. Modern photodetectors include an amplifier; the amplification photodetector simultaneously amplifies and absorbs the incoming light.
    a. **Photodiode:** Photodiodes are light detecting devices fabricated in semiconductor materials, when a photon of sufficient energy strikes a diode, it creates an electron-hole pair. This mechanism is also known as the inner photoelectric effect. If the absorption occurs in the junction's depletion region, or one diffusion length away from it, these carriers are swept from the junction by the built-in electric field of the depletion region. Thus holes move toward the anode, and electrons toward the cathode, and a photocurrent is produced. The total current through the photodiode is the sum of the dark current (current that is generated in the absence of light) and the photocurrent, so the dark current must be minimized to maximize the sensitivity of the device. Photodiodes are similar to regular semiconductor diodes except that they may be either exposed or packaged with a window or optical fiber connection to allow light to reach the sensitive part of the device (Tavernier, 2011). The most common photodiodes are the P-N photodiode, PIN photodiode, avalanche photodiode, waveguide photodiode and amplification photodiode (Syms, 1992; Piprek, 2003).
    b. **Infrared-Detectors:** This detector is sensible to infrared radiation (IR). Thermal and photonic photodetectors are the main types of infrared-detectors. The effects of the incident IR radiation can be followed through many temperature dependent phenomena; a thermal detector type has no wavelength dependence. Major characteristics indicating infrared detector performance are photo sensitivity, the photodetector is wavelength dependent. In comparison with visible and ultraviolet rays, infrared radiation has small energy. To increase infrared detection efficiency, the detector

should be cooled. The signal output from a detector is generally quite small and needs to be amplified. When designing a preamplifier, it is considered the impedance match between the detector and the amplifier, if the incident light is modulated by a chopper, the use of a tuned amplifier is necessary to reduce noise. If the detector is cooled, it is also practical to cool the amplifier. IR photodetectors are classified in intrinsic type and extrinsic type (Ge:Au, Ge:Hg, Ge:Cu, Ge:Zn, Si:Ga, Si:As), the intrinsic type is classified on Photoconductive type (PbS, PbSe, InSb, HgCdTe) and Photovoltaic type (Ge, InGaAs, Ex. InGaAs, InAs, InSb, HgCdTe) (Rogalski, 2002). The most common infrared detectors are InGaAs PIN photodiode and linear image sensor, PbS and PbSe photoconductive detector, InAs and InSb photovoltaic detector, MCT (HgCdTe) and InSb photoconductive detector, and two color detector (McClure, 2003; Tribolet, 2008).

c. **Charge Coupled Devices (CCD):** CCD is the abbreviation for charge-coupled device, also known as CCD image sensor. CCD sensors are typically built with silicon technology. Silicon based integrated circuits (ICs), consisting of a dense matrix of photodiodes or photogates that operate by converting light energy, in the form of photons, into electronic charges. When a photon strikes a silicon atom in or near a CCD photosite, the photon usually produces a free electron and a hole via the photoelectric effect.

Figure 1. Photodetectors classification

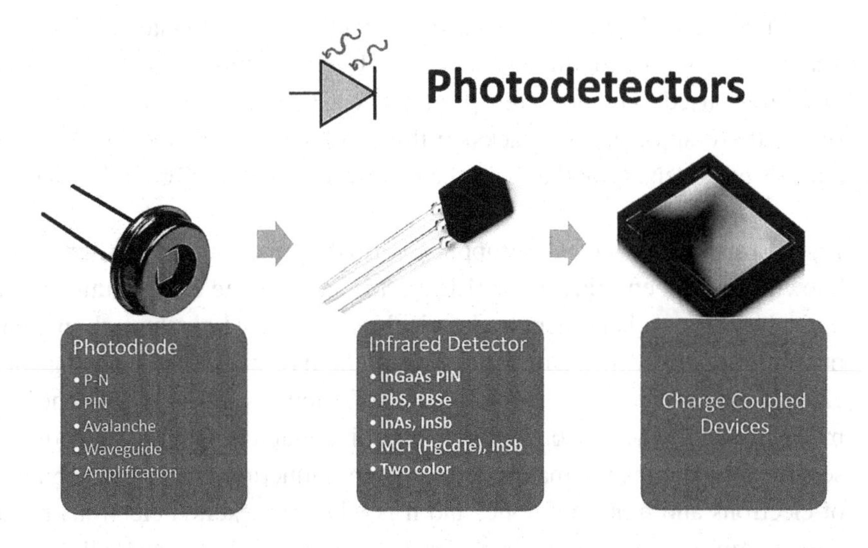

The primary function of the CCD is to collect the photogenerated electrons in its potential wells (or pixels) during the CCD's exposure to radiation, and the hole is then forced away from the potential well and is eventually displaced into the Si substrate. While more light that is incident on a particular pixel, the higher the number of electrons that accumulate on that pixel. The output signal is then transferred to the computer for image regeneration or image processing (Li, 2006).

The pixels in a CCD sensor can be arranged in various configurations, the linear and area array are the classical architectures. A linear or linescan CCD sensor consists of a single line of pixels, adjacent to a CCD shift register that is required for the read-out of the charge packets. The isolation between the pixels and the CCD register is achieved by a transfer gate. Typically, the pixels of a linear CCD are formed by the photodiodes. Two-dimensional imagining is possible with area array CCD sensors; the entire image is captured with one exposure, eliminating the need for any movement by the sensor or the scene. As charge coupled device (CCD) technology matures it dominates the digital imaging, for commercial and industrial manufacturing applications (Janesick, 2001).

Conventional CCDs are designed for the front illuminated mode of operation. Front illuminated CCDs are quite economical to manufacture by using standard wafer fabrication procedures, and are popular in consumer imagining applications, as well as industrial-grade applications. However, front-illuminated CCDs are inefficient at short wavelengths (e.g., blue and UV) due to the absorption of photons by the polysilicon layers if photogates are used as the pixel element. Typically, conventional front-illuminated CCDs are adequate for low-end applications and for consumer electronics. But, for large, professional observatories and high-end industrial inspection systems that demand that require extremely sensitive detectors, conventional thick frontside-illuminated chips are rarely used. Thinned back-illuminated CCDs offer a more compatible solution for these applications. Backside-thinned back-illuminated chips are rarely used. Thinned back-illuminated CCDs offer a more compatible solution for these applications. Backside-thinned back-illuminated CCDs exhibit a superior responsively, remarkably in the shorter wavelength region (Li, 2003).

2. **Light Emitting Devices:** Other application of the p-n junction is as a light source, known as a light emitting diode (LED). It is based in the electroluminescence optical-electrical phenomenon, electrical generation of light through spontaneous emission. It is when a material emits light in response to the passage of an electric current, a radiative recombination of electrons and holes are done in the material, usually a semiconductor. In a semiconductor electrons and holes are separated by doping the material to form a p-n junction, then a recombination of electrons and holes are done, and it is when the excited electrons release their energy as photons (light). As an isolated lump of material will not emit

significant quantities of light, in thermal equilibrium (room temperature) the number of downward electron transitions is extremely small. To improve the optical output, the material should be moved far from equilibrium, so that the rate of spontaneous emission is considerably increased. This might be done by taking (for example) a p-type material, which already contains a large holes density, and pouring electrons into it by a forward-biased p-n junction. Spontaneous emission generates photons that travel in random directions, so the emission is isotropic, the emission is also un-polarized, and the output is incoherent, with a spectrum consisting of a broad range of wavelengths, consequently, the LED cannot be used as a source for high-speed, long-distance optical communications; the dispersion caused by such an extended spectrum would be far too large. LED transmitters are therefore restricted to short-haul applications.

A single photon is able to generate an identical second photon by stimulating the recombination of an electron-hole pair. This photon multiplication is the key physical mechanism of lasing. The second photon exhibits the same wavelength and the same phase as the first photon, doubling the amplitude of their monochromatic wave. Subsequent repetition of this process leads to strong light amplification. However, the competing process is the absorption of photons by the generation of new electron-hole pairs. Stimulated emission prevails when more electrons are present at the conduction band level than at the valence band level. This carrier inversion is the first requirement of lasing and it is achieved at pn-junctions by providing conduction band electrons from the n-doped side and valence band holes from the p-doped side. The photon energy is given by the band gap, which depends on the semiconductor material. Continuous current injection into the device leads to continuous stimulated emission of photons, but only if enough photons are constantly present in the device to trigger this process. Therefore, optical feedback and the confinement of photons in an optical resonator is the second basic requirement of lasing.

The absorption and simulated emission process produced in a LED is the principle of a light amplification by stimulated emission of radiation (LASER), the exponentially growing wave produce an amplification ought to offer a method of light generation. To achieve this, the electron density is raised far above the equilibrium level by the injection of electrons across a forward-biased diode. The laser is a device based on the stimulated emission of electromagnetic radiation. It emits light through a process of optical amplification; it emits spatial coherently light as a tight spot, achieving a very high irradiance, or they can have very low divergence in order to concentrate their power at a great distance. The laser enables applications such as cutting, welding and lithography for optical disk drives, laser printers, barcode scanners, fiber-optic, optical communication, surgery, skin treatments, device enforcement for marking targets and measuring range and speed, and lighting display.

The first semiconductor lasers consisted of heavily doped p-n junctions with cleaved facets. This construction is inefficient, as there is no defined region in which recombination can take place. Carriers can be lost to diffusion before recombination occurs. As a result, these early devices required a lot of current to reach threshold. Threshold currents were reduced with the developments of the double-heterostructure laser, which has a thin region of semiconductor with a smaller energy gap sandwiched between two oppositely doped semiconductors with a wide bandgap energy. When forward biased, carriers flow into the active region and recombine more efficiently because of the potential barriers of the heterostructure confine the carriers to the active region. There is an added advantage of guided the laser light, because the refractive index of the cladding layers is less than the active region. See Figure 2.

a. **Burrus-Type LED:** In place of the simple p-n junction, a more complicated structure is used for the Burrus-type LED. Typically, this might be a P-n-N double heterostructure, containing three layers: a p-type layer with a wide energy gap, a narrow-gap n-type layer, and a wide-gap n-type layer. A single heterojunction provides potential barriers of different heights for electrons and holes; the double heterostructure provide a high barrier at different positions for electrons and holes. In this way, the recombination region may be limited to a defined region of space, resulting in an increase in the radiative recombination efficiency. This is a high-efficiency device, suitable for use with multimode fiber. To fabricate the device, a double heterostructure is first grown on a substrate, and a SiO_2

Figure 2. Light emitting devices classification

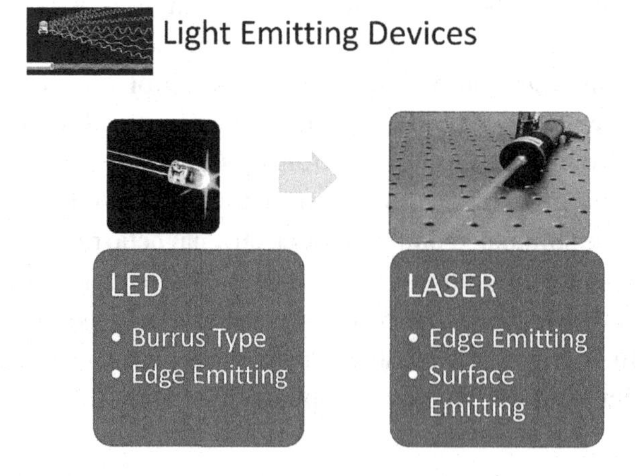

isolation layer is then deposited on the lower layer of the heterojunction. This layer is etched, to expose the hetero- junction over a small window. Metallization layers are then added to the upper and lower surfaces; clearly, the lower of these makes contact with the heterojunction only in the region of the window. A deep well, aligned with this window, is then etched right through the substrate to the upper layer of the heterojunction, and a multimode optical fiber is epoxied into the well in contact with the LED surface. When a forward current is passed through the LED, it flows mainly through the region immediately below the fiber. The recombination region (or active region) is therefore confined vertically by the double heterojunction and laterally by the distribution of current flow.

b. **Edge Emitting LED:** This Edge Emmiting LED also called superluminescent LED presents an alternative geometry to the Burrus-type LED (Fukuda, 1999). This is also based on a double heterostructure, but now the emission is taken from the edge of the junction rather than its surface. Generally, the current is forced to flow through a narrow strip down the device center line by the introduction of a current-blocking silica layer. As a result, the active volume is constrained vertically by the double heterostructure and laterally by the current flow. Most importantly, light generated by spontaneous emission is also confined vertically to a certain extent; the refractive index difference at the top and bottom of the active layer results in total internal reflection. This tends to channel a large fraction of the light towards the emission windows at relatively shallow angles.

c. **Edge Emitting Laser:** In this Laser the light propagates in a direction along the wafer surface of the semiconductor chip and is reflected or coupled out at a cleaved edge. These are the original and still very widely used form of semiconductor lasers. Within the edge-emitting laser structure, the laser beam is guided in a waveguide structure. Typically, one uses a double heterostructure, which restricts the generated carriers to a narrow region and at the same time serves as a waveguide for the optical field. This arrangement leads to a low threshold pump power and a high efficiency. Depending on the waveguide properties, particularly its transverse dimensions, it is possible either to obtain an output with high beam quality but limited output power, or an output with high output power but with poor beam quality.

d. **Surface Emitting Laser:** In the surface-emitting laser, the light propagates in the direction perpendicular to the semiconductor wafer surface. These are further subdivided into monolithic and external-cavity devices: Monolithic means that the laser resonator is realized in the form of two semiconductor Bragg mirrors with the quantum well section in between. Such a device is called VCSEL (vertical cavity surface-emitting laser) and is electrically pumped in most cases (Lorenser, 2003).

Sensors and Data Fusion

Data fusion refers to the synergistic use of information from different sensors for proper quality and integrity of the decision making to accomplish a task required by a system. All tasks that demand any type of parameter estimation from multiple sources can benefit from the use of data/information fusion methods. The terms information fusion and data fusion are typically employed as synonyms; but in some scenarios, the term data fusion is used for raw data (obtained directly from the sensors) and the term information fusion is employed to define already processed data. In this sense, the term information fusion implies a higher semantic level than data fusion. Other terms associated with sensor fusion that typically appear in the literature include decision fusion, data combination, data aggregation, multi-sensor data fusion, and data fusion (Castanedo, 2013).

Systems with only one sensors are not recommended for those with non-lineal behavior, the observations made by one sensor could be uncertain and occasionally incorrect, a single sensor could not detect all the critical characteristics of the system as could be done with multiple sensors, which could provide different information each one, in special if they are from different technologies and can provide different information regarding the system. Besides if a unique sensor is used, and it fails the system is shutdown while this sensor it is not working. The redundant information can reduce uncertainty and increases the accuracy with which the characteristics are perceived by the system. Several sensors delivering redundant information increase reliability in case of errors or failure of a sensor. Further information from various sensors to characterize the environment perceived in a way that would be impossible using only receive the information for each sensor separately.

The goal of using sensor fusion in multi-sensor environments is to obtain a lower detection error probability and a higher reliability by using data and information from multiple distributed sources, as shown in Figure 3.

1. **Sensors and Data Fusion Classification:** Fusion techniques and methods are difficult to classify due to they involve a big variety of application fields. A general classification is based on data criteria, as described at Figure 4.
   a. **Relation between the Input Data Sources (Sensors):** Complementary: Each sensor provides complementary information that can be combined together, the information provided by the sensors represents different

Figure 3. Sensors and data fusion applied to systems with non-lineal behavior

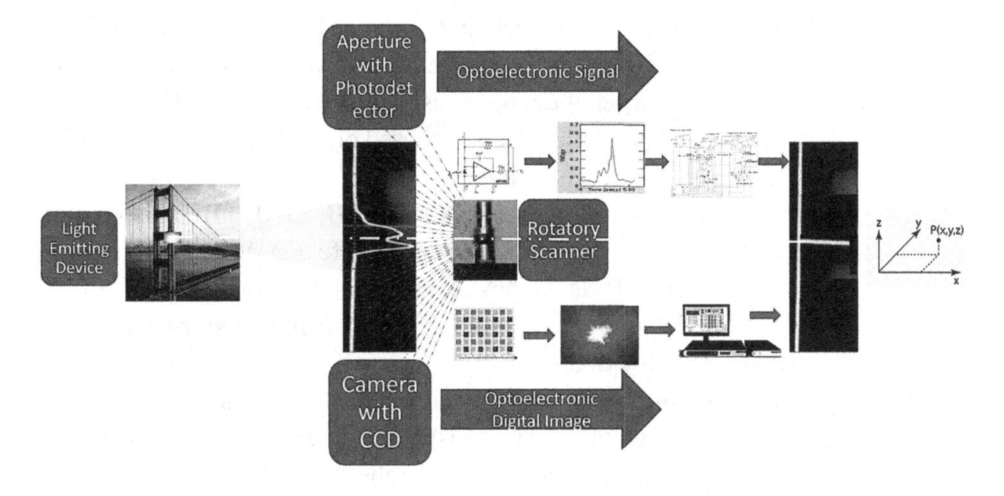

parts of the scene and could thus be used to obtain more complete global information (Vauhkonen, 2014).

i. **Redundant:** Various sensors providing information about the same target or scene (Abidi, 1992).
ii. **Cooperative:** Two or more sensors are independently used to perform detection, providing information that is combined into new information that is typically more complex than the original information (Labayrade, 2005).

b. **Input/Output Data Sensors Types and Their Nature:** Low Level Data in-Data out (DAI-DAO): The raw data are directly provided as an input to the data fusion process. This type of data fusion process inputs and outputs raw data; the results are typically more reliable or accurate. Data fusion at this low level is conducted immediately after the data are gathered from the sensors. The algorithms employed at this level are based on signal and image processing algorithms.
   i. **Low Level Data In-Feature Out (DAI-FEO):** At this low level, the data fusion process employs raw data from the sources to extract features or characteristics that describe an entity in the environment;
   ii. **Medium Level Feature In-Feature Out (FEI-FEO):** At this medium level, both the input and output of the data fusion process are features or characteristics (shape, texture, and position), the data fusion process addresses a set of features with to improve, refine or obtain new features. This process is also known as feature fusion, symbolic fusion, information fusion or intermediate-level fusion.

iii. **High Level Feature In-Decision Out (FEI-DEO):** Takes symbolic representations as sources and combines them to obtain a more accurate decision, this high level obtains a set of features as input and provides a set of decisions as output. Most of the classification systems that perform a decision based on a sensor's inputs fall into this category of classification.

iv. **High Level Decision in-Decision out (DEI-DEO):** This type of classification is also known as decision fusion. It fuses input decisions to obtain better or new decisions (Dasarathy, 1997).

v. **Multiple Level:** It is a combination of data provided from different levels of abstraction.

c. **Data Abstraction Level:** It is typically classified in three levels of abstraction: measurements, characteristics, and decisions. Other possible classifications of data fusion based on the abstraction levels are as follows:

i. **Signal Level:** Directly addresses the signals that are acquired from the sensors.

ii. **Pixel Level:** Operates at the image level and could be used to improve image processing tasks.

iii. **Characteristic:** Employs features that are extracted from the images or signals (i.e., shape or velocity),

iv. **Symbol:** At this level, information is represented as symbols; this level is also known as the decision level (Luo, 2002).

d. Data fusion levels defined by the JDL (Joint Directors of Laboratories).

i. **First Layer Sources of Information:** Local and distributed sensors linked to a data fusion system, and or a priori information, as databases.

ii. **First Layer Human-Computer Interaction (HCI):** It is an interface that allows inputs to the system from the operators and produces outputs to the operators. HCI includes queries, commands, and information on the obtained results and alarms. It incorporates not only multimedia methods for human interaction (graphics, sound, tactile interface, etc.), but also methods to assist humans in direction of attention, and overcoming human cognitive limitations.

iii. **First Layer Source Preprocessing:** An initial process allocates data to appropriate processes and performs data pre-screening. Source preprocessing reduces the data fusion system load by allocating data to appropriate processes. The database management system stores the provided information and the fused results. This system is a critical component because of the large amount of highly diverse information that is stored.

   iv. **Second Layer Level 1 Processing (Object Refinement):** Transforms sensor data into a consistent set of units and coordinates, refines and extends in time estimates of an object's position, kinematics, or attributes, assigns data to objects to allow the application of statistical estimation techniques, and refines the estimation of an object's identity or classification.
   v. **Second Layer Level 2 Processing (Situation Refinement):** Formal and heuristic techniques are used to examine, in a conditional sense, the meaning of Level 1 processing results.
   vi. **Second Layer Level 3 Processing (Threat Refinement):** It is an impact assessment, at this level the impact of techniques used in level 2 are evaluated to obtain a proper perspective. And a future projection is performed to identify possible risks, vulnerabilities, and operational opportunities. This level includes an evaluation of the risk or threat and a prediction of the logical outcome.
   vii. **Second Layer Level 4 Processing (Process Refinement):** It is a process concerned about other processes, monitors the data fusion process performance to provide information about real-time control and long-term performance, identifies what information is needed to improve the multilevel fusion product (inferences, positions, identities, etc.), determines the source specific requirements to collect relevant information (i.e., which sensor type, which specific sensor, which database), and allocates and directs the sources to achieve mission goals (Luo, 2002).

2. **Sensors and Data Fusion Methods:** Most current sensors data fusion methods algorithms can be broadly classified as illustrate at Figure 5.
   a. **Estimation Methods:** These methods take the weighted average of redundant information coming from a group of sensors and use it as the value of the fusion sensors. These methods determine the state of the target under movement (typically the position) given the observation or measurements. State estimation techniques are also known as tracking techniques with a final goal of obtain a global target state from the observations. The estimation problem involves finding the values of the vector state (e.g., position, velocity, and size) that fits as much as possible with the observed data. From a mathematical perspective, we have a set of redundant observations, and the goal is to find the set of parameters that provides the best fit to the observed data. In general, these observations are corrupted by errors and the propagation of noise in the measurement process. State estimation methods fall under level 1 of the JDL classification and could be divided into two broader groups:

i. **Linear Dynamics and Measurements:** Here, the estimation problem has a standard solution. Specifically, when the equations of the object state and the measurements are linear, the noise follows the Gaussian distribution, do not refer to it as a clutter environment; in this case, the optimal theoretical solution is based on the Kalman filter;

ii. **Nonlinear Dynamics:** The state estimation problem becomes difficult, and there is not an analytical solution to solve the problem in a general manner. In principle, there are no practical algorithms available to solve this problem satisfactorily. This method allows real-time processing of data. Some of them are: Maximum Likelihood and Maximum Posterior, Particle Filter, The Kalman Filter, The Distributed Kalman Filter, Distributed Particle Filter, and Covariance Consistency Methods: Covariance Intersection/Union. Being The Kalman filter the most popular.

b. **Classification Methods:** A data association is performed to determine the set of measurements that correspond to each target. The multidimensional feature space can be partitioned into distinct regions, each representing an identification or identity class. The location of a feature vector is compared to pre-specified locations in feature space. A similarity measure must be computed, and each observation is compared to a priori classes. A feature space may be partitioned by geometrical or statistical boundaries. Therefore, the templating approach may declare a unique identity or an identity with an associated uncertainty. The implementation of parametric templates is computationally efficient for multisensor fusion systems. The most common methods are Nearest Neighbors and K-Means.

c. **Inference Methods:** Bayesian inference allows multisensor information to be combined according to the rules of probability theory. Bayes' formula provides a relationship between the a priori probability of a hypothesis, the conditional probability of an observation given a hypothesis, and a posteriori probability of the hypothesis. Bayesian inference updates the probabilities of alternative hypotheses, based on observational evidence. New information is used to update the a priori probability of the hypothesis.

d. **Artificial Intelligence Methods:** In these methods decisions are typically taken based on the knowledge of the perceived situation, a high-level inference is required, as the human reasoning, for pattern recognition, planning, induction, deduction, and learning.

Figure 4. Sensors and data fusion classification

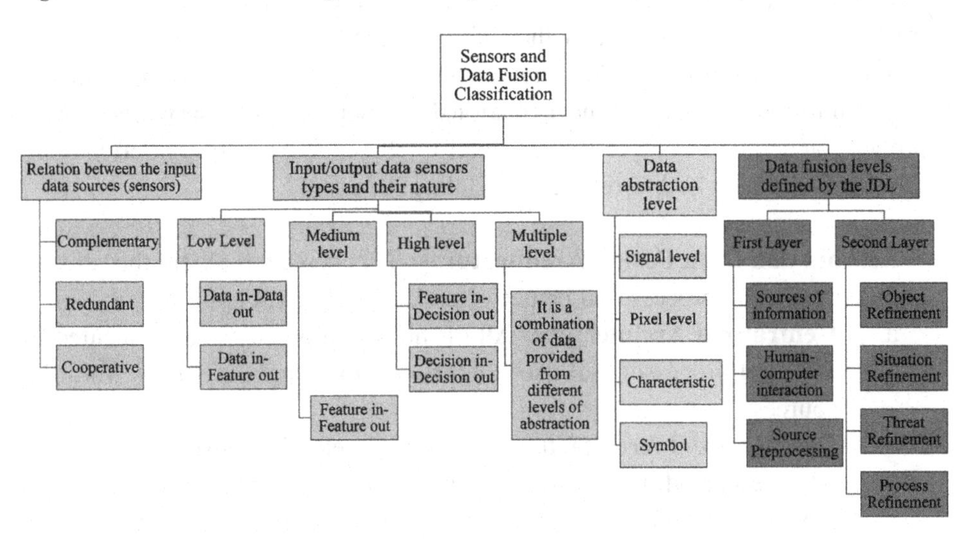

Figure 5. Sensors and data fusion methods

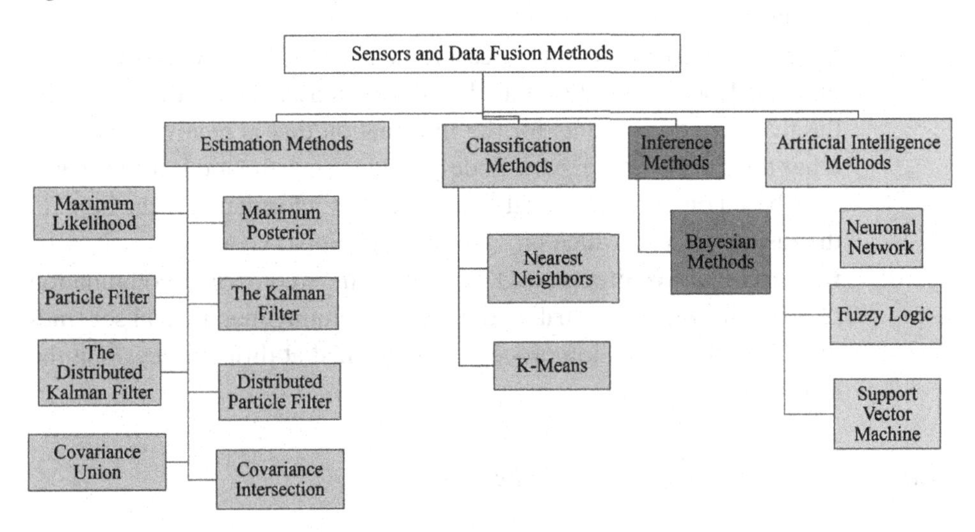

The artificial intelligence methods are based on an initial data set and the rule-based knowledge base comprising rules, frames and scripts. The inference process uses the a priori data set, and searches the complete set of rules to identify applicable rule. The rule selection strategies from among multiple applicable rules include refraction, actuation, rule ordering, specificity, and random choice. Examples of these methods are Neuronal Network, Fuzzy Logic, Support Vector Machines, etc.

Artificial intelligence methods can be seen as the advanced versions of the estimation, the classification, and the inference methods. Artificial intelligence methods can be model-free, rather than model-specific, and have sufficient degree of freedom to fit complex nonlinear relationships, with the necessary precautions to properly generalize. As a result, artificial intelligence methods can effectively conduct sensor fusion at different levels (Hall, 1997).

3. **Sensors and Data Fusion Architectures:** This describes where the sensors and data fusion process will be performed, as shown in Figure 6.
    a. **Centralized Architecture:** All of the fusion processes are executed in a central processor that uses the provided raw measurements from the sources.
    b. **Decentralized Architecture:** It is composed of a network of nodes in which each node has its own processing capabilities and there is no single point of data fusion. Therefore, each node fuses its local information with the information that is received from its peers. Data fusion is performed autonomously, with each node accounting for its local information and the information received from its peers.
    c. **Distributed Architecture:** Measurements from each source node are processed independently before the information is sent to the fusion node; the fusion node accounts for the information that is received from the other nodes. Therefore, each node provides an estimation of the object state based on only their local views, and this information is the input to the fusion process, which provides a fused global view.
    d. **Hierarchical Architecture:** Other architectures comprise a combination of decentralized and distributed nodes, generating hierarchical schemes in which the data fusion process is performed at different levels in the hierarchy.

Figure 6. Sensors and data fusion architectures

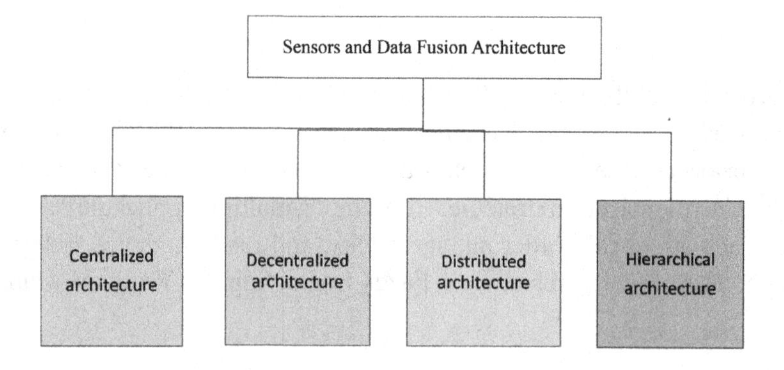

MACHINE VISION FOR SPATIAL COORDINATE MEASUREMENT

Machine vision methods to provide spatial coordinates measurement has developed in a wide range of technologies, the computer vision has guide the machine vision to the tendency of duplicate the abilities of human vision by electronically perceiving and understanding an image with a high a dimensional data, increasing the data storage requirement, and the time processing due to the complexity of algorithms to extract important patterns and trends and understand what data says (learning data). For some applications it is not necessary to acquire the whole vision of a scene as makes the human vision to extract the features of interest (Luo, 2002), it is possible to only extract the significant characteristics from the scene by the use of a positioning laser (PL), a rotatory mirror scanner and dynamic triangulation as shown at Figure 7, for detailed system overall see (Basaca-Preciado, 2014; Rodriguez-Quiñonez, Sergiyenko, Preciado, et al, 2014; Flores-Fuentes, 2014a).

Machine Vision System Overall

The main sensor is a light scanning system, it can be a coherent light emitter source such as a laser or an incoherent light source such as a bulb like the ones used in motor vehicles. We assume that for any light emitter source there is only one energy centre that represents its point position.

Figure 7. 3D laser scanning system for spatial coordinates

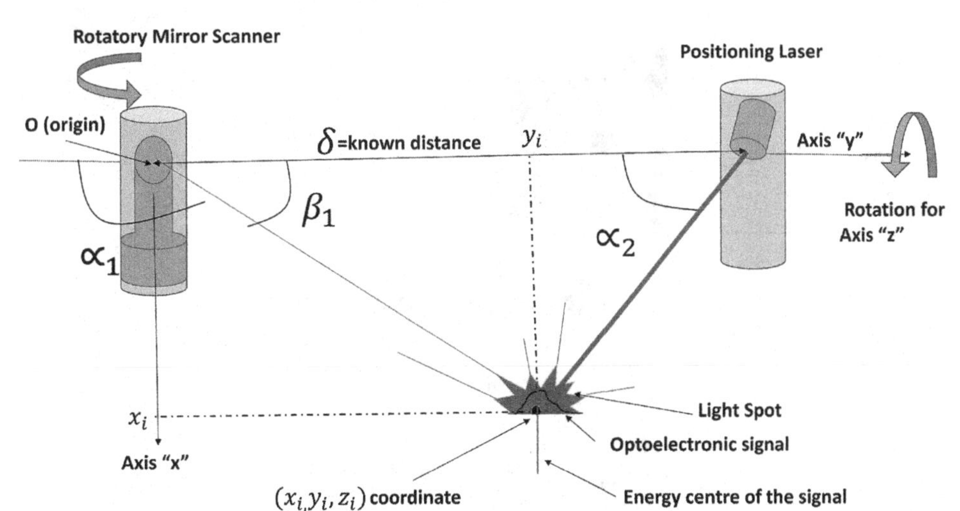

A rotating optical aperture, designed as a 45° slanted mirror surface attached to a cylindrical rod that deflects the light beam into a double convex lens provided with an interference filter and a photodiode. The cylindrical rod is mounted onto the shaft of a dc electrical motor, and as it rotates, an electronic signal is generated. Figure 8 shows the main elements of the optical scanning aperture. When the mirror starts spinning, the sensor "s" is synchronized when the origin generates a pulse indicating the 0° position and the starting of a cycle of 360° that finishes immediately before the "s" sensor generates the next beginning pulse. These pulses are used to calculate the scanning frequency and the zero reference, which is used to measure the angle where the light emitter source is positioned.

The signal timing diagram in Figure 9 shows the starting pulse and the optoelectronic signal relation used to calculate the light emitter energy signal centre position as described in the equations below. The interval $T_{2\pi}$ is the time for one motor revolution, defined as the time between m_1 and m_1 as in (1), which are expressed by the code $I_{2\pi}$ as defined in (2). As for m_1, it is the reference of the starting of one motor revolution (0°).

Figure 8. Optical scanning aperture

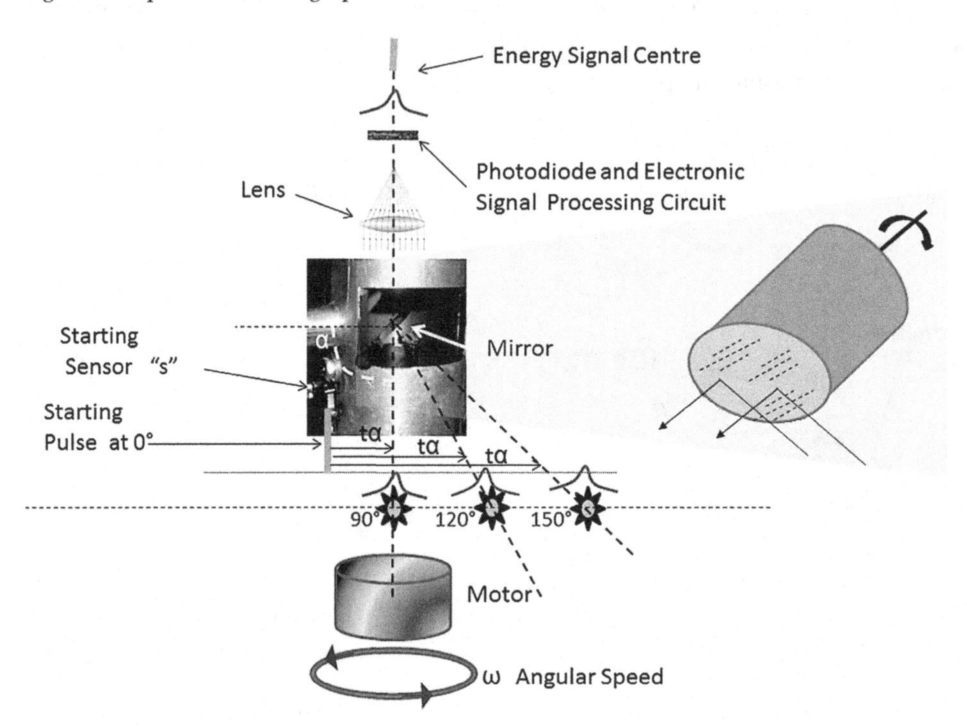

$$T_{2\pi} = \frac{2\pi}{\omega} = \frac{2\pi}{2\pi f} = \frac{1}{f} \tag{1}$$

$$I_{2\pi} = T_{2\pi} f \tag{2}$$

The time t_α is the time between the reference of the starting of one motor revolution and the energy signal centre, that is, the interval between m_1 and m_2. This can be expressed by the code I_α as defined in (3). As for m_2, it is the energy signal centre. The angle under measurement is then calculated by (4).

$$I_\alpha = t_\alpha f \tag{3}$$

$$\alpha = \frac{2\pi I_\alpha}{I_{2\pi}} = \frac{2\pi t_\alpha f}{T_{2\pi} f} = \frac{2\pi t_\alpha f}{\frac{2\pi}{\omega} f} = t_\alpha \omega = t_\alpha 2\pi f \tag{4}$$

Figure 9. Optical scanning aperture timing

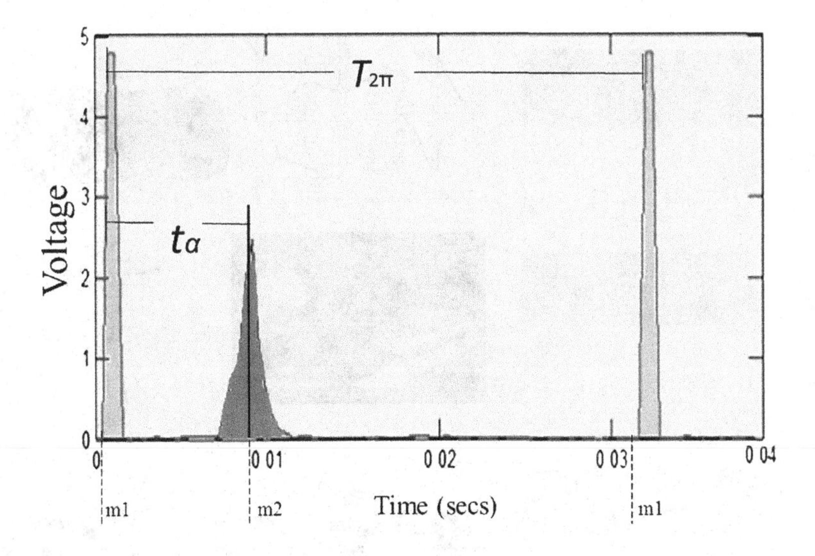

where:

m_1 is the reference of the starting of one motor revolution.
m_2 is the energy signal centre.
$T_{2\pi}$ is the time interval of one cycle, from m_1 to m_1.
ω is the angular speed
f is the scanning frequency (cycles per second).
$I_{2\pi}$ is the interval code of one cycle, from m_1 to m_1.
I_α is the range code from the starting of one cycle to the energy signal centre.
t_α is the interval from the starting of the signal to the energy signal centre, from m_1 to m_2.

Applications of Machine Vision for Spatial Coordinate Measurement.

The present chapter surged in the research continuity of a 3D Vision System for mobile robot navigation application, a 3D medical laser scanner, and a structural health monitoring system, shown at Figure 10.

Figure 10. Applications of machine vision for spatial coordinate measurement

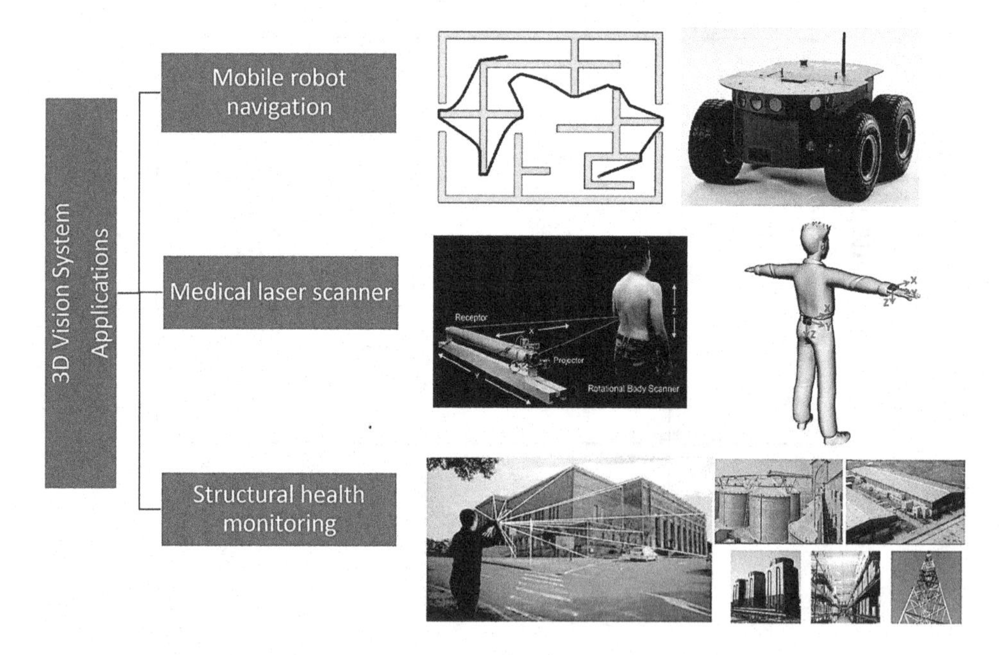

1. **Mobile Robot Navigation:** Mobile Robot Navigation, also known as Navigation System (NS) utilizes Technical Vision System (TVS) to obtain the characteristics of obstacles appearing in the field of view (FOV) of a mobile robot; with that information the NS calculates a new trajectory to reach its programmed goal. The mobile robot needs to have a goal programmed to begin operation, then an initial trajectory is calculated by calculating the shortest path between current position and goal. The mobile robot will follow the initial trajectory until an obstacle is detected by an array of infrared (IR) sensors. The IR sensors signal will trigger the TVS to perform a scan and obtain the characteristics (3D coordinates) of the detected obstacles. If no obstacles are detected, the mobile robot will reach its goal and stop advancing.

This NS is based in the Pioneer 3-AT (Figure 11) which is a small, four-wheel, four-motor skid-steer intelligent robotics platform, with all the characteristics to be a very suitable research development platform to implement the navigation and 3D vision system for autonomous mobile robots (Basaca-Preciado, 2014).

2. **3D Medical Laser Scanning:** For Medical Scanning System, the 3D measurements of the human body surface or anatomical areas have gained importance in many medical applications. Medical professionals widely use size and shape to assess nutritional status and developmental normality and to calculate the requirements of drug, radiotherapy, and chemotherapy dosages, as well as the production of prostheses. This system has the capability to realize precise measurements by 3D point clouds sampled from the surface of human body, used to obtain the 3D coordinates of the body that are in the FOV in front of the system as shown in Figure 12. This system is based in medical professional

Figure 11. Pioneer 3-AT mobile robotic platform

traditionally manual measurement of the body size and shape to assess health status and guide treatment (Rodriguez-Quiñonez, Sergiyenko, Preciado, et al, 2014).

3. **Structural Health Monitoring System:** Important engineering constructions as shown at Figure 13, require geometrical monitoring to predict their structural health during their lifetime. Structure experiences deterioration and damage due to environmental conditions and excessive load conditions. Optical Scanning Systems (OSS) provides position measurements for structural health monitoring (SHM) task by a spatial coordinates measurements. This system is based on an initial prototype consisting of an incoherent light source emitter (non-rotating) mounted on the structure under monitoring, a passive rotating optical aperture sensor designed with a 45°-sloping mirror, and embedded into a cylindrical micro rod. A photodiode to capture the beam of light while the cylindrical micro rod mounted on a dc electrical motor shaft is rotating, and a signal energetic centre detector by digital and analog processing (Flores-Fuentes, 2014a), (Flores-Fuentes, 2014b).

Figure 12. 3D rotational body scanner

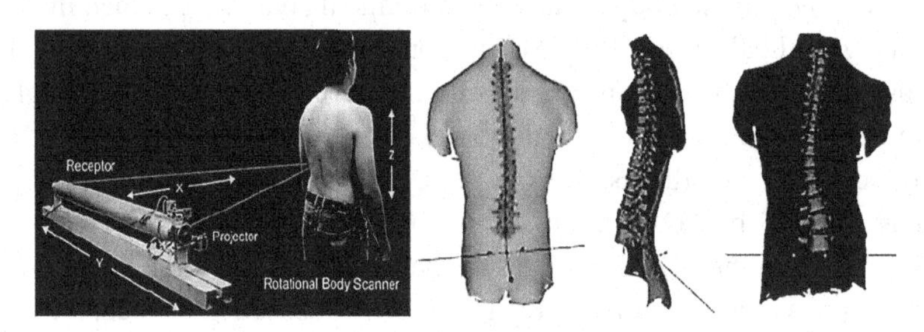

Figure 13. Structures (historical center, skyscraper, suspension bridge, dam water retention, antenna, and industrial warehouses)

Optoelectronic Scanning and Digital Image Processing through the Energetic Center of Laser Spot

Optoelectronic scanning systems generate different shapes of signals, depending on the light source and the sensor of the scanner. A typical position measuring process includes an emitter source of light, as a laser diode or an incoherent light lamp and the position sensitive detector like photodiode and CCD as a receiving device, which collects a portion of the back-reflected light from the target.

When a photodiode is used as a sensor on a scanner with rotating mirror it can also originate a similar Gaussian-like shape with some noise and deformation, the spatial distribution function of light has an Airy-function-like shape, by the other side the CCD produces output currents related to the "centre of mass" of light incident on the surface of the device, CCD use the light quantity distribution of the entire beam spot entering the light receiving element to determine the beam spot centre or centroid and identifies this as the target position.

Concluding that, the photodiode signal originates a similar function to a CCD, consequently, it is possible to enhance the accuracy measurements in optoelectronic scanners with a rotating mirror, using a method for improving centroid accuracy by taking measurement in the energy centre of the signal (Sonka, 2014), (Li, 2014).

1. **Energetic Center of Laser Optoelectronic Signal Detected by Photodiode:** The photodiode is the sensor, which is used for laser-scanning (Beiser, 1995). As SHM is an upcoming tendency of determining the integrity of structures and development of strategies to prevent undesirable damage, the OSS was designed, it has the necessity off detect a light emitter mounted on the structure under monitoring and calculate its energy centre localization. The OSS generates the targeted signals to be analyzed by the proposed methods previously assessed (Rivas-Lopez, 2014).

These six Energy Centre Localization Methods (ECLM), (Geometric Centroid, Power Spectrum Centroid, Analog Processing by Electronic Circuit, Saturation and Integration, Rising Edge, and Peak Detection), Figure 14 shows MATLAB code for Power Spectrum Centroid y graphical results of six ECLM. ECLM were assessed based on the assumption that the OSS signal from light emitter scanning is a Gaussian-like shape signal, the light emitter is an incoherent light, considered a punctual light source and due to the fact that with distance the light source expands its radius a cone-like or an even more complex shape is formed depending on the properties of the medium through which the light is travelling. And to reduce errors in position measurements, the best solution is taking the measurement in the energy centre of the signal generated by the OSS.

Figure 14. ECLM graphic representation

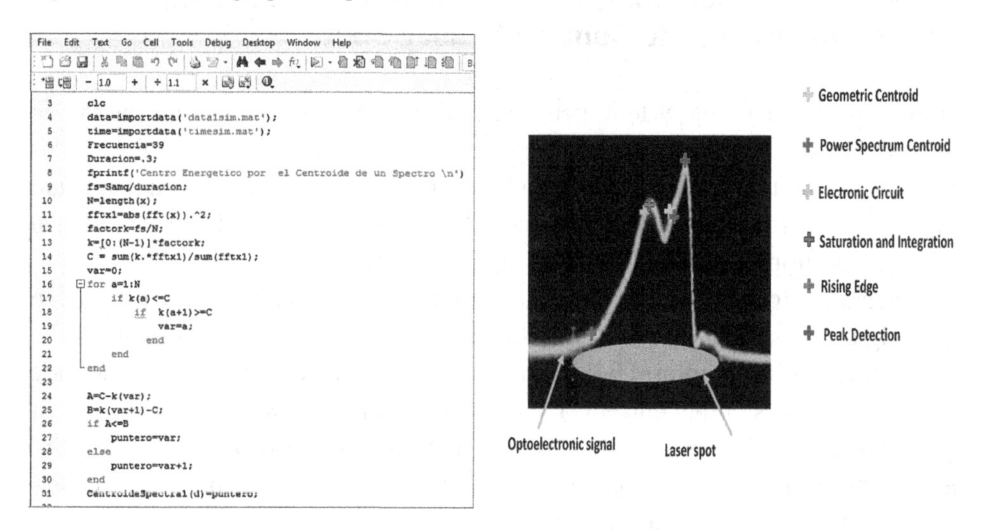

According with the results from previous assessment made with incoherent light source the most accurately ECLM are the Geometric Centroid and the Power spectrum centroid, by this reason the Geometric Centroid ECLM was selected to perform the energetic centre localization of a laser spot to measure laser beam deviation.

2. **Energetic Center of Laser Image Signal Detected by CCD:** The CCD is the sensor who gives place to the close-range photogrammetry (Franca). Photogrammetry gives information about the geometry of an object or surface through the use of photographs, provides less dense surface information but with high quality, especially along object space discontinuities, edges and borders, it can be defined as the automation of information extraction from digital images, based on image analysis methods.

A camera is located parallel to the OSS photodiode to obtain a second source of information regarding angle β_1, by using a CCD sensor to acquire an image of the FOV at the same time as the OSS photodiode. The use of two methods has been proposed, one focused in the determination of edges and the shape centroid and other focused in the density of points measured on the surface of laser reflection. Both of them calculate the energetic center (image matrix centroid) applying morphological operations. Mathematical morphology is a tool for extracting image components that are useful in representation and description of region shape, it works directly on spatial domain, is a non-linear approach for detection of edges, by image transformations based on simple expanding and shrinking operations.

Edge detection identifies and locates discontinuities in an image. An edge may be regarded as boundary between two dissimilar regions in an image, classical methods of edge detection involve the convolution of an image with an operator (like Canny and Prewitt), which is developed to be sensitive to gradients in the image and returning values of zero in uniform areas.

In the method focused in the determination of edge and the shape centroid, a laser light beam is projected in a desired coordinate to create an edge, then the laser light scattered reflection is used to generate a minimum illumination difference (MID), to isolate only the laser light reflection from the image and to detect the edges between the two images (with/without laser light beam), a continuity the edge location is performed by an operator like Canny or Prewitt. Second, the laser light reflection energetic center is localized by the application of several morphological operations as dilation, erosion, boundaries location, fill image regions and remove open area (Li, 2013). MATLAB code and results in Figure 15.

In the method focused in the density of points of laser reflection, the first step the image is captured, and the red color spectrum is obtained, a Gaussian blur filter (5x5 averaging matrix) is applied to smooth the captured image. By calculating the convolution of the original image and a 5x5 averaging matrix, is obtained the desired smoother image which allows to find the center of the laser spot more precisely.

Figure 15. Graphic image processing stages by method focused in the determination of edge and the shape centroid

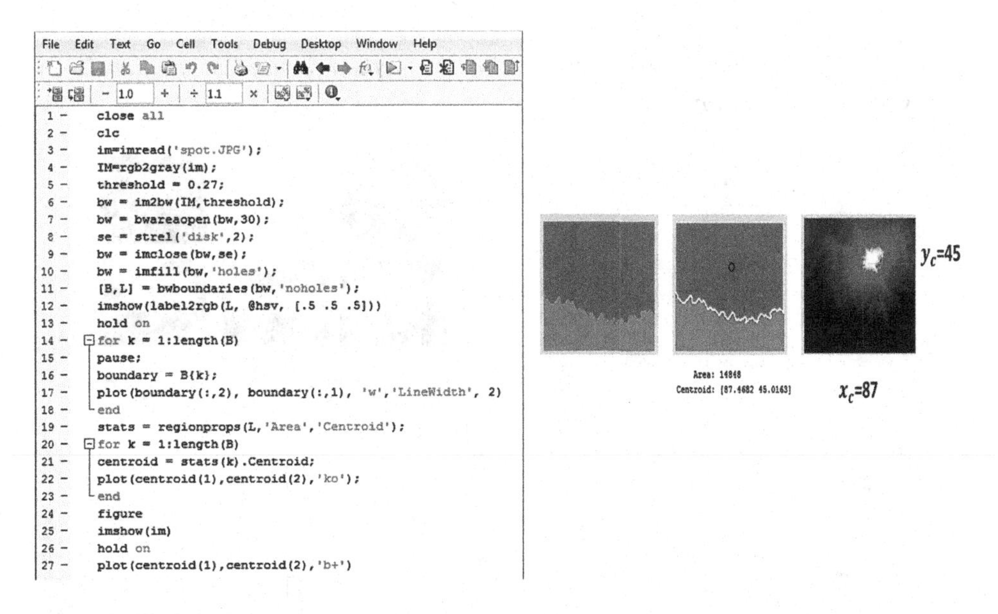

Straightforward formal implementation of the two-dimensional convolution equation in spatial form is used, calculated for finite intervals, defined by image frame limits.

After the image has been blurred the maximum values of red color are calculated for each row (x) and column (y) of the 2D matrix representing the image; the brighter the red color in the image, the higher the red value.

The final results are the x and y coordinates of the center of the laser spot inside the captured image (i.e. these coordinates instantly represented in pixels and further, converted to X and Y) (Damelin, 2012; Zhang, 2012). MATLAB code and graphical results are shown in Figure 16.

These two Energy Centre Localization Methods (ECLM) for CCD Images have not been formerly evaluated to compare them, due to the measurement quality depends on surface reflection properties and lighting conditions, the surface reflection properties are dictated by a number of factors, as a) angle of the laser ray hitting, b) surface material, and c) roughness. Although for this fusion sensor purpose, the method focused in the density of points of laser reflection has been selected to be used during the experimentation.

Photodiode and CCD Fusion for Machine Vision

The photodiode and CCD sensors fusion is the base for those systems with non-lineal behavior that require the use of information from various sensors for pattern

Figure 16. Graphic image processing stages by method focused in the density of points of laser reflection

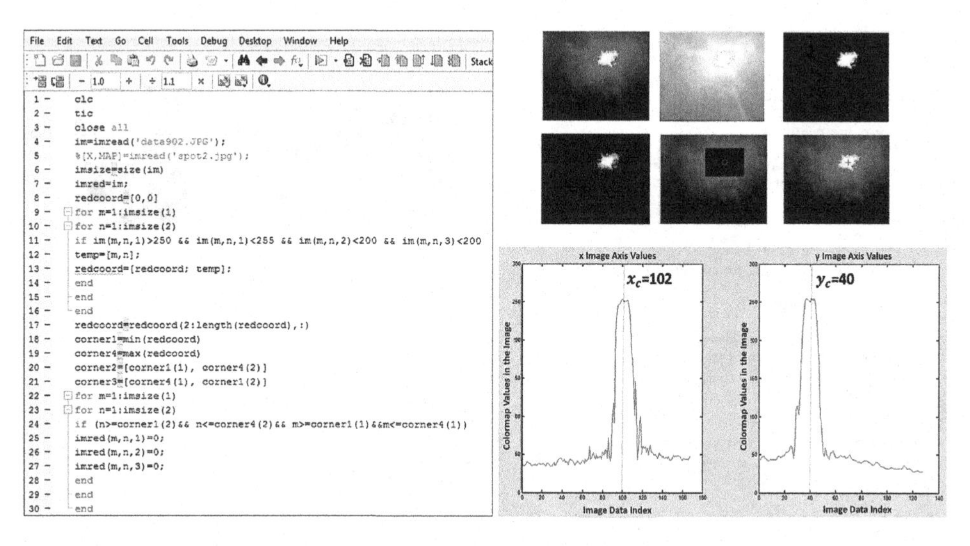

recognition to produce an enhanced output. The goal of using data fusion in multisensor environments is to obtain a lower detection error probability and a higher reliability by using data from multiple distributed sources. Artificial intelligence methods for sensor and data fusion require human reasoning such as pattern recognition, planning, deduction and learning. The processes of inference used by expert systems begin with a group of initial data (data prior) and some basic rules. Machine Learning algorithms are investigated to fuse the sensors data, such as Support Vector Machine (SVM). The best model generated by each algorithm is called estimator. It is shown that the employment of estimators based on artificial intelligence can improve significantly the performance achieved by each sensor alone (Faceli, 2004; Banerjee, 2012; Waske, 2007).

In previous OSS experimentation angle β_1 measurements with photodiode has been done through an optical table, to observe the measurement error through the FOV, a total of 6020 measurements built the dataset that was used to train and to test an SVM algorithm in the error prediction, to measurement corrections of angle β_1, obtaining a satisfactory improvement, increasing from 53.39% to 97.60% the percentage of measurements with error lower than 2.9°. Corrected angles β_1 were used to calculate the spatial coordinates by dynamic triangulation. At the same time Navigation and Medical Scanning Systems experimentations were applying the Feed Forward back propagation neural networks (NN) Levenberg–Marquardt algorithm with Widrow-Hoff Height/bias learning function to predict the error of spatial coordinates previously calculated with measurement of angle β_1, NN provided an acceptable approximation of the measurement error up to 99.97%, however, the validation of the method shows only a 96% of performance.

For sensors and data fusion experimentation the designed process explained at Figure 17 was followed. A total of 80 measurements built the new experimentation database of 80 instances (*n* rows) and 6 variables (*p* columns), the database values corresponding for column *p*=1 are the real objective angle under measurement $\left(\beta_{1R}\right)$, for *p*=2 are the real objective distance under measurement, for *p*=3 are the system scanning motor frequency, for *p*=4 are the angle calculation from photodiode output $\left(\beta_{1photodiode}\right)$, for *p*=5 are the angle calculation from CCD output $\left(\beta_{1CCD}\right)$, and for *p*=6 are the error measurement $\left(E\right)$ calculated by (5).

$$E = \beta_{1R} - \beta_{1M} \tag{5}$$

Figure 17. Sensors and data fusion experimentation flowchart

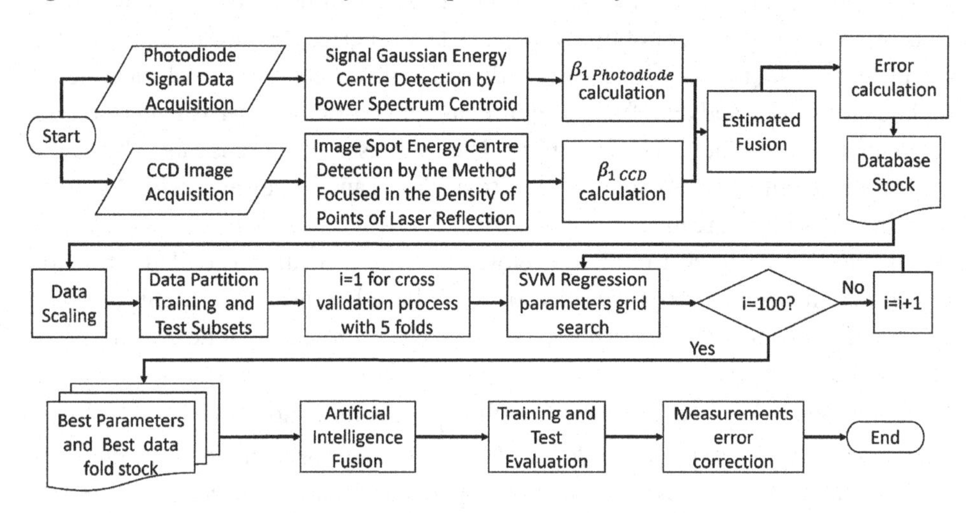

where:

E is the measurement error, representing how far the measurement from the real value is.

β_{1R} is the angle real value.

β_{1M} is the angle measured by the system (estimated fusion of photodiode and CCD by weighted average).

A scaling step on each column took place (from -1 to 1). The scaled database was partitioned in two sets, 80% training dataset and 20% test dataset, obtaining a training dataset of 64 rows with 6 columns and a test dataset of 16 rows with 6 columns. The 5 first columns are indicated as attributes and the last column as target.

In order to fusion data by artificial intelligence, a properly validate the SVM regression performance was executed, the well-known k-fold cross-validation method was executed, repeating 100 times a 5-fold and in each run performing grid search parameters, as illustrated at Figure 18.

The following kernel parameters were analyzed:

- **Kernel Type:** [Polynomial, RBF],
- **Gamma:** [2^{-5}, 2^{-4}, 2^{-3}, 2^{-2}, 2^{-1}, 2^{0}, 2^{1}, 2^{2}, 2^{3}, 2^{4}, 2^{5}],
- **C:** [2^{-5}, 2^{-4}, 2^{-3}, 2^{-2}, 2^{-1}, 2^{0}, 2^{1}, 2^{2}, 2^{3}, 2^{4}, 2^{5}],
- **NU:** [0.015, 0.1, 0.2, 0.3, 0.4, 0.5, 0.6, 0.7, 0.8, 0.9],
- **Polynomial Degree in Polynomial Kernel:** [1, 2, 3].

Figure 18. Flowchart critical steps illustration

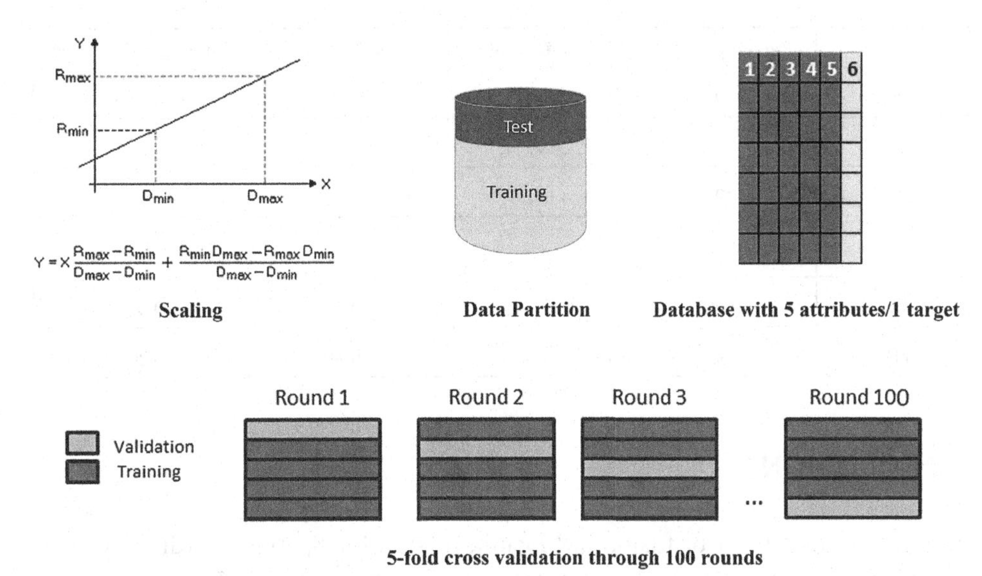

EXPERIMENTAL RESULTS

By implementing the method of finding the energetic center of the laser spot by CCD, a second value of the β_1 angle was obtained, the average of that value and the previously corrected angle by artificial intelligence algorithm generated a new value for the β_1 angle, being the first fusion sensor experimentation.

Since the first experimentations for β_1 angle error correction included the known values of frequency (scanner rotation), angle and distance of the scanning aperture (SA) and the angle of incidence β_1 of the laser light on the phototransistor as input data (attributes) for the intelligence algorithm, this new experimentation, utilizes the intelligence algorithm SVMR as a fusion sensor technique, to include as attributes for the training process the β_1 angle measured for both sensors, photodiode and CCD in order to obtain the best sensors characteristics fusion. Showing results illustrated on Table 1.

The SVMR algorithm has demostrated been an efficient intelligent tool for prediction, the change of attributes for the target prediction to include the CCD sensor, has improved the β_1 angle measurement accuracy (with error lower than 1°), for the spatial coordinate dynamic triangulation calculation.

Table 1. Photodiode and CCD fusion sensor

Target	Photodiode Result	CCD Result	SVM Prediction	Fusion Result Error
90°	89.4°	91.1°	90.35°	0.35°
95°	98.1°	99.9	95.60°	0.60°
100°	99.4°	101.0°	100.08°	0.08°
⋮	⋮	⋮	⋮	⋮
105°	103.2°	104.0°	104.60°	0.40°
110°	108.0°	108.0°	109.80°	0.20°

CONCLUSION

The goal of using sensor fusion in Machine Vision for Spatial Coordinates Measurement is to obtain a lower measurement error probability and a higher reliability by using data from multiple distributed sources (signal energetic center of sensors, previously corrected by artificial intelligence). This method implements the use of a second value of a measured variable to be used as an atributte in SVMR prediction, which directly affects the overall system accuracy and consequently the monitoring, navigation and scanning systems performance.

Due to the measurement quality depends on surface reflection properties and lighting conditions, the surface reflection properties are dictated by angle of the laser ray hitting, and surface roughness material, further experimentation will be realized scanning over different surface materials like wood, skin, paper, paperboard, egg carton, etc, to build a database to train the system to be able to operate in almost any environment, installed in an optimal hardware architecture in order to further increase the system accuracy and expand the optical scanning applications.

The author's project consists of three main research stages (Laboratory, Outdoor, and Field). The design of experiment and measurements in this chapter describes the results of the Laboratory research stage, and takes them as background for Outdoor research stage, where the fusion of 2D/3D Image and Range Scanner Datasets will be done with data from real environment measurements, while the Field research stage summarize future challenges for accomplish the project scope of SHM for heritage buildings in Autonomous University of Baja California at Mexicali city. In the Laboratory Research Stage, we started with the build of the one OSS for 3D measurements, the use of one OSS provides of polar coordinates in a plane, the use of two OSS separated at known distances, provide of a spatial coordinate in a plane by dynamic triangulation calculations, while the use of three OSS provides the full scope of 3D spatial coordinates measurement. In this stage one OSS was built and characterized at laboratory environment. During tests has been observed that it presents a non-linear behavior throughout all scanning scene, and in order to obtain an accuracy increase the requirement of computational intelligence techniques support surged, for pattern recognition to produce an enhanced output in a modern approach to the prediction of the non-linear measurement error of the OSS to compensate it. The outcome of our project has been a prototypal system for supporting an expert user in conducting structural analyses on building by extracting spatial coordinate information from light emitters mounted over a structure as indicator, in order to evaluate any structural displacement. Users may thus monitor the state of previous interventions, reason about the stability and integrity of the structure, identify potential risks, and plan reinforcement activities accordingly. Besides the specific scenario considered here, such a system may find application in several areas, as for instance in a museum scenario, bridge, damp, industrial structures or any other historical building monitoring projects. Concluding that this chapter outlines the proposed strategy by Artificial Intelligent Algorithms, and the capability to understand images and range scanner data through their fusion.

Future challenges for accomplish the project scope of SHM for buildings are:

- Incorporate a third aperture to the OSS, the use of three apertures in the OSS provides the full scope of 3D spatial coordinates measurement.
- Perform real time measurements by the integration of the signal processing stages (acquisition, energy signal centre calculations, error compensations, fusion and triangulation).
- Integrate the OSS signal processing by hardware, such as Field Programmed Gate Array (FPGA) to satisfy the monitoring system requirements to continue their tasks with safety and high accuracy on temporary occlusion condition.
- Include adaptive elements to the signal processing task in order to enhance the system to be capable of performance at an unknown environment.
- Implement a communication system for measurements record by wireless.
- Evaluate the requirement of time processing reduction.
- Reduce the energy consumption of the OSS.
- Perform measurement analysis to identify structures natural patterns from damages.

REFERENCES

Abidi, M. A. (1992). *Data fusion in robotics and machine intelligence*. Academic Press Professional, Inc.

Banerjee, T. P., & Das, S. (2012). Multi-sensor data fusion using support vector machine for motor fault detection. *Information Sciences*, *217*, 96–107. doi:10.1016/j.ins.2012.06.016

Basaca-Preciado, L. C.-Q.-L.-B., Sergiyenko, O. Y., Rodríguez-Quinonez, J. C., García, X., Tyrsa, V. V., Rivas-Lopez, M., & Starostenko, O. et al. (2014). Optical 3D laser measurement system for navigation of autonomous mobile robot. *Optics and Lasers in Engineering*, *54*, 159–169. doi:10.1016/j.optlaseng.2013.08.005

Beiser, L. (1995). Fundamental architecture of optical scanning systems. *Applied Optics*, *31*(34), 7307–7317. doi:10.1364/AO.34.007307 PMID:21060601

Castanedo, F. (2013). A review of data fusion techniques. *The Scientific World Journal*, 1–19. PMID:24288502

Dagenais, M. a. (1995). *Integrated optoelectronics*. Academic Press.

Damelin, S. B. (2012). *The mathematics of signal processing*. Cambridge University Press.

Dasarathy, B. V. (1997). Sensor fusion potential exploitation-innovative architectures and illustrative applications. *Proceedings of the IEEE*, *1*(85), 24–38. doi:10.1109/5.554206

Elfes, A. (1992). Multi-source spatial data fusion using Bayesian reasoning. *Data fusion in robotics and machine intelligence*, 137-163.

Faceli, K., de Carvalho, A. C. P. L. F., & Rezende, S. O. (2004). Combining intelligent techniques for sensor fusion. *Applied Intelligence*, *3*(20), 199–213. doi:10.1023/B:APIN.0000021413.05467.20

Flores-Fuentes, W. L.-N.-C.-B.-Q., Rivas-Lopez, M., Sergiyenko, O., Gonzalez-Navarro, F. F., Rivera-Castillo, J., Hernandez-Balbuena, D., & Rodríguez-Quiñonez, J. C. (2014a). Combined application of power spectrum centroid and support vector machines for measurement improvement in optical scanning systems. *Signal Processing*, *98*, 37–51. doi:10.1016/j.sigpro.2013.11.008

Flores-Fuentes, W. L.-Q.-B.-C., Rivas-Lopez, M., Sergiyenko, O., Rodriguez-Quinonez, J. C., Hernandez-Balbuena, D., & Rivera-Castillo, J. (2014b). Energy Center Detection in Light Scanning Sensors for Structural Health Monitoring Accuracy Enhancement. *Sensors Journal, IEEE*, *7*(14), 2355–2361. doi:10.1109/JSEN.2014.2310224

Franca, J. G. (2005). A 3D scanning system based on laser triangulation and variable field of view. *Image Processing,2005. ICIP 2005. IEEE International Conference on*. IEEE.

Fukuda, M. (1999). *Optical semiconductor devices*. John Wiley and Sons.

Gonzalez-Aguilera, D. L.-G. (2010). *Camera and laser robust integration in engineering and architecture applications*. INTECH Open Access Publisher. doi:10.5772/9959

Hall, D. L., & Llinas, J. (1997). An introduction to multisensor data fusion. *Proceedings of the IEEE, 1*(85), 6–23. doi:10.1109/5.554205

Hou, B. (2011). Charge-coupled devices combined with centroid algorithm for laser beam deviation measurements compared to a position-sensitive device. *Optical Engineering (Redondo Beach, Calif.), 50*(3), 033603–033603. doi:10.1117/1.3554379

Janesick, J. R. (2001). *Scientific charge-coupled devices*. SPIE Press.

Klein, L. A. (2003). Sensor and data fusion: a tool for information assessment and decision making. Bellingham, WA: SPIE Press.

Labayrade, R., Royere, C., Gruyer, D., & Aubert, D. (2005). Cooperative fusion for multi-obstacles detection with use of stereovision and laser scanner. *Autonomous Robots, 2*(19), 117–140. doi:10.1007/s10514-005-0611-7

Li, F. a. (2006). *CCD image sensors in deep-ultraviolet: degradation behavior and damage mechanisms*. Springer Science and Business Media.

Li, F. M. (2003). Degradation behavior and damage mechanisms of CCD image sensor with deep-UV laser radiation. *Electron Devices. IEEE Transactions on, 12*(51), 2229–2236.

Li, L., Ma, G., & Du, X. (2013). Edge detection in potential-field data by enhanced mathematical morphology filter. *Pure and Applied Geophysics, 4*(179), 645–653. doi:10.1007/s00024-012-0545-x

Li, X., Zhao, H., Liu, Y., Jiang, H., & Bian, Y. (2014). Laser scanning based three dimensional measurement of vegetation canopy structure. *Optics and Lasers in Engineering, 54*, 152–158. doi:10.1016/j.optlaseng.2013.08.010

Lorenser, D. A. (2003). Towards wafer-scale integration of high repetition rate passively mode-locked surface-emitting semiconductor lasers. *Applied Physics. B, Lasers and Optics, 8*(79), 927–932.

Luo, R. C.-C., & Yih, . (2002). Multisensor fusion and integration: Approaches, applications, and future research directions. *Sensors Journal, IEEE*, *2*(2), 107–119. doi:10.1109/JSEN.2002.1000251

Marston, R. M. (1999). *Optoelectronics circuits manual*. Butterworth-Heinemann.

McClure, W. F. (2003). 204 years of near infrared technology: 1800-2003. *Journal of Near Infrared Spectroscopy*, *6*(11), 487–518. doi:10.1255/jnirs.399

Piprek, J. (2003). *Semiconductor optoelectronic devices: introduction to physics and simulation*. Academic Press.

Rivas-Lopez, M. C.-F.-Q.-B.-B. (2014). Scanning for light detection and Energy Centre Localization Methods assesment in vision systems for SHM. *2014 IEEE 23rd International Symposium on* (pp. 1955-1960). Industrial Electronics (ISIE).

Rodriguez-Quiñonez, J. B.-L.-F.-P. (2014). Improve 3D laser scanner measurements accuracy using a FFBP neural network with Widrow-Hoff weight/bias learning function. *Opto-Electronics Review*, *4*(22), 224–235.

Rodriguez-Quiñonez, J. C., Sergiyenko, O. Y., Preciado, L. C. B., Tyrsa, V. V., Gurko, A. G., Podrygalo, M. A., & Balbuena, D. H. et al. (2014). Optical monitoring of scoliosis by 3D medical laser scanner. *Optics and Lasers in Engineering*, *54*, 175–186. doi:10.1016/j.optlaseng.2013.07.026

Rogalski, A. (2002). Infrared detectors: An overview. *Infrared Physics & Technology*, *3*(43), 187–210. doi:10.1016/S1350-4495(02)00140-8

Shih, H.-C. P. (2015). High-resolution gravity and geoid models in Tahiti obtained from new airborne and land gravity observations: Data fusion by spectral combination. *Earth, Planets, and Space*, *1*(67), 1–16.

Snyder, W. E. (2010). *Machine vision*. Cambridge University Press. doi:10.1017/CBO9781139168229

Sonka, M. A. (2014). *Image processing, analysis, and machine vision*. Cengage Learning.

Syms, R. R. (1992). *Optical guided waves and devices*. McGraw-Hill.

Tavernier, F. A. (2011). *High-speed optical receivers with integrated photodiode in nanoscale CMOS*. Springer Science and Business Media. doi:10.1007/978-1-4419-9925-2

Tribolet, P. A. (2008). Advanced HgCdTe technologies and dual-band developments.*SPIE Defense and Security Symposium*. International Society for Optics and Photonics.

Vauhkonen, J. a. (2014). Tree species recognition based on airborne laser scanning and complementary data sources. In Forestry applications of airborne laser scanning (pp. 135-156). Springer. doi:10.1007/978-94-017-8663-8_7

Waske, B. a. (2007). Fusion of support vector machines for classification of multisensor data. *Geoscience and Remote Sensing. IEEE Transactions on, 122*(45), 3858–3866.

Weckenmann, A. D.-R., Jiang, X., Sommer, K.-D., Neuschaefer-Rube, U., Seewig, J., Shaw, L., & Estler, T. (2009). Multisensor data fusion in dimensional metrology. *CIRP Annals-Manufacturing Technology, 2*(58), 701–721. doi:10.1016/j.cirp.2009.09.008

Yallup, K. A. (2014). *Technologies for smart sensors and sensor fusion*. CRC Press.

Zhang, F.-M. A.-H.-H. (2008). Multiple sensor fusion in large scale measurement. *Optics and Precision Engineering*, (7), 18.

Zhang, Q., Su, X., Xiang, L., & Sun, X. (2012). 3-D shape measurement based on complementary Gray-code light. *Optics and Lasers in Engineering, 4*(50), 574–579. doi:10.1016/j.optlaseng.2011.06.024

Chapter 2
CMOS Image Sensor:
Analog and Mixed-Signal Circuits

Arjuna Marzuki
Universiti Sains Malaysia, Malaysia

ABSTRACT

In the image sensing optical receiver, such as CMOS Image Sensors (CIS), in general, comprise of photodiode and analog-mixed -signal circuits to amplify small photocurrent into digital signals. The CMOS image sensors are now the technology of choice for most imaging applications, such as digital video cameras, scanner and numerous other. Even though their sensitivity does not reach the one of the best actual CCD's (whose fill factor is about 100%), they are now commonly used because of their multiple functionalities and their easy fabrication. CMOS analog-mixed -signal circuits play a very important role for CMOS image sensors. This chapter will present the recent works on the CMOS analog-mixed -signal circuit for image sensing application, particularly for device related to machine vision application. A review of designing aspect of CIS are done. A proposed light integrating CIS was developed and measured.

INTRODUCTION

The application of CMOS image sensor (CIS) for image sensing device requires a solid understanding of the CMOS image sensor itself (Edmund, 2014). While the understanding of the analog-mixed- signal circuits (basic photodiode, amplifier, ADC, I/O and power supply) are essentially important for the use (and design) the

DOI: 10.4018/978-1-5225-0632-4.ch002

CMOS image sensor. With these understandings, it will help in the design of an image sensing device such as a scanner (Yoshida, 2005). There is so far no detail study on process selection advantages and disadvantages from the mixed-signal circuit design or CMOS Image sensor perspective. There is also so far no detail study on the methodology/approach in designing the CMOS image sensor. Latest good review on the CMOS image sensor is more than 8 years ago (Bigas, Cabruja, Forest & Salvi, 2006). The latest basic analog circuit for CIS was reviewed about 4 years ago (Marzuki, Abd. Aziz, & Abd. Manaf, 2011).

The objective of this chapter:

- To explain some of the most important parameters such as noise, dynamic range, sensitivity and resolution in relation to the analog-mixed- signal circuit design.
- Specification of CMOS Image sensor IC in the market.
- To explain technology option/type of CMOS process.
- To explain some of design methodologies used in the designing the CMOS Image sensor.
- To update some of well-known circuit technique/architecture.
- To contribute a new concept of light integrating which is suitable for CMOS image sensor.
- To highlight the future research direction.

REVIEW OF THE STATE OF THE ART OF CMOS IMAGE SENSOR

This section will review:

1. System Requirement,
2. CMOS Image sensor IC specifications,
3. Technology,
4. Design Methodology, and
5. Circuit Technique.

System Requirement

Table 1 shows the requirement of two machine visions. System 1 (Yoshida, Kamaruzzaman, Jewel, & Sajal, 2007) is about a currency counterfeit scanner while system 2 (Erlank & Steyn, 2014) is stars tracker.

Table 1. Systems requirement

	System 1-Counterfeit Scanner	System 2-Star Tracker
Processor	A microcontroller (Image processing engine embedded in the microcontroller)	ARM Cortex M3 processor (limited memory)
Capture technique	Sensor controller to grab the image frame	FPGA, SRAM Module (As frame buffer)
Programmable	Not Available	Can be programmed after assembly (via TTL/I2C/UART)
Type of power supply	Single power supply	Single 3.3 V supply
Power Consumption	Not Available	Average power consumption is < 500 mW
Processor frequency	Processor operated at 20 MHz	Processor operated at 48 MHz
Object distance	Object must be near to the sensor	Object is very far
Optical technique	Vertical angle reading, Employing lens	Large field of view (FOV), Employing lens for large FOV
Detection	Infrared carbon ink detection (need IR LED)	To detect three stars
System Accuracy	Not Available	Accuracy: 0.01 degree RMS (Root Mean Square)

The captured image in System 1 is processed by a microcontroller PIC-16F648A or ATMega88 (AVR). The microcontroller then determines the validity of the note based on an OCR technique by looking for the characters ‘B’, ‘A’ and ‘N’ in the scanned image. The success-rate of the counterfeit detection with properly captured image is 100% and the average processing time is 250 milliseconds with above mentioned microcontroller.

The System 2 was implemented as three separate printed circuit boards (PCBs) which stack on top of one another. Each PCB is approximately 3 x 4.5 cm and was designed to be as small as possible to minimize the system’s volume. The top PCB contains the Melexis image sensor (Melexis, n. d.) and acts as a mount for the lens. The middle PCB contains an FPGA and an SRAM module, which together act as a frame buffer. The bottom PCB contains an ARM Cortex M3 processor and various supporting electronics.

All image processing, star matching and attitude determination algorithms run on the processor. The processor, clocked at 48MHz, is fast enough to output attitude updates at 1Hz. The processor also monitors the current consumption of the SRAM module and performs a power cycle if abnormal currents are measured. The system could be reprogrammed in orbit to track the earth’s horizon or the moon, instead of the stars. The system is supplied by a single 3.3V supply and has I2C and TTL Serial interfaces.

Table 2. CMOS image sensor specification

Parameters	Specifications	Comments
Resolution(pixels), x	20k≤x≤500k	The resolution is quite low to medium size.
Pixel size (μmxμm)	4-5x4-6	Optimized for resolution. Large pixel provides high sensitivity.
Dark current (ADC/s)	¼ ADC/s	ADC is analog to digital converter
Sensitivity (nW/cm^2) at Signal to Noise Ratio (SNR) of 10	25	Fill factor*Quantum Efficiency (QE), QE is the ratio of photon-generated electrons that the pixel captures to the photons incident on the pixel area
Responsivity (V/lux-s)	7-40	Lux is light intensity, luminous exposure. 20 V/lux-s is considered high sensitivity (Xu et al., 2014).
Power consumption (mW)	250	Portable device should use low power consumption
Supply voltage (V)	1.8-3.3	1.8 for digital, 3.3 V for analog
Control	SPI, TTL, I2C, UART	Method to control the system
Data	Parallel, fast serial output	Basic format of data for further processing. Separate clock is required for fast output signal.
Clock (MHz)	20-50	Built in clock, PLL is optional
Frame rate (fps)	60	High speed frame is for high speed market.

For counterfeit scanner, the 'detection' mechanism (from hardware point of view) provided by the CMOS image sensor is by frame or row/column, while for star tracker is pixel by pixel checking.

CMOS Image Sensor Specification

Table 2 is a summary of CMOS image sensor ICs specification as derived from (Yoshida et al., 2007; Melexis, n. d.; Omnivision, n. d.; Bigas et al., 2006; Xu et al., 2014). The CIS chips from Omnivision and Melexis are found to be suitable for machine vision system described in Table 1.

Resolution can be described as number of pixels. While the sensitivity (as indicated in Table 2) indicates the required power to achieve SNR of 10. The Signal to Noise (SNR) which can be defined from noise and output signal is a realistic parameter which indicates the performance of CIS in an application. Noise is a dominant parameter for low illumination application. Dynamic Range is the ratio of maximum output signal over noise.

Shutter (rolling or global) feature is normally required for machine vision application, and it is available in CIS chips (Melexis, n. d.; Omnivision, n. d.) for the application. Medium *image quality* and high *frame rate* are also main requirements for the CIS (Innocent, n. d.). In summary, the requirement for the CIS is not stringent as compared to high speed digital camera. Low-medium resolution is probably is enough for a certain machine vision application.

Technology

This section describes CMOS technology, Back side Illumination (BSI) technology and photo devices which applicable to CIS design.

1. **General Comments on Technology or Process for CMOS Image Sensor:** Generally, 4 type of process are used, standard CMOS, Analog-Mixed Signal CMOS, Digital CMOS and CMOS Image Sensor process. The latter is the process developed specifically for CMOS Image sensor. There are many foundries available for the development of CMOS Image sensor. The most obvious different between this process and other process is the availability of photo devices such as pinned photodiode. Advantages of smaller dimension technology are smaller pixel, high spatial resolution and lower power consumption. A technology lower than 100 nm requires modification to fabrication process (not following the digital road map) and pixel architecture (Wong, 1997). Fundamental parameters such as leakage current (will affect the sensitivity to the light) and operation voltage (will affect dynamic range, i.e., the saturation, pinned photodiode most likely not going to work at low voltage (Wong, 1997) are very important when a process is selected for CIS development. Because of this limitation, new circuit technique is introduced
    a. Old circuit such as standard pixel circuit cannot be used when using 0.1 micron and lower (Bigas et al., 2006). This is due to topology requires high voltage; because the maximum supply voltage is now lower.
    b. Calibration circuit and cancellation circuit are normally employed to reduce noises.

In order to increase the resolution into multi mega pixel and hundreds of frame rate lower dimension technology is normally chosen. Evidently it has been reported that 0.13 micron (Takayanagi et al., 2013) and 0.18 micron (Xu et al., 2014) are good enough to achieve good imaging performance.

These modifications of CMOS process have started at 0.25 micron and below to improve their imaging characteristics –As process scaling going to much lower than 0.25 micron and below several fundamental parameters are degraded namely

Figure 1. Pixel size vs. CIS technology node

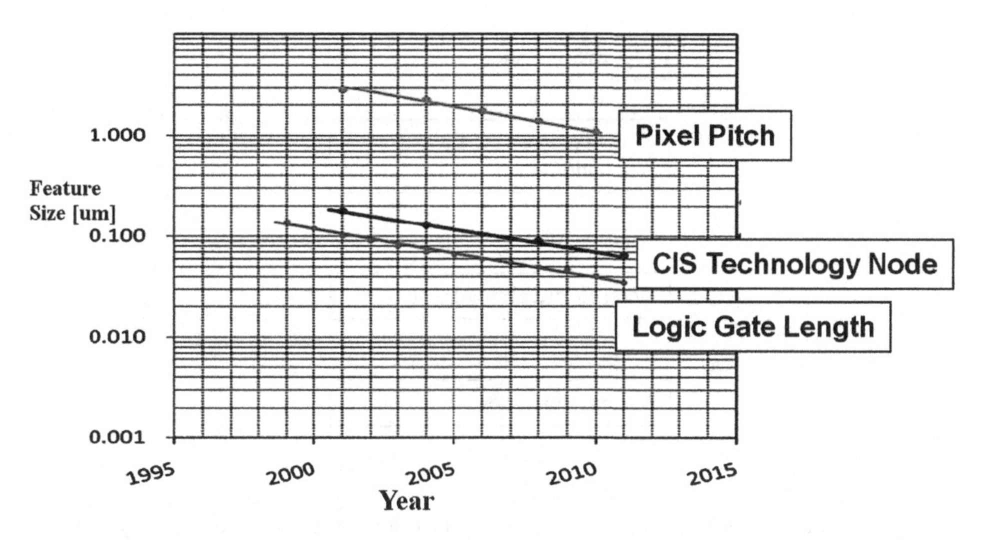

photo responsibility and dark current. Therefore, the modifications are focused to mitigate these parameters degradation (Gamal, 2002), (Bigas et al., 2006). System requirement (such as supply voltage, temperature) is also one of the criteria in selecting suitable process. Price of tool and development cost will also determine the process selection. A description of Pixel size vs. CIS Technology is shown in Figure 1(Hwang, 2012).

2. **Backside Illumination (BSI):** BSI technology eliminates the need to push light through the layers of metal interconnections. With this, high Quantum efficiency (QE) can be achieved. However, this technology incurs additional costs due to extra process, such as stacking and Through Silicon Via (TSV). Pixel of 1.1 micron seems to be the tipping point advantage over Front side illumination (FSI) (Aptina, 2010). A work in (Sukegawa, 2013) uses BSI to improve the resolution through the pixel size. As the BSI is targeted for very small pixel, process such as 90 nm is preferred, of course this is very expensive process, and thus this CMOS Image sensor is meant for expensive application (e.g. high end camera).
3. **Photo Devices:** The typical photo detector devices are photodiode and phototransistor. Typical photodiode devices are N+/Psub, P+/N_well, N_well/Psub and P+/N_well/Psub (back to back diode) (Ardeshirpour, 2004). Phototransistor devices are P+/n_well/Psub (vertical transistor), P+/N_well/P+ (Lateral transistor) and N_well/gate (tied phototransistor) (Ardeshirpour, 2004).

Figure 2. Design flow of CMOS image sensor

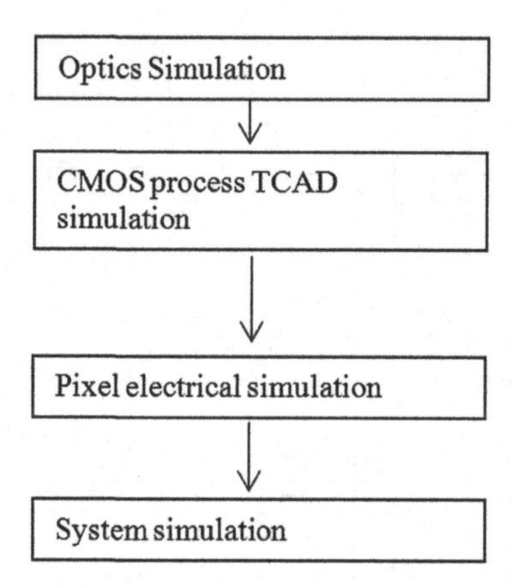

This standard photo device still requires micro lens and color filter array (CFA). Quantum efficiency (QE) of photodiodes in standard CMOS is usually below 0.3 (Scheffer, Dierickx, & Meynants, 1997).

The devices which are normally developed for modified CMOS process are photo gate, pinned photodiode and amorphous silicon diode. These devices will improve the sensitivity of CIS. Pinned photodiode which has low dark current offers good imaging characteristic for CIS (Lulé, 2000).

Design Methodology

A typical design flow of CMOS Image sensor is shown in Figure 2.

Wave propagation simulation can be done for optics simulation (Agranov, 2007). Image systems evaluation tools is available to conduct the optics simulation as well as system simulation (Farrell, Xiao, Catrysse, & Wandell, 2004). Commercially avaibale Technology Computer Aided Design (TCAD) tools such as from Synopsys and Silvaco can be used to simulate the process/technology of the photo devices. There is a work (Passeri, 2002) (mixed-mode simulation) which combine the TCAD and pixel level simulation. There are many Electronic Design Automation (EDA) tool availabe for pixel electrical simulation, these EDA tools are similar to any IC design tool such as spectre, spice, verilog-A and verilog. These tools could be time consuming if the number of pixels are large. Indeed if large pixels together with deep submicron process is required, more capital has to be provided (cost of tools are more expensive for very deep submicron, especially below 90 nm).

Even though CMOS foundry provides the models for supported design tools, sometimes designer still have to model the sub block on their own to suit the CIS specification. This can speed up the pixel electrical simulation time, however, this will degrade the accuracy. For system simulation, VHDL-AMS, System-C or matlab can be used to predict overall function and performance.

Circuit Technique

This section discusses the architecture or topology for the analog-mixed-signal circuit design and several techniques to reduce power consumption and noise.

1. **CMOS Image Sensor Architecture:** Data converter is one of major sub blocks in CIS, its function is to convert analog signal into digital signal. This approach will improve signal robustness and provide for further processing. Analog to Digital Converter (ADC) is a typical circuit used as the data converter. Other type is 'oscillator' circuit concept.
    a. **Pixel Level ADC:** Digital pixel sensor offers wide dynamic range (Trépanier, Sawan, Audet, & Coulombe, 2002). Processing can also be done at the pixel level. This concept is shown in Figure 3. The topology in Figure 3 is similar to memory architecture. Major disadvantage is low fill factor.
    b. **Column Level ADC:** This topology is shown in Figure 4. This approach offer the best performance trade-off. A Correlated doubled sampling (CDS) circuit is normally used to reduce noises (Feng, 2014). This circuit is employed prior the column ADC.
    c. **Chip Level ADC:** Chip level ADC or sometimes called matrix level ADC is depicted in Figure 5. The ADC for this topology has to be very fast (David, 1999), this topology would also consume very high current. ADC type suitable for CIS topology is pipelined ADC. However, SAR (Deguchi et al., 2013) and Flash type ADC (Loinaz, 1997) have also been reported in this type of CIS design.

Table 3 is a summary of ADC choice of location for CMOS Image sensor (CIS) (Feng, 2014). Due to shutter requirement, column level ADC is suitable for CIS architecture.

2. **ADC Type:** This section discusses major ADC architecture namely slope ADC, SAR ADC, flash ADC, pipelined DAC and Delta Sigma ADC. The dynamic range of ADC should be higher than the final CIS dynamic range.

Figure 3. Pixel level ADC CIS topology

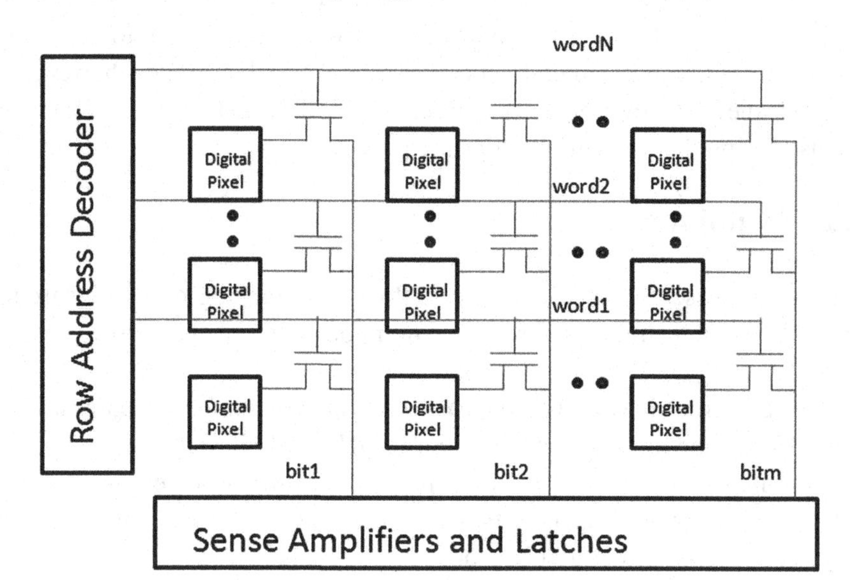

Table 3. ADC choice of location in CIS

Parameter	Pixel Level ADC	Column Level ADC	Chip Level ADC
Power	Low (Trépanier et al., 2002)	Medium (Takayanagi et al., 2013)	High
Area	Fill factor low	Area medium	High
Speed (Frame Rate)	Highest	Medium	Limited by ADC
Noise	Elimination of temporal noise (Gamal, 2002)	Medium	Low
Global Shutter	Implementable	Implementable	Not Implementable

a. **Slope ADC:** Dual slope ADC type has greater noise immunity than the other type ADCs. While single slope ADC is sensitive to the switching error. However, a single slope ADC approach was still used in BSI application due to its small area (Suzuki, 2015). In (Kleinfelder, Lim, Liu, & Gamal, 2001), slope ADC is also employed. This concept is shown in Figure 6.

b. **SAR ADC:** This is a popular ADC type for column level ADC (Takayanagi et al., 2013), also for chip level ADC (Deguchi et al., 2013). A 3-bit SAR ADC has been developed for pixel level ADC (Zhao et al., 2014). The conversion in this type of ADC always starts with MSB decision follow with LSB (Song, 2000). The technique is shown in Figure 7. An

Figure 4. Column level ADC CIS topology

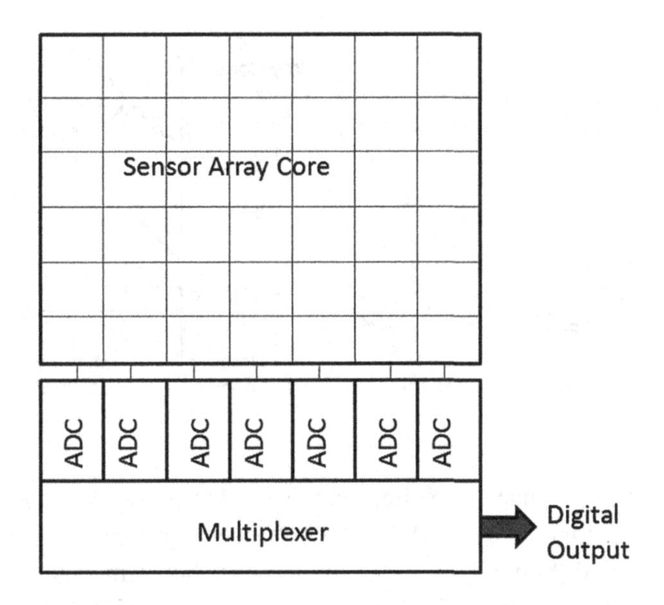

Figure 5. Chip level ADC CIS topology

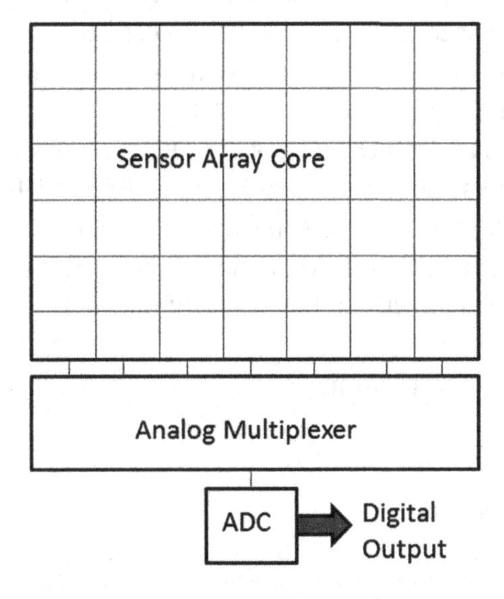

algorithm is also depicted. The SAR algorithm is important (to control DAC output with progressing dividing the range by 2) in SAR ADC.

c. **Flash ADC:** This is the straightest forward way of designing an ADC. Another improvement of flash ADC is folding ADC. The Flash ADC probably loses importance in CMOS Image sensor since it is limited to 8

Figure 6. Single slope ADC topology

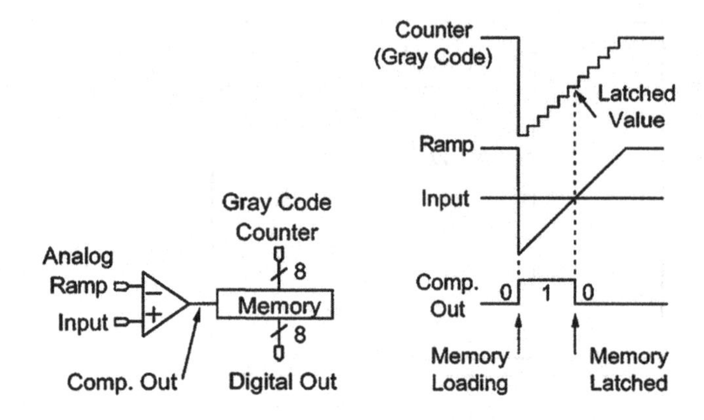

or 10 bit (Loinaz, 1998; Innocent, n. d.). However, this ADC topology is suitable for chip level ADC. The ADC technique is shown in Figure 8.

d. **Pipelined ADC:** This ADC topology as shown in Figure 9 is concept suitable for chip level ADC (Hamami, Fleshel, Yadid-pecht, & Driver, 2004). 1.5-bit stage is normally used in designing pipelined ADC (Abd Aziz & Marzuki, 2011), thus requires only trilevel DAC (Song, 2000), this is described in Figure 10. Error correction is required to produce the correct output. Further improvement to pipelined ADC topology are algorithmic, cyclic or recursive ADC.

e. **Delta Sigma ADC:** Oversampled ADCs have the advantage of filtering temporal noise (Norsworthy, Schreier, & Temes, 1997). The idea is similar to synchronous analog voltage to frequency converter. This ADC has been employed at pixel (Mahmoodi & Joseph, 2008) level and column level (Chae et al., 2011). Basic idea of the ADC is shown in Figure 11.

Table 4 shows the ADC performances.

Table 4. ADC performances

Topology	Latency	Speed	Accuracy	Area
Flash	Low	High	Low	High
SAR	Low	Low-medium	Medium-high	Low
Delta-sigma	High	Low	High	Medium
Pipeline	High	Medium-high	Medium-high	Medium
Slope	Low	Low	High	Low

Figure 7. SAR ADC technique

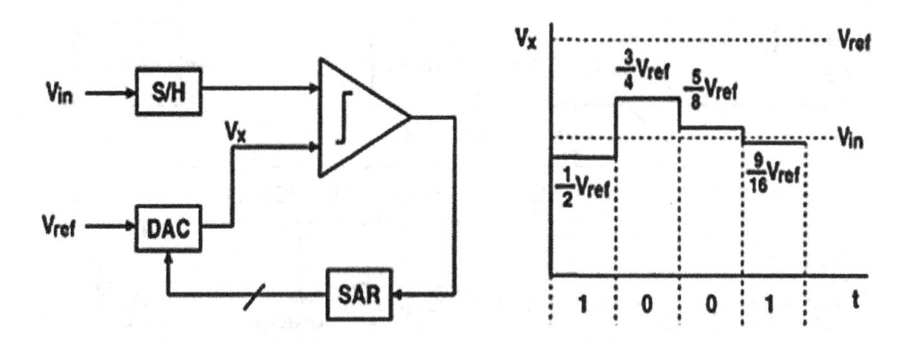

Figure 8. FLASH ADC technique

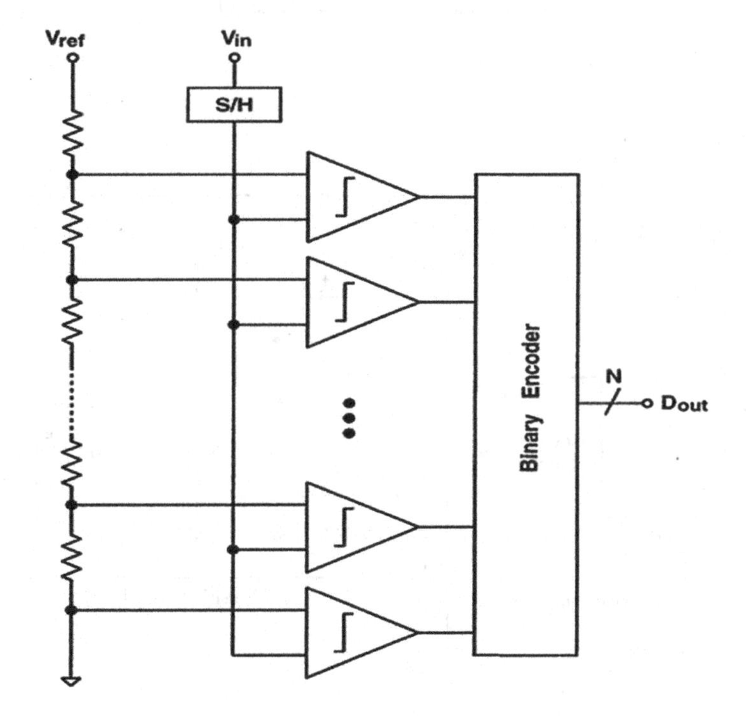

In summary, the column-level ADC topology is popular choice for CMOS image sensor due to good tradeoff between read out speed, silicon area and power consumption (Lyu, Yao, Nie, & Xu, 2014).

3. **Pixel Circuitry:** A low fixed pattern noise (LFPN) capacitive transimpedance amplifier (CTIA) for active pixel CMOS image sensors (APS) with high switchable gain and low read noise is shown in Figure 12. The LFPN CTIA

Figure 9. Pipelined ADC

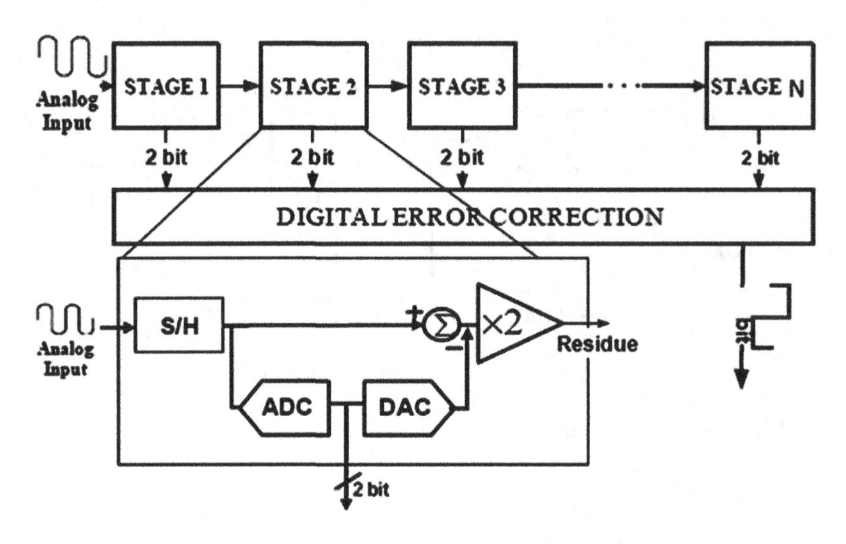

Figure 10. 1.5bit stage

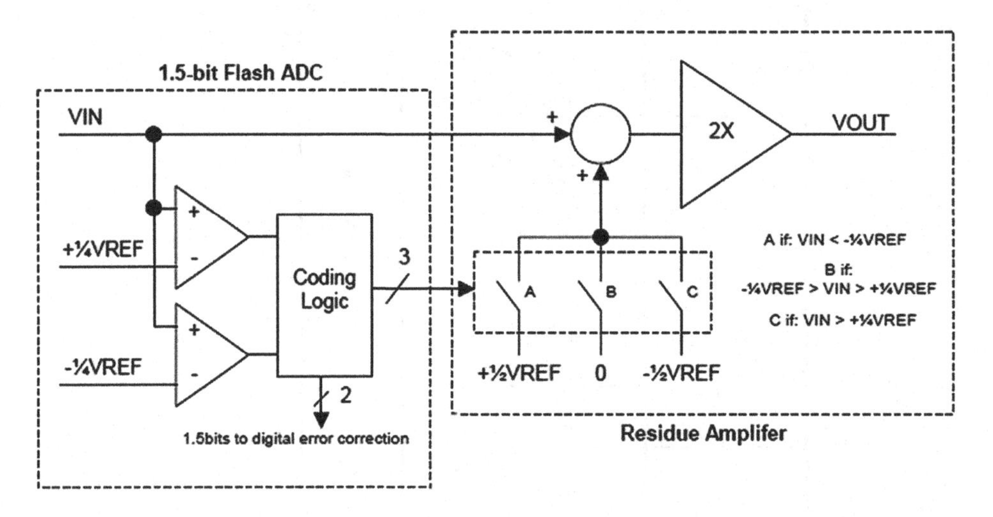

Figure 11. Delta Sigma ADC

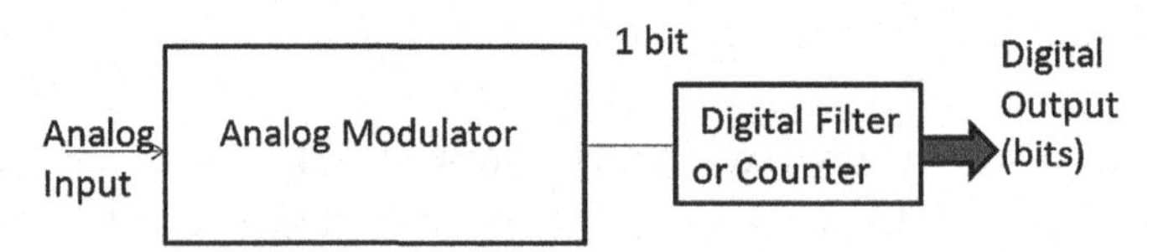

Figure 12. CTIA
Fowler, Balicki, How, & Godfrey, 2001.

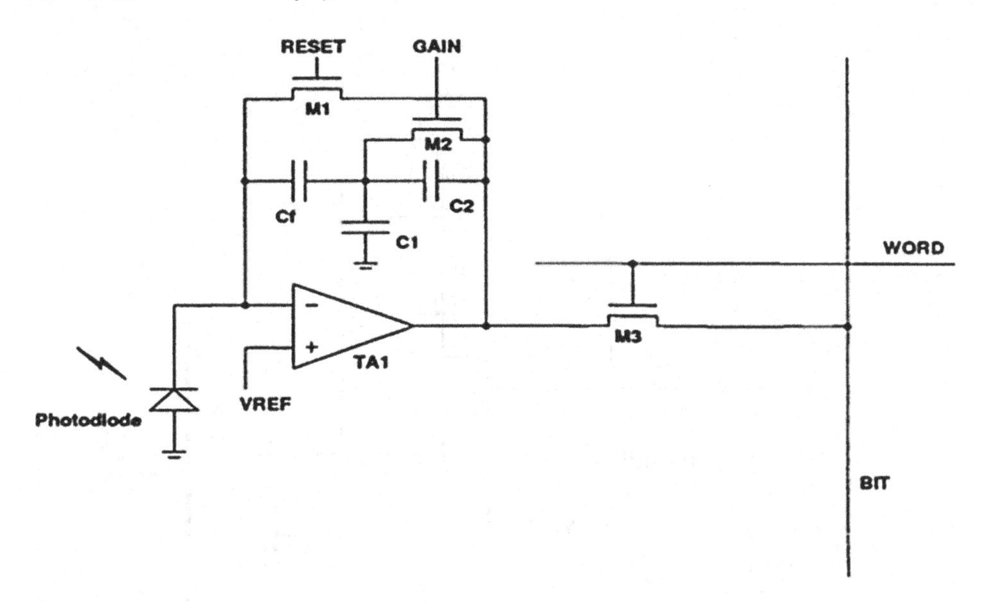

APS uses a switched capacitor voltage divider feedback circuit to achieve high *sensitivity*, low gain FPN, and low read noise. The circuit consists of a transconductance amplifier TA1, a photodiode, a network of feedback capacitors and switches (Cl, C2, Cf, Ml, M2), and a bit line select transistor M3. WORD is used to select each row of pixels, BIT is the output bus for each column in the sensor, RESET and GAIN are used to reset the pixel and control the pixel gain, and VREF is the pixel bias voltage.

Another circuit is shown in Figure 13. The integration capacitor, Cint is used as a feedback component. The photo current now is coming from Cint and Vdiode remains constant throughout the integration period.

A pixel image sensor (Massari, Gottardi, Gonzo, Stoppa, & Simoni, 2005), implementing real-time analog image processing is shown in Figure 14. The pixel consists of different functional blocks: *Sensing Block, Computing Unit, Bank of Switches, Memories*. The pixel is capable to process edge detection, motion detection, image amplification and dynamic-range boosting by means of a highly interconnected pixel architecture based on the absolute value of the difference among the neighbor pixels.

Another example of processing at pixel level is Matrix Transform Imager Architecture (Bandyopadhyay, Lee, Robucci, & Hasler, 2006). The circuit is shown in Figure 15. Photodiode is placed as current source for a differential pair; this

Figure 13. Pixel schematic
Goy, Courtois, Karam & Pressecq, 2001.

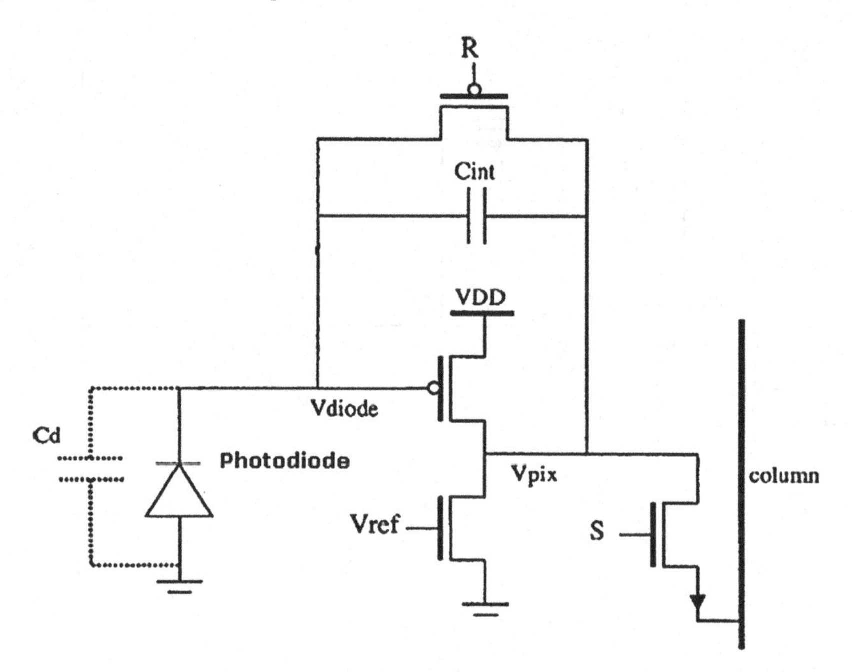

Figure 14. Pixel schematic
Massari, Gottardi, Gonzo, Stoppa, & Simoni, 2005.

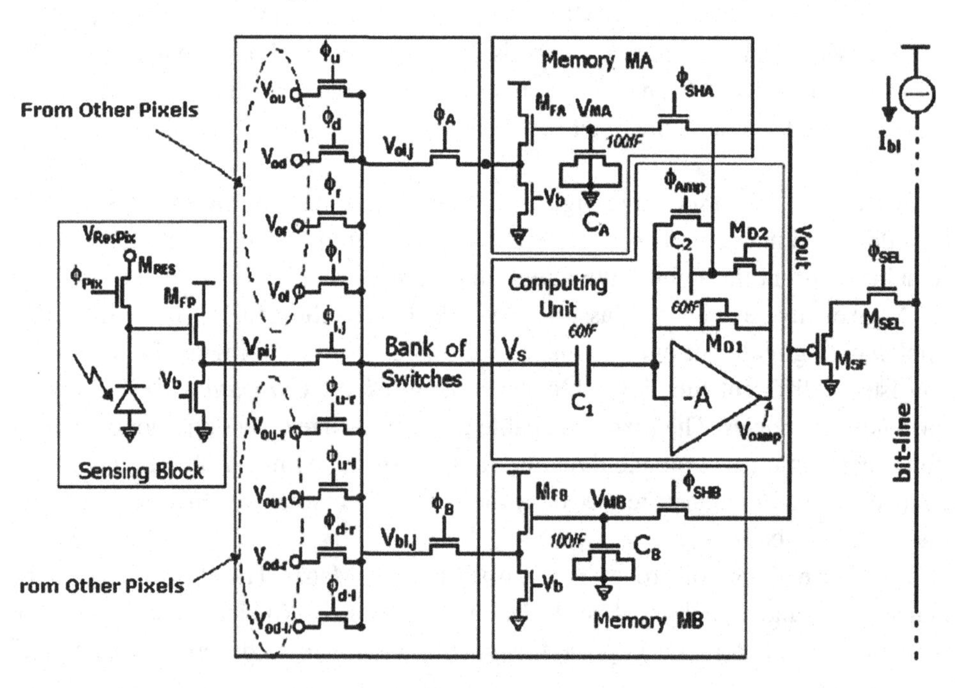

Figure 15. Multiplication of photcurrent and input voltages
Bandyopadhyay, Lee, Robucci, & Hasler, 2006.

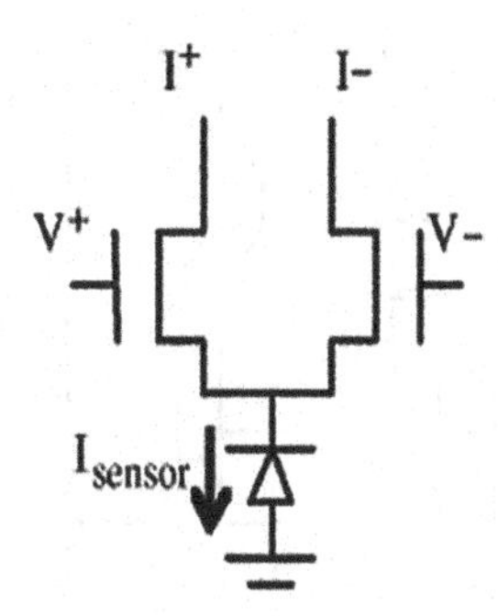

produces output currents in the form of multiplication of photocurrent and input voltages.

Digital Pixel Sensor (DPS) concept is similar to the solution used in CMOS neuron-stimulus chip. DPS in number is found useful for on-chip compression. Figure 16 shows an example of DPS (Zhang & Bermak, 2010) used for comprehensive acquisition concept. The photodiode is used to discharge the input capacitance of the comparator and photodiode itself.

A common 3T pixel with 2D metal grating structure are proposed as new color pixel (Fu, Zhan, Lin, & Wu, 2009), however it has low transmittance due to light reflection. Another solution is to Transverse Field Detector (TFD) (Langfelder, Longoni, & Zaraga, 2009), however common 3T pixel cannot be used. Figure 17

Figure 16. Analog circuit of DPS
Zhang & Bermak, 2010.

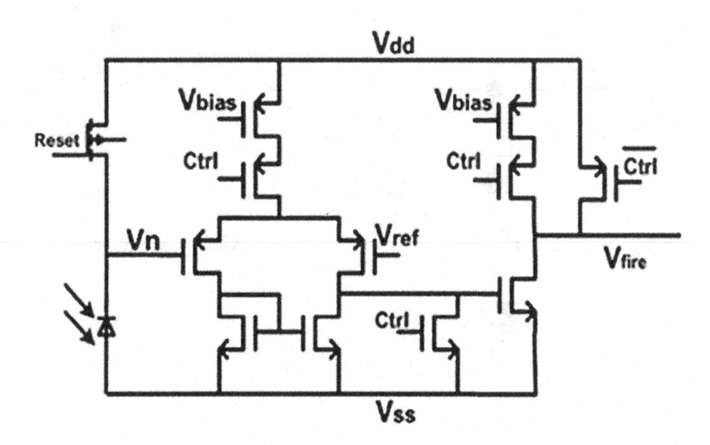

Figure 17. Pixel circuit for TFD
Langfelder, Longoni, & Zaraga, 2009.

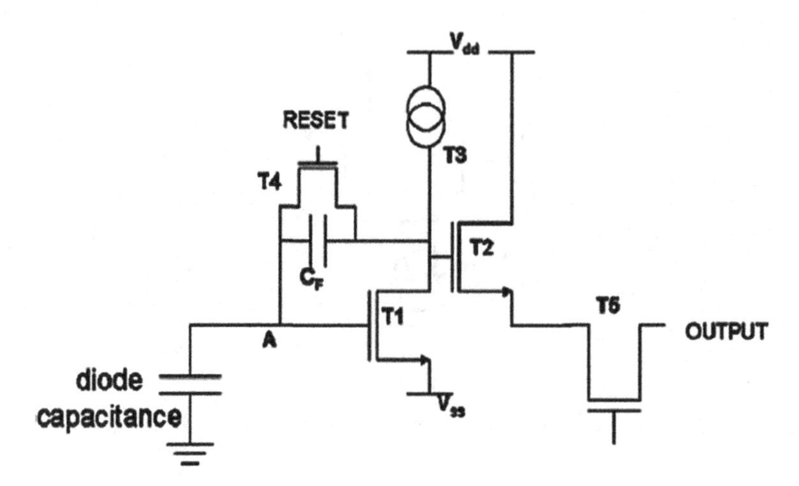

shows of suitable readout circuit for the TFD. With the circuit, there will be no voltage variation at the photodiode output.

To improve the flicker artifacts, off-chip solution can be implemented (Poplin, 2006). The solution employs a discrete Fourier transform to identify the flicker frequency. Another solution (Hurwitz et al., 2001) is to employ two chips, with automated gain control (AGC) circuit. The pixel circuit with AGC is shown in Figure 18.

Figure 18. Pixel circuit with AGC
Hurwitz et al., 2001.

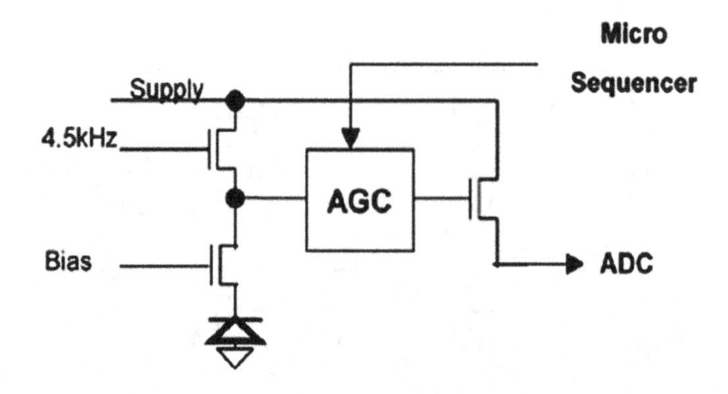

4. **Input/Output (I/O):** I/O could be very simple circuit with just protection devices (Barr, 2007). Figure 19 shows a slow analog signal which has slow rise time, is applied into TTL or CMOS input without a Schmitt trigger circuit. For a typical inverter there no clear different between threshold to indicate hi to low and vice versa (They are the same). The output of the inverter can be low or high for both conditions. Therefore, the solution is to have the circuit with different threshold voltage for low-to high and high to low. One of the solutions is the Schmitt trigger circuit (Camenzind, 2005).
   a. **Schmitt Trigger Circuit:** As shown in Figure 20, when Input is low, ML is on, output is low. Now the voltage at source of M1 and M2 is approximately equal to R2/R3. When Input is high, M1 is ON, output is high. Now the voltage at the source of M1 and M2 is approximately equal to R1/R3. Another Schmitt trigger circuit (Barr, 2007) is shown in Figure 21. P3 and N3 force output to settle/approach very fast. By vary the size of these transistors; it will lead to a hysteresis (i.e. different threshold voltage).
   b. **Bidirectional Pad:** Sharing pad (Figure 22) between input and output is common for pin or I/O limited device. If OE is low, M1 and M2 are off, thus PAD can be used as input PAD. If OE is high, M2 and M1 would work as normal output driver, thus PAD is now configured as output PAD. A Schmitt trigger circuit can be added after the basic protection circuits. Some foundries would offer very good guidance in implementing the I/O.
   c. **Serial to Parallel interface (SPI):** This circuit is normally used for data control. There should be 'synchronization' between clock, data and control/latch. Obviously for output, parallel configuration is preferred due to its fast operation. However, if the I/O is limited, high speed serial output such as low voltage differential signaling (LVDS) (Innocent, n. d.) interface can be used.

Figure 19. Slow analog signal into inverter

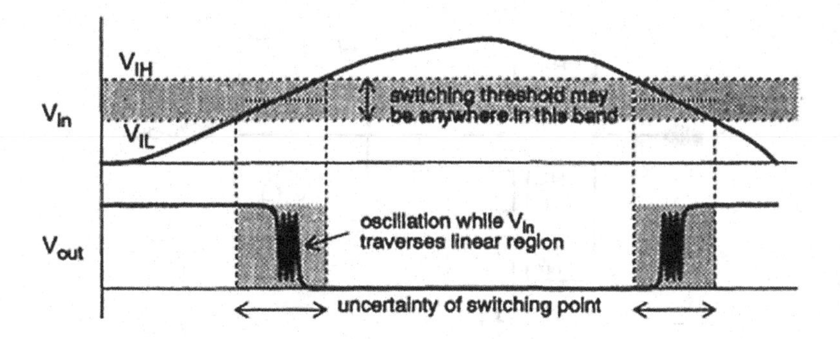

Figure 20. Simple Schmitt trigger concept

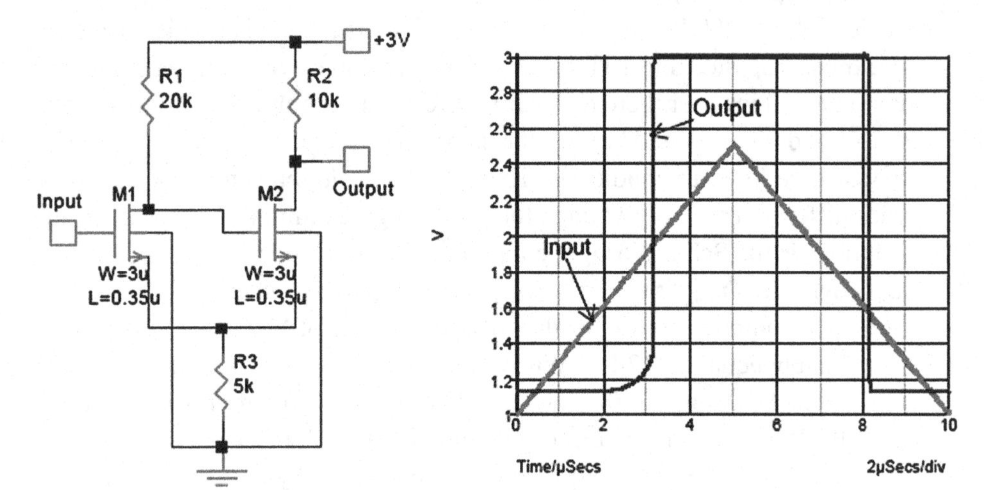

An example of 3 ‘wire’ SPI control is shown in Figure 23. Table 5 describes terminology of the SPI control.

Figure 24 shows the basic logic diagram of SPI, the first column of D flip flop is configured as shift register while the second column of flip-flop is configured as latches.

5. **Voltage Reference and Clock Generation:** Dual supply voltage is normally used in designing CMOS image sensor (Gao & Yadid-Pecht, 2012). Analog block supply voltage is normally higher than the digital supply voltage.

Figure 21. CMOS Schmitt trigger circuit concept

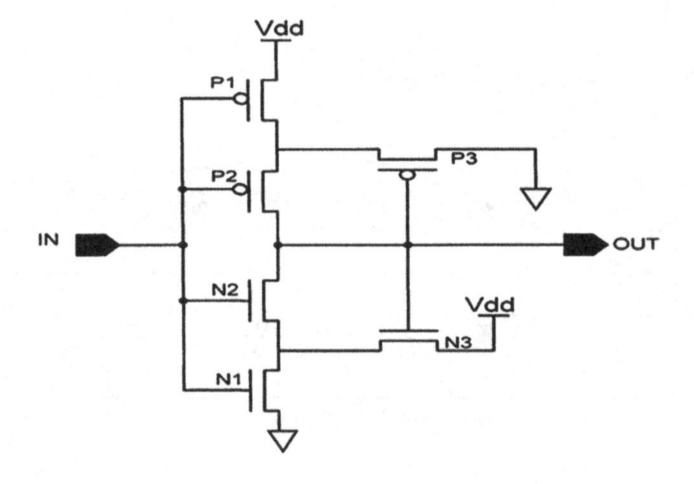

Figure 22. A three-state bidirectional output buffer

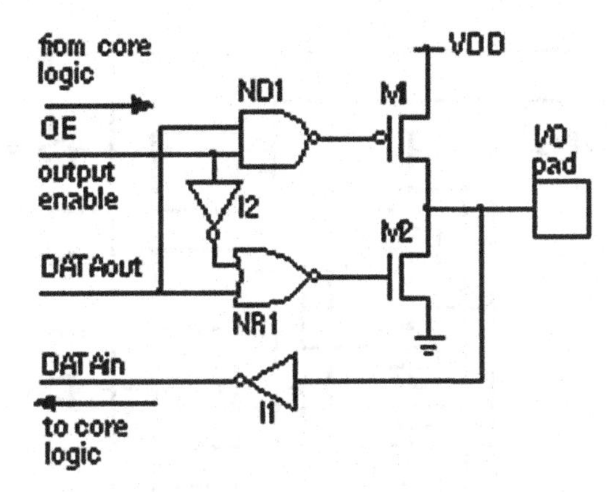

Figure 23. Setup and hold condition, and chip select (latch control)

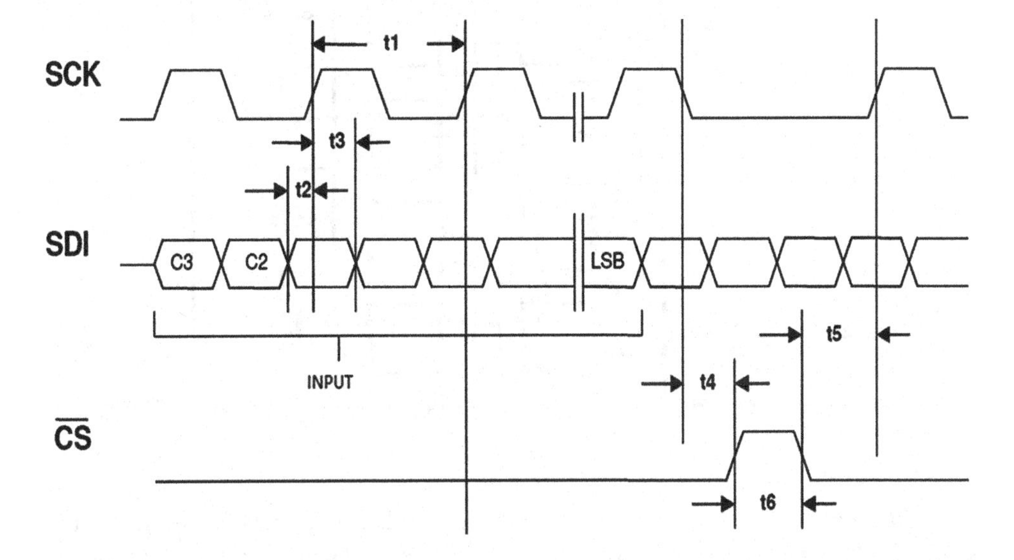

Table 5. SPI terminology

Symbol	Parameter
t1	Clock cycle time
t2	Data setup time
t3	Data Hold time
t4	SCK Falling edge to CS Rising edge
t5	CS Falling Edge to SCK Rising edge
t6	CS pulse width

Figure 24. Basic logic diagram of SPI

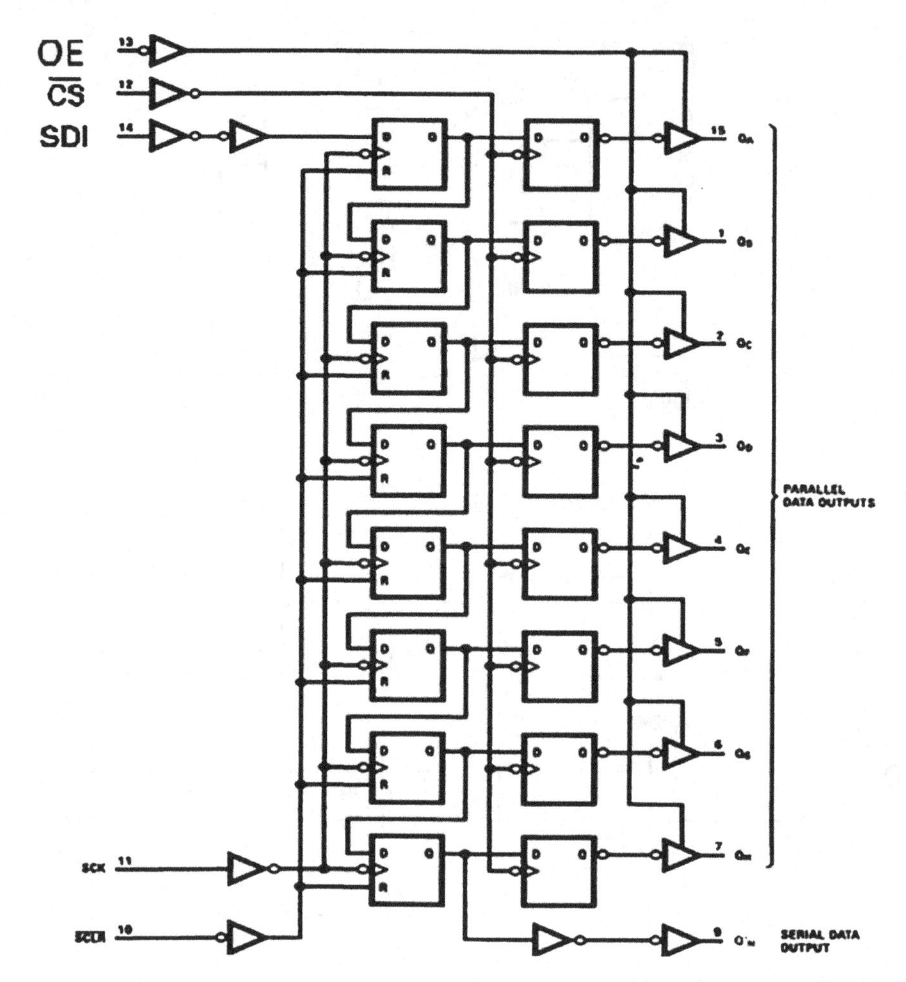

Voltage references are therefore essential (Subbiah, Suss, Kravchenko, Hosticka, & Krautschneider, 2014), a bandgap circuit is usually developed (Wen, Zhong, & Ke, 2012), as they are used for ADC reference and amplifier. There is also work on PLL (Chen et al., 2013) to provide for clock generator. Single distributed clock is employed in (Meynants, 2010). For fast output, a separate clock is employed, as in (Lindgren et al., 2005).

6. **Low Power and Low Noise Technique:** Selection of low power and low noise process or technology is a typical engineering decision in ensuring low power and low noise design respectively. The methods which can be used to lower down the power consumption of CIS can be categorized as biasing method, circuit techniques and advanced power management circuit.

In biasing method, sub threshold region or weak inversion biasing is one the approaches to achieve low current consumption (Mahmoodi & Joseph, 2008). This technique can be applied to an OTA or an amplifier for an ADC. Triode region biasing can also be used to further reduce the power consumption (Tang, Cao, & Bermak, 2010).

In circuit technique, regenerative latch (Mahmoodi & Joseph, 2008) can be used to reduce the digital power consumption. Reducing/scaling the capacitors in the pipeline stages (for ADC) can also reduce the power consumption (Cho & Gray, 1995).

In advanced power management technique, a 'smart' approach such as harvesting solar energy can also be employed to reduce the power consumption (Cevik, Huang, Yu, Yan, & Ay, 2015). The readout circuit can be selectively ON if only required. Pixels can also be periodically activated to further reduce power consumption (Zhao et al., 2014).

The noise of CIS can be reduced at different level. At pixel level, thermal noise can be reduced by CDS and oversampling. The flicker noise is reduced by using large device, periodically biasing the transistor and proper PMOS substrate voltage biasing (Yao, 2013). At column level, calibration can be used to reduce Fixed Pattern Noise (FPN) as discussed in work (Xu et al., 2014). At ADC level, kT/C noise is reduced by selecting suitable value for Cf and Cs of S/H Circuit and buffer (Hamami, Fleshel, Yadid-pecht, & Driver, 2004). At photodiode level, high conversion gain helps reduce referred-to-input noise (Wong, 1996).

SIMPLE ANALOG AND MIXED-SIGNAL CIRCUITS FOR CIS

The objective of this section is to discuss a simple analog mixed-signal circuit front-end circuit suitable for CMOS Image sensor. The approach is based on light integrating technique. A Standard 0.35µm Analog technology was chosen for this front-end circuit while a standard Cadence Analog design flow was used in the designing of these circuitries.

The proposed work could have similar idea to any CTIA based image sensor (Xui, 2014).This section is organized as follows.

1. **Design:** This section discusses the design aspect. It covers concept, RGB Photodiode, Integrator and Capacitor array, single to differential and sample/hold (S/H) buffer amplifier and sensor implementation.
   a. **Concept Overview:** The design objective is to integrate 1. Analog function (including gain stage functions), 2. RGB sensor and 3. Analog to digital converter function, into single chip solution. Figure 25 shows basic block diagram of proposed sensor with a device controller. Due to this

Figure 25. Block diagram of RGB CIS

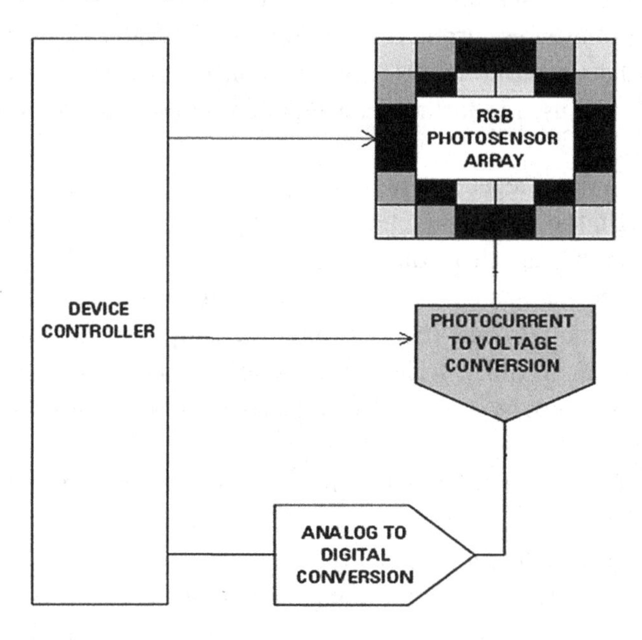

chosen architecture, a global shutter technique cannot be implemented. Figure 26 shows the block diagram of the integrated sensor with digital output employing switched capacitor circuit techniques. The sub blocks will be explained in next sub-sections. The basic concept was derived from previous work (Marzuki *et al.*, 2007). The proposed control signals for the sensor are shown in Figure 27. From Equation (1),

Figure 26. RGB photodiode and analog mixed-signal circuit block diagram

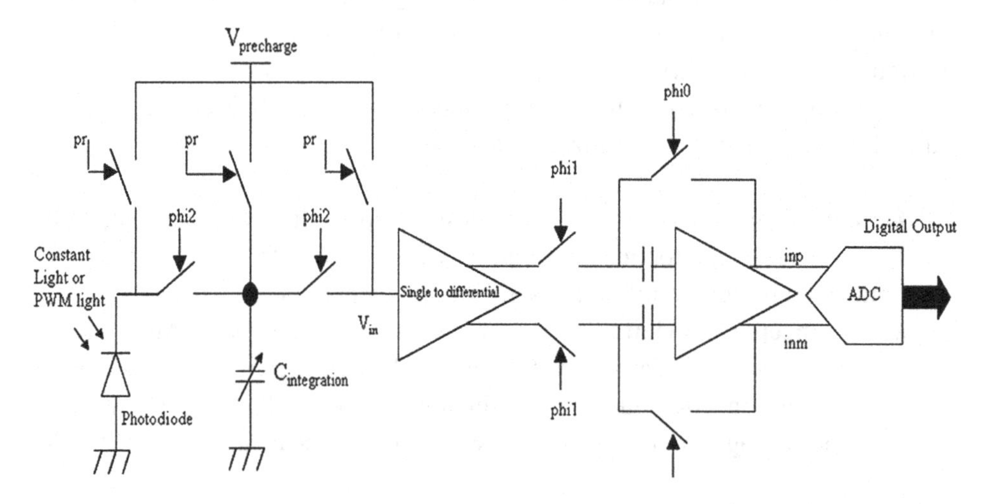

Figure 27. Control signals

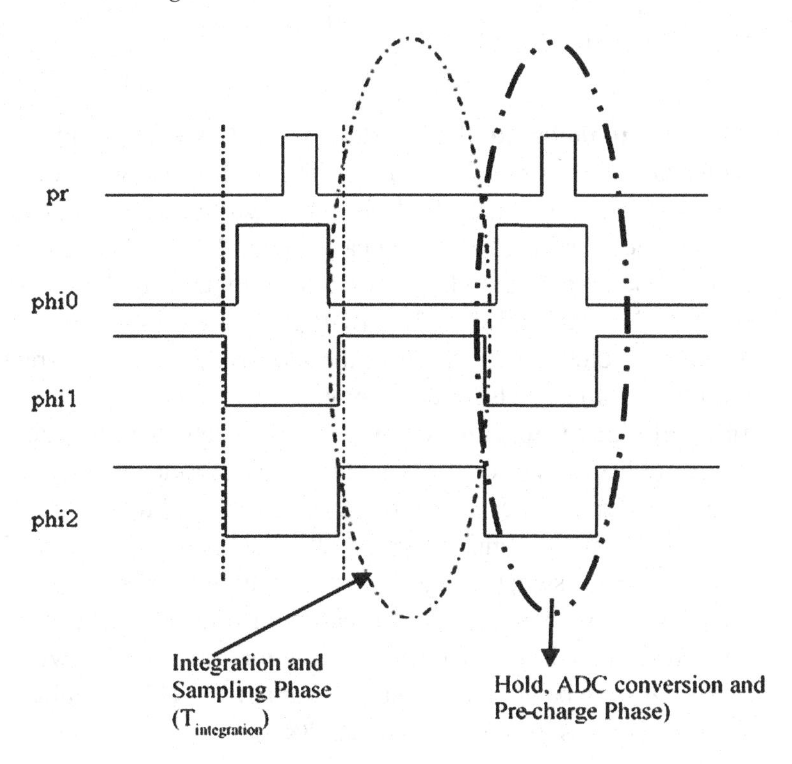

$$V_{in} = V_{precharge} - \frac{I_{photodiode} \times T_{integration}}{C_{integration}} \quad (1)$$

When the light incident on the photodiode and the integration phase (see Figure 27) is activated, then the voltage V_{in} is inversely proportional to the light intensity. A Single to differential amplifier is then used to produce differential voltages for differential input ADC, these values are later sampled by the S/H amplifier. The sampled values are hold for analog to digital conversion. At the same time, the $C_{integration}$ is charged back to $V_{precharge}$ value. A pipeline ADC with 8-bit resolution is used for analog to digital conversion. The ADC can receive ± 1.2 V, i.e. differential voltages with the nominal voltage (common mode voltage or VCM) of 1.2 V. For output of 0 DEC, the differential voltages is -1.2 V (e.g. inp = 0. 6 V, inm = 1.8 V) while for output 256 DEC (2^8), the differential voltage is 1.2 V. Equation (2) describes the relationship of voltages and ADC output. Both Equation (1) and (2) have shown the concept capability of integrating the three mentioned functions into single silicon.

$$ADC\ Output(DEC) = \left(\frac{(inp - inm) + VCM}{2VCM} \right) \times 256 \tag{2}$$

b. **RGB Photodiode:** RGB photodiodes are N-Well type photodiodes with color photo array (CFA) filters. The RGB filters were arranged in common centroid pattern. Light is converted to photocurrent by these photodiodes. An interface circuit called PhotoMux is designed to select pixels (photodiode and channels (RGB)). Figure 28 shows the RGB photodiodes. The total size of the RGB photodiodes is 400 μm×400 μm. These photodiodes will determine the sensitivity and dark current. This prototype has 12 pixels for each color.

c. **Integrator and Capacitor Array:** Figure 29 shows the integrator circuit with capacitor array of 4×8 pF. The value of the capacitance can be selected through capsel. The capacitor block is pre-charged to ~1.8V ($V_{precharge}$) when 'pr' signal is high. Before 'phi2' signal is high 'pr' signal is first low. The 'phi2' signal is high when integrating is selected. During the integration period, the pre-charged capacitor voltage starts to decrease as described by the Equation (1). The capacitor array together with photodiode sizes can be used to adjust the required V_{in}. This is similar to gain stage function as in work (Lim *et al.,* 2006).

When all capacitors are selected (32pF), if the photodiode current (I) is 125 nA (assumed) and using Equation (3), the slope or dV/dT = 3906.25. This is not much different compared to the simulated slope (3824). Figure 30 shows integrator output simulation results.

Figure 28. RGB photodiodes

Figure 29. Integrator and capacitor array

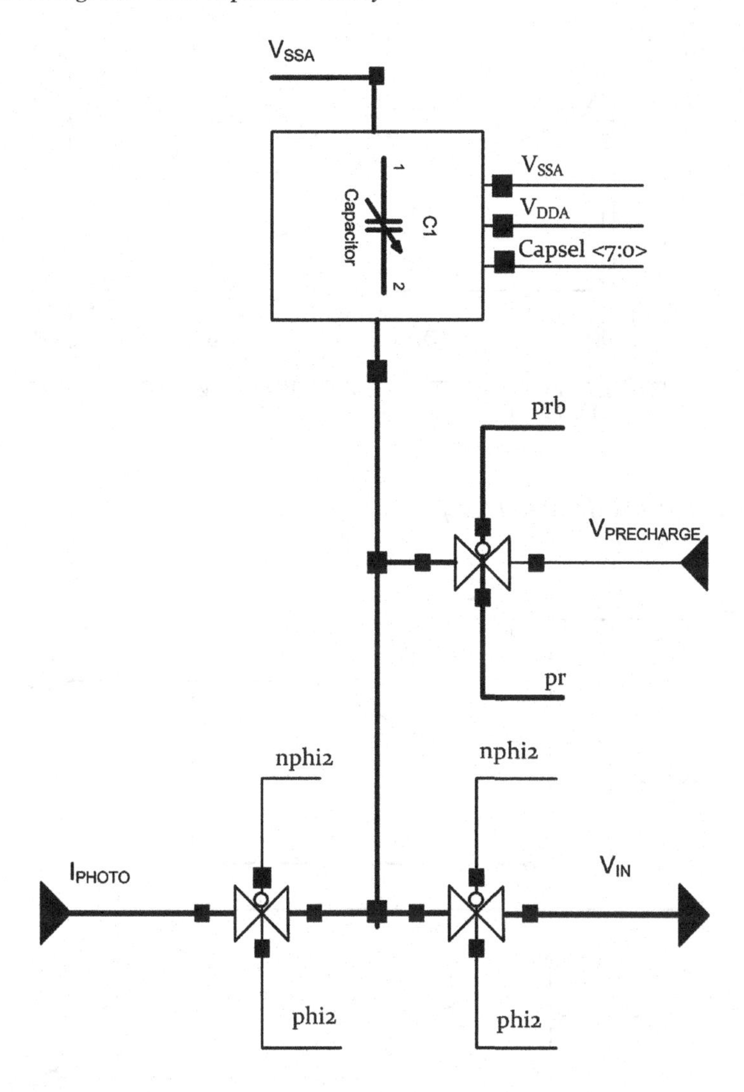

$$I = C\frac{dV}{dT} \tag{3}$$

Obviously, the integrator will set the dynamic range and sensitivity of the CIS.

d. **Single to Differential Amplifier:** In Figure 31, the outputs, OUTP and OUTN are the differential signals from single-ended input IN (V_{in} of integrator circuit). The VCM is set to 1.2 V. Two single-end output amplifiers are used to produce the differential signal.

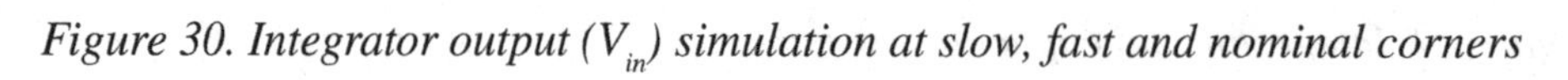

Figure 30. Integrator output (V_{in}) simulation at slow, fast and nominal corners

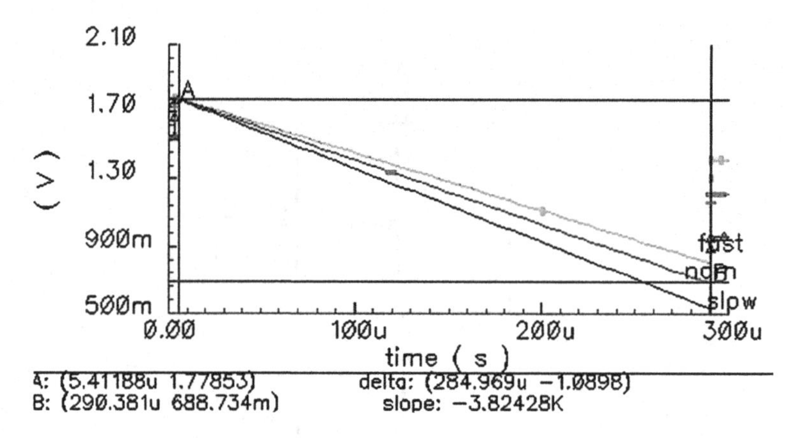

Figure 31. Single to differential amplifier

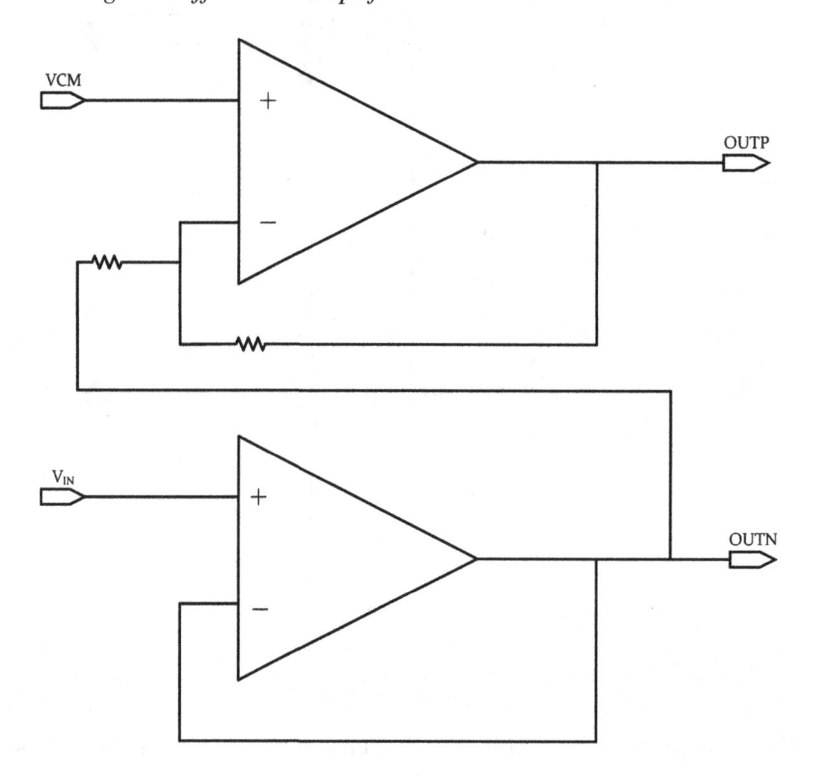

Figure 32. Non inverting output (OUTN)

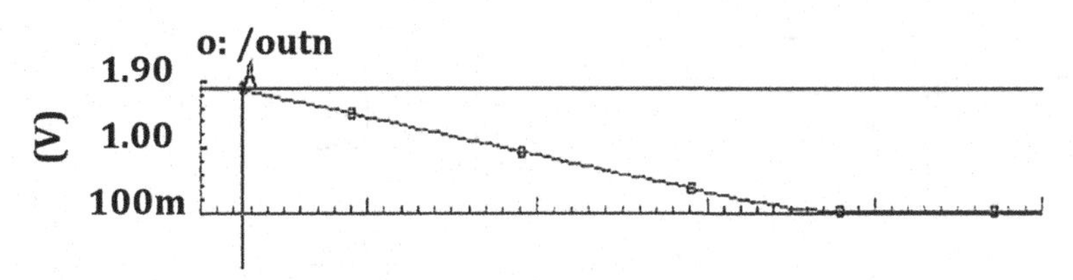

The output of the non-inverting amplifier (OUTN = inm) will be cut-off at 100 mV as shown in Figure 32. From Equation (2), this corresponds to maximum digital sensor output of 245 DEC if OUTP (equal to inp) is maintained at 1.2 V (a pseudo differential signal to ADC).

An S/H buffer amplifier as shown in Figure 33 consists of an operational amplifier, 2 capacitors (4 optional capacitors), 12 switches, 2 transistors (act as switches). Bottom plate sampling technique is used as this can reduce the substrate noise (Baker, 2010).

The S/H buffer amplifier is used prior to the ADC block due to the ADC that is used in the design is a free running ADC type. Two capacitors with value of 100 fF are used as the S/H capacitors. The value of the capacitor is chosen to achieve the best kT/C noise.

Figure 33. S/H buffer amplifier

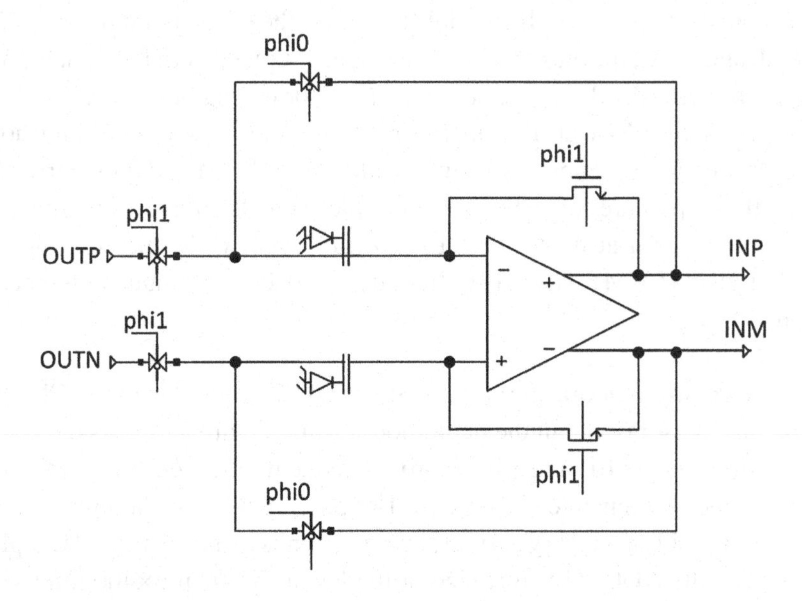

Figure 34. Operational amplifier with a common mode feedback circuit

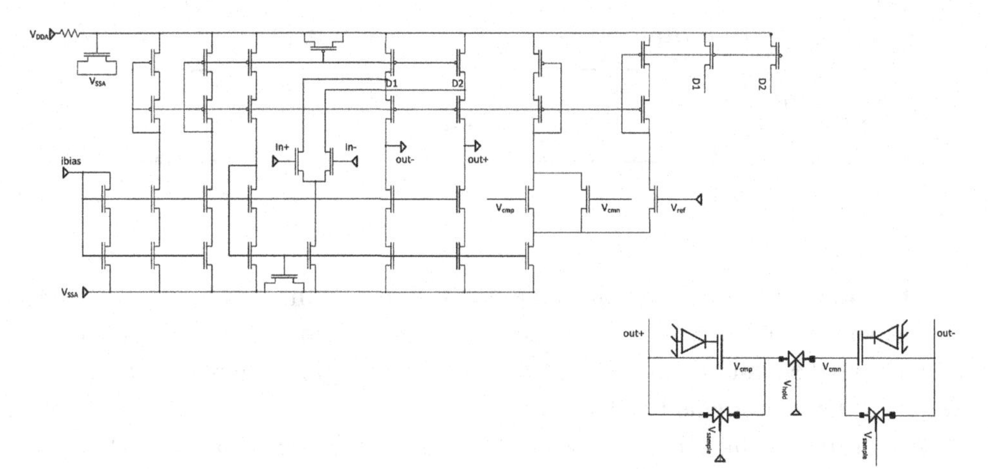

During the sampling (phi0 is low) of the differential voltages, the output of the operational amplifier (as shown Figure 33) is shorted to its input and the DC voltage is set at VCM by a common mode feedback circuit in the operational amplifier (Gray *et al.*, 2010). During holding phase (phi1 is low) the output is shorted to the left/bottom plate of the S/H capacitor. By employing the common mode feedback circuit (CMFB), the output of this S/H buffer amplifier is always re-centered at VCM. The CMFB employs capacitive sensing technique. Two capacitors (as shown in Figure 34) are used to average out the differential output voltage, the averaged output is connected to the CMFB amplifier input, the CMFB amplifier compares with VCM, and adjust the biasing current until the averaged output is equal to VCM. Some switches are added at the input and output for testing purposes.

Figure 35 shows S/H buffer amplifier outputs during sampling and holding periods. During the sampling, both outputs are tied to VCM and only during holding phase, the outputs are connected to the sampled signals. Figure 35 also elaborates the differential signal at output of 1.19 V is achieved whilst the common mode voltage is 1.19 V (VCM). The S/H buffer amplifier seems immune to temperature variation.

e. **CIS Implementation:** Figure 36 shows the floor plan of the CMOS image sensor (without the photodiode) while Figure 37 shows the layout of implemented design in 0.35 μm CMOS technology. A Non-Overlap Block is used to generate phi0 and phi1 or phi2. A Biasing Circuitries are used to provide necessary current sources or sinks to the operational amplifiers and the ADC. The total size of the layout is 800 μm×400 μm (without

Figure 35. S/H buffer amplifier output at fast corner and vs temperature

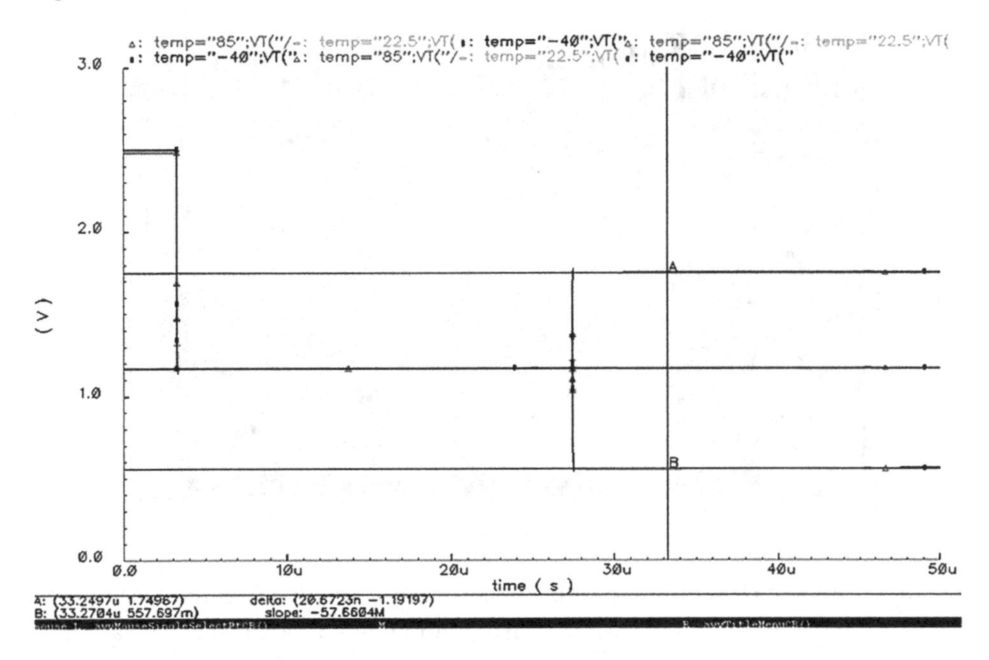

the RGB photodiodes). PhotoMux is used to select the pixel (photodiode and channel). A pipelined ADC type is used for the CIS implementation.

2. **Development and Measurement Results:** This section describes the development of the sensor and the measurement results.

Figure 36. Floor plan of the novel CMOS RGB sensor

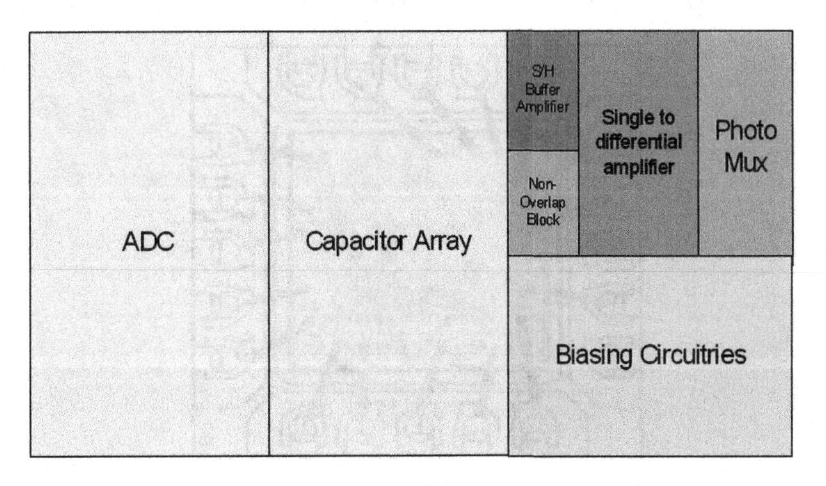

Figure 37. Layout of the implemented design

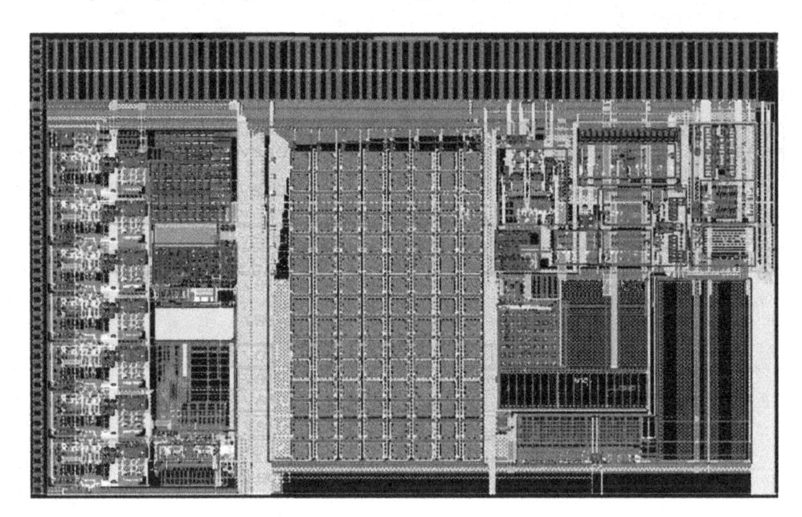

a. **Sensor Development and Measurement Results:** Figure 38. is microphotograph of the CMOS sensor with Digital block inside a clear QFN package.

Figure 39 shows cross section of the packaged sensor. Depicted on left and top of Figure 38 is the analog and the ADC (two mentioned functions) which are covered with metal to protect from incident light. More than half of the layout is consumed by the digital part and I/O.

Figure 38. Microphotograph of the novel CMOS RGB

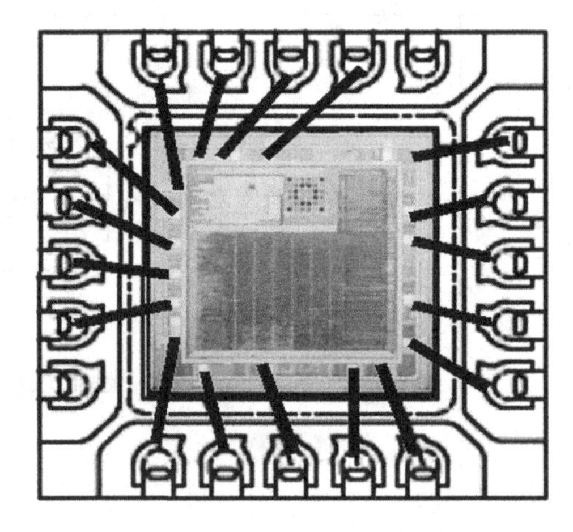

Figure 39. 5 x 5 mm QFN clear package, 0.75 thickness, 20-L pin out, NiPdAu selective plating

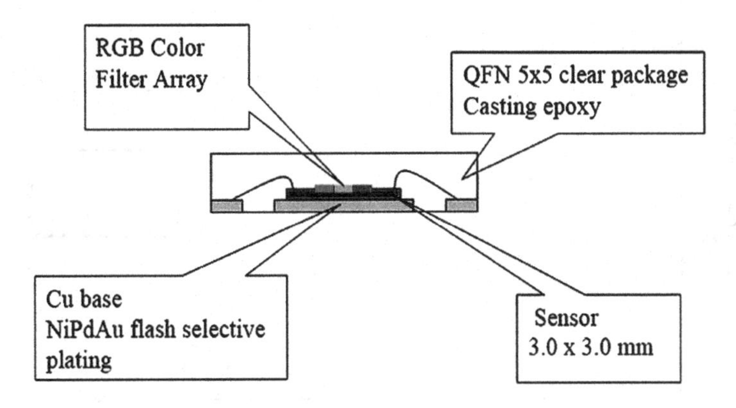

Table 6. Current consumption (VDD =2.6V)

States	Digital	Analog	Digital and Analog
Static	0.24uA	1.883mA	~2 mA
Sleep	0.22uA	0.02uA	~0.2uA
Normal operation	4.104mA	1.891mA	6.127mA

Table 6 is current consumption of the CMOS Image sensor integrated with peripheral digital circuit at different states.

Table 7 shows integrator designed capacitor value versus measured value. Low value of capacitance, especially for 4 pF and 8 pF as shown in Table 7 might not be accurate due to parasitic component associated with switches.

Table 7. Measured vs. designed

Measured (pF)	Designed (pF)
4	5.76
8	10.6
12	13.5
16	16.9
20	20.8
24	25.5
28	28.9
32	32.7

Figure 40. Single channel novel CMOS RGB color

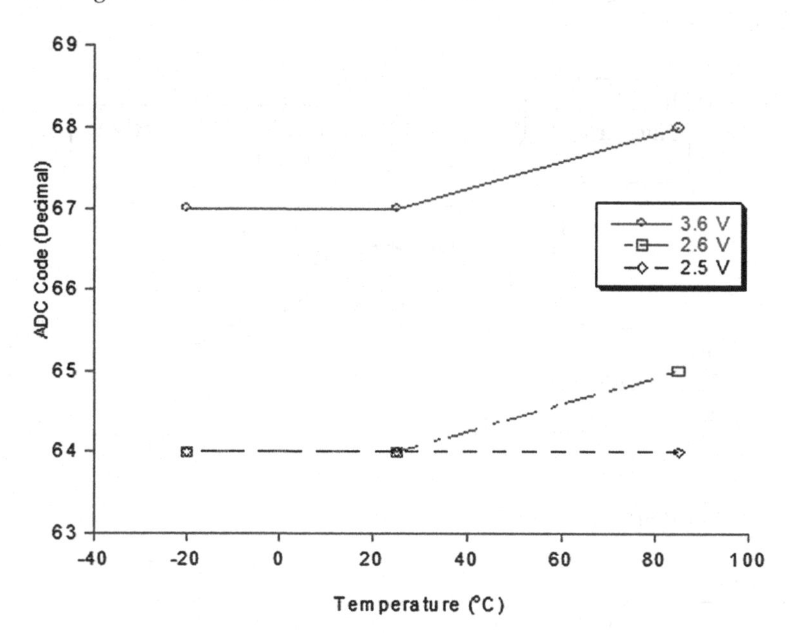

Figure 40 shows the ADC and S/H buffer amplifier quarter code in DEC (1/4 of 256) (at dark condition) vary proportionally with temperature. The measurement is conducted inside dark room and when ADC is configured to receive a pseudo differential signal. From five unit tested over temperature, the worst variation with temperature is +/- 1 LSB. This is due that VREFP (positive voltage reference for the ADC) is varied by +/- 10 mV w.r.t. room temperature. It has also shown that the average of dark code variation over temperature of the same unit is 1 LSB. The measurement is also conducted at different VDDs (2.5 V, 2.6 V and 3.6 V).

Figure 41 shows the maximum output code which is 245 DEC. This is similar to the simulated results (see previous section). For this measurement, the ADC is still configured to receive a pseudo differential signal or voltage. The Dynamic range in terms of ADC output would be from 1/4 of 256 to 245 DEC. It was also found that detection range (by using standard Luxeon DCC strip) for this light integrating sensor is from 0.2 to 6 klux. This parameter indicates the sensitivity of the sensor.

Figure 41. Single channel CMOS image sensor output vs. light (light intensity)

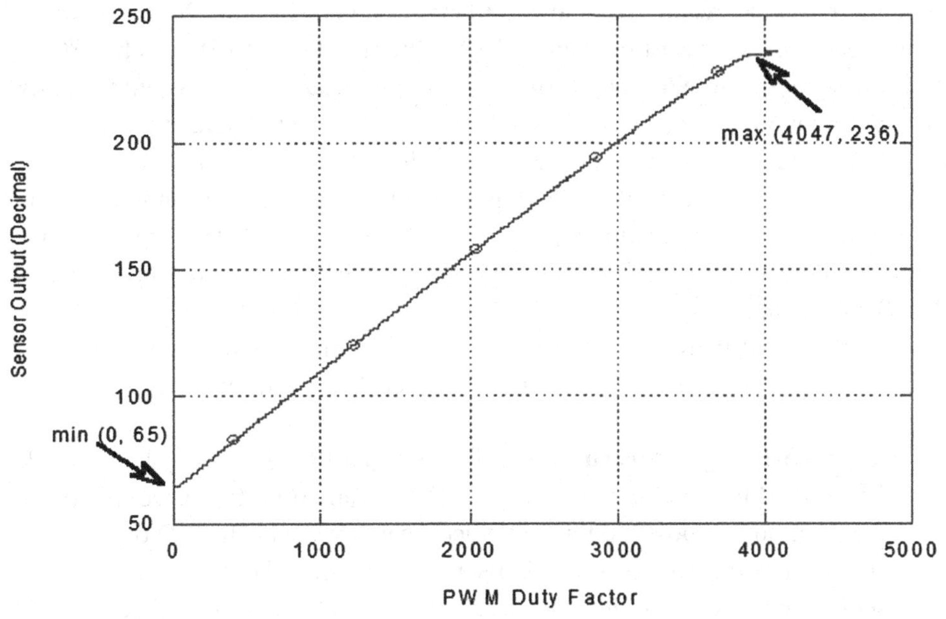

FUTURE RESEARCH DIRECTION

This section discusses the future research direction of CMOS Image sensor for machine vision application.

1. **High Sensitivity and High Dynamic Range CIS with High Resolution:** New circuit techniques would be required to achieve high sensitivity in CIS applicable to machine vision. The focus area is to work on new circuit technique (Xu et al., 2014). Selection of technology would also improve the sensitivity. Together with high speed requirement, it will be suitable for fast moving object application.

Advanced driver assistance systems (ADAS) are systems developed to automate/ enhance vehicle systems for safety and better driving based upon vision/camera systems, sensor technology and vehicle communication systems. ADAS applications utilize image sensors to provide enhanced safety features such as parking assistance, lane departure warnings and collision avoidance systems. However, performances of cameras degrade in wide dynamic range scenarios (e.g. low-angle strong sun-light, the headlights of oncoming vehicles, shadows in summer), because in these situations, dynamic range exceeds the capabilities of the conventional CMOS/CCD image sensors. Dynamic range is the ratio of the highest (lightest) signal which an imaging sensor can record to the lowest (darkest) signal (Li et al., 2015). A typical nature scene has a contrast ratio around 10,000:1 (Li et al., 2015), which is 80 dB of dynamic range. In (Xu et al., 2014), the reported dynamic range is only 50.1dB. Special care has to be made if ADAS going to employ the CIS (Xu et al., 2014). A commercial available CIS chip (Melexis, n. d.) should be able to provide the required dynamic range. Obviously, the CIS chip is able to support the dynamic range but not the high sensitivity requirement. In (Brockherde et al., 2004), a high dynamic range and high sensitivity CIS has been developed, however the resolution is less than 0.5 million.

The main area of research is to create new circuit without employing latest technology in order to achieve a resolution higher than 1 million.

2. **CIS for 3-D Object Measurement:** For a standard 2-D sensing system, a single CIS should be capable to achieve any 2-D required performance. However, this is almost impossible for 3-D object measurement. The 3-D object measurement is required to use ‘multisense’ technique. Thus requires multiple sensors approach. A single CIS (Lindgren, 2005) has been developed for 3-D measurement. The CIS area can be separated into 4 parts. The first part of the CIS is to extract the 3-D shape. The second part is to cater internal material light scatter measurement. The third part is to capture the color information and the final part to grey image with high resolution. The direction of research is to reduce the so called sensor area part to less than 4 or any complexity of the photodiodes front end circuit.
3. **Vision System on Chip:** The idea of ‘system on chip’ allures the CIS research and development into the concept of ‘vision system on chip’. Previous work (Ginhac, 2009) has introduced processor at pixel level. This is an attempt to develop vision system on chip. The challenge here is to develop a very fast low power small area ADC for each pixel.

CONCLUSION

It is founded the pixel circuitry does not have to be modified due to the available multiple VDD CMOS technology. At the moment the CIS is considered a slave device, the CIS still does not work as system on chip. There are many works on the circuit technique of CMOS Image sensor and the technology or process. The latest trend for the technology is to use the BSI approach in order to increase the resolution. While for the circuit technique, it appears the pixel circuitry does not evolve much for at least 5~10 years, this could be due to technology selection (due to cost

& market) to be bigger than 100 nm. The approach of using non-ADC for the CMOS Image sensor architecture is also enticing for new type of application. The work on design tool is very limited. Effort has to be made to have a single solution of designing tool for CMOS Image sensor. For the light integrating approach for CIS. The gain stage component is eliminated by using selectable capacitors (capacitor array) but the function remains. The technique has also made the integration of the sensor straight forward. An ADC is integrated together with the circuits into single silicon in order to produce digital outputs (a chip-level ADC approach). The digital output can be stored and processed in a processor for a complete machine-vision based device.

REFERENCES

Abdul Aziz, Z. A., & Marzuki, A. (2010). Residual Folding Technique adopting switched capacitor residue amplifiers and folded cascode amplifier with novel pmos isolation for high speed pipelined ADC applications. *Proceedings of the 3rd AUN/SEED-Net Regional Conference in Electrical and Electronics Engineering:International Conference on System on Chip Design Challenges (ICoSoC 2010)* (pp. 14–17).

Agranov, G., Mauritzson, R., Barna, S., Jiang, J., Dokoutchaev, A., Fan, X., & Li, X. (2007). Super Small, Sub 2μm Pixels for Novel CMOS Image Sensors. *Proceedings of the Extended Programme of the 2007 International Image Sensor Workshop* (pp. 307–310).

Aptina. (2010). *An Objective Look at FSI and BSI.* Aptina White Paper.

Ardeshirpour, Y., Deen, M. J., Shirani, S., West, M. S., & Ls, O. N. (2004). 2-D CMOS based image sensor system for fluorescent detection. *Proceedings of the Canadian Conference on Electrical and Computer Engineering* (pp. 1441–1444). IEEE.

Baker, R. J. (2010). *CMOS: Circuit Design, Layout, and Simulation* (3rd ed.). Wiley-IEEE Press. doi:10.1002/9780470891179

Bandyopadhyay, A., Lee, J., Robucci, R., & Hasler, P. (2006). MATIA: A Programmable 80 μW/frame CMOS Block Matrix Transformation Imager Architecture. *IEEE Journal of Solid-State Circuits, 41*(3), 663–672. doi:10.1109/JSSC.2005.864115

Barr, K. (2007). *ASIC Design in the Silicon Sandbox: A Complete Guide to Building Mixed-Signal Integrated Circuits.* McGraw Hill Professional.

Bigas, M., Cabruja, E., Forest, J., & Salvi, J. (2006). Review of CMOS image sensors. *Microelectronics Journal, 37*(5), 433–451. doi:10.1016/j.mejo.2005.07.002

Brockherde, W., Bussmann, A., Nitta, C., Hosticka, B. J., & Wertheimer, R. (2004). High-Sensitivity, High-Dynamic Range 768 x 576 Pixel CMOS Image Sensor. Proceedings of the ESSCIRC (pp. 411–414).

Camenzind, H. (2005). *Designing Analog Chips*. Academic Press.

Cevik, I., Huang, X., Yu, H., Yan, M., & Ay, S. (2015). An Ultra-Low Power CMOS Image Sensor with On-Chip Energy Harvesting and Power Management Capability. *Sensors (Basel, Switzerland), 15*(3), 5531–5554. doi:10.3390/s150305531 PMID:25756863

Chae, Y., Cheon, J., Lim, S., Kwon, M., Yoo, K., Jung, W., & Han, G. (2011). A 2.1 M Pixels, 120 Frame/s CMOS Image Sensor ADC Architecture with Column-Parallel Delta Sigma ADC Architecture. *IEEE Journal of Solid-State Circuits, 46*(1), 236–247. doi:10.1109/JSSC.2010.2085910

Cho, T., & Gray, P. R. (1995). A 10b 20MSamples/s 35mW Pipeline A/D Converter. *IEEE Journal of Solid-State Circuits, 30*(3), 166–172. doi:10.1109/4.364429

David, Y. (1999). *Digital pixel cmos image sensors*. Stanford University.

Deguchi, J., Tachibana, F., Morimoto, M., Chiba, M., Miyaba, T., & Tanaka, H. K. T. (2013). A 187.5μVrms -Read-Noise 51mW 1.4Mpixel CMOS Image Sensor with PMOSCAP Column CDS and 10b Self-Differential Offset-Cancelled Pipeline SAR-ADC. Proceedings of the ISSCC 2013 (pp. 494–496)

El Gamal, A. (2002). Trends in CMOS Image Sensor Technology and Design. In *Electron Devices Meeting*, (pp. 805–808). doi:10.1109/IEDM.2002.1175960

Erlank, A. O., & Steyn, W. H. (2014). Arcminute Attitude Estimation for CubeSats with a Novel Nano Star Tracker. *Proceedings of the19th World Congress The international Federation of Automatic Control* (pp. 9679–9684). Cape Town: IFAC. doi:10.3182/20140824-6-ZA-1003.00267

Farrell, J. E., Xiao, F., Catrysse, P. B., & Wandell, B. A. (2004). A simulation tool for evaluating digital camera image quality. In *Electronic Imaging 2004* (pp. 124–131). International Society for Optics and Photonics.

Feng, Z. (2014). *Méthode de simulation rapide de capteur d'image CMOS prenant en compte les paramètres d'extensibilité et de variabilité*. Ecole Centrale de Lyon.

Fowler, B., Balicki, J., How, D., & Godfrey, M. (2001). Low FPN High Gain Capacitive Transimpedance Amplifier for Low Noise CMOS Image Sensors.*Proceedings of the SPIE*(Vol. 4306). doi:10.1117/12.426991

Fu, Q., Zhan, W., Lin, Q., & Wu, N. (2009). A Novel CMOS Color Pixel for Vision Chips. *Proceedings of the IEEE Sensors 2009 Conference*. doi:10.1109/ICSENS.2009.5398508

Ginhac, D., Dubois, J., Heyrman, B., & Paindavoine, M. (2009). A high speed programmable focal-plane SIMD vision chip. *Analog Integrated Circuits and Signal Processing*, *65*(3), 389–398. doi:10.1007/s10470-009-9325-7

Goy, J., Courtois, B., Karam, J. M., & Pressecq, F. (2001). Design of an APS CMOS Image Sensor for Low Light Level Applications Using Standard CMOS Technology. *Analog Integrated Circuits and Signal Processing*, *29*(1/2), 95–104. doi:10.1023/A:1011286415014

Gray, P. R., Hurst, P. J., Lewis, S. H., & Meyer, R. (2010). *Analysis and Design of Analog Integrated Circuits (5th ed.).*Wiley.

Hamami, S., Fleshel, L., Yadid-pecht, O., & Driver, R. (2004). CMOS Aps Imager Employing 3.3V 12 bit 6.3 ms/s pipelined ADC. *Proceedings of the 2004 International Symposium onCircuits and Systems ISCAS'04.*

Hurwitz, J., Smith, S., & Murray, A. A. (2001). *A Miniature Imaging Module for Mobile Applications*. ISSCC. doi:10.1109/ISSCC.2001.912559

Hwang, S. H. (2012). *CMOS image sensor: current status and future perspectives.* Retrieved from http://www.techonline.com

Innocent, M. (n. d.). *General introduction to CMOS image sensors*. Academic Press.

Kleinfelder, S., Lim, S., Liu, X., & El Gamal, A. (2001). A 10 000 Frames/s CMOS Digital Pixel Sensor. *IEEE Journal of Solid-State Circuits*, *36*(12), 2049–2059. doi:10.1109/4.972156

Langfelder, G., Longoni, A., & Zaraga, F. (2009). Further developments on a novel Color sensitive CMOS detector.*Proc. of SPIE*(Vol. 7356). doi:10.1117/12.822291

Li, Y., Qiao, Y., & Ruichek, Y. (2015). Multiframe-Based High Dynamic Range Monocular Vision System for Advanced Driver. *IEEE Sensors Journal*, *15*(10), 5433–5441. doi:10.1109/JSEN.2015.2441653

Lim, K., Lee, J. C., Panotopulos, G., & Heilbing, R. (2006). Illumination and Color Management in Solid state lighting. *Proceedings of the 41th IAS Annual Meeting of theIEEE Industry Applications Conference*. doi:10.1109/IAS.2006.256908

Lindgren, L., Melander, J., Johansson, R., & Möller, B. (2005). A Multiresolution 100-GOPS 4-Gpixels/s Programmable Smart Vision Sensor for Multisense Imaging. *IEEE Journal of Solid-State Circuits*, *40*(6), 1350–1359. doi:10.1109/JSSC.2005.848029

Loinaz, M. J., Singh, K. J., Blanksby, A. J., Member, S., Inglis, D. A., Azadet, K., & Ackland, B. D. (1998). A 200-mW, 3.3-V, CMOS color Camera IC producing 352x288 24-b Video at 30 Frames/s. *IEEE Journal of Solid-State Circuits*, *33*(12), 2092–2103. doi:10.1109/4.735552

Lulé, T., Benthien, S., Keller, H., Mütze, F., Rieve, P., Seibel, K., & Böhm, M. (2000). Sensitivity of CMOS Based Imagers and Scaling Perspectives. *IEEE Transactions on Electron Devices*, *47*(11), 2110–2122. doi:10.1109/16.877173

Lyu, T., Yao, S., Nie, K., & Xu, J. (2014). A 12-bit high-speed column-parallel two-step single-slope analog-to-digital converter (ADC) for CMOS image sensors. *Sensors (Basel, Switzerland)*, *14*(11), 21603–21625. doi:10.3390/s141121603 PMID:25407903

Mahmoodi, A., & Joseph, D. (2008). Pixel-Level Delta-Sigma ADC with Optimized Area and Power for Vertically-Integrated Image Sensors. *Proceedings of the 51st Midwest Symposium on Circuits and Systems MWSCAS '08* (pp. 41–44). doi:10.1109/MWSCAS.2008.4616731

Marzuki, A., Abdul Aziz, Z. A., & Abd Manaf, A. (2011). *A Review of CMOS Analog Circuits for Image Sensing Application. Proceedings of the 2011 IEEE international conference on Imaging systems and techniques*. Penang: IEEE.

Marzuki, A., Pang, K. L., & Lim, L. (2007). *Method and Apparatus for Integrating a Quantity of Light.* U.S. Patent 2007/0235632 A1.

Massari, N., Gottardi, M., Gonzo, L., Stoppa, D., & Simoni, A. (2005). A CMOS Image Sensor with Programmable Pixel-Level Analog Processing. *IEEE Transactions on Neural Networks*, *16*(6), 1673–1684. doi:10.1109/TNN.2005.854369 PMID:16342506

Melexis. (n. d.). *MLX75412BD, Avocet HDR Image sensors.* Retrieved from www.melexis.com

Meynants, G. (2010). *Global Shutter Image Sensors for Machine*. ISE.

Norsworthy, S. R., Schreier, R., & Temes, G. C. (Eds.), (1997). *Delta-sigma data converters: theory, design, and simulation* (Vol. 97). New York: IEEE press.

Omnivision. (n. d.). *OV7740, 1/5" CMOS VGA (640 x 480) CameraChip™ sensor*. Retrieved from http://www.ovt.com

Optic, E. (2014). *Imaging Electronics 101: Understanding Camera Sensors for Machine Vision Applications.* Retrieved from http://www.edmundoptics.com/technical-resources-center/imaging/understanding-camera-sensors-for-machine-vision-applications/

Passeri, D., Placidi, P., Verducci, L., Pignatel, G. U., Ciampolini, P., Matrella, G., & Marras, A., and G. M. B. (2002). Active Pixel Sensor Architectures in Standard CMOS Technology for Charged-Particle Detection Technology analysis.*Proceedings of PIXEL 2002 International Workshop on Semiconductor Pixel Detectors for Particles and X-rays.*

Poplin, D. (2006). An Automatic Flicker Detection Method for Embedded Camera Systems. *IEEE Transactions on Consumer Electronics*, *52*(2), 308–311. doi:10.1109/TCE.2006.1649642

Scheffer, D., Dierickx, B., & Meynants, G. (1997). Random addressable 2048x2048 active pixel image sensor. *IEEE Transactions on Electron Devices*, *44*(10), 1716–1720. doi:10.1109/16.628827

Song, B. (2000). Nyquist-Rate ADC and DAC. In E. W. Chen (Ed.), *VLSI Handbook*. VLSI.

Sukegawa, S., Umebayashi, T., Nakajima, T., Kawanobe, H., Koseki, K., Hirota, I., & Fukushima, N. (2013). A 1/4-inch 8Mpixel Back-Illuminated Stacked CMOS Image Sensor. Proceedings of ISSCC 2013 (pp. 484–486). IEEE.

Suzuki, A., Shimamura, N., Kainuma, T., Kawazu, N., Okada, C., Oka, T., & Wakabayashi, H. (2015). A 1/1.7 inch 20Mpixel Back illuminated stacked CMOS Image sensor for new Imaging application. Proceedings of ISSCC '15 (pp. 110–112). IEEE.

Takayanagi, I., Yoshimura, N., Sato, T., Matsuo, S., Kawaguchi, T., Mori, K., & Nakamura, J. (2013). A 1-inch Optical Format, 80fps, 10. 8Mpixel CMOS Image Sensor Operating in a Pixel-to-ADC Pipelined Sequence Mode.*Proc. Int'l Image Sensor Workshop* (pp. 325–328).

Tang, F., Cao, Y., & Bermak, A. (2010). *An ultra-low power current-mode CMOS image sensor with energy harvesting capability*. ESSCIRC. doi:10.1109/ESSCIRC.2010.5619822

Trépanier, J., Sawan, M., Audet, Y., & Coulombe, J. (2002). A Wide Dynamic Range CMOS Digital Pixel Sensor. *Proceedings of the 2002 45th Midwest Symposium on Circuits and Systems MWSCAS '02*. IEEE. doi:10.1109/MWSCAS.2002.1186892

Wong, H. P. (1997). *CMOS Image sensors - Recent Advances and Device Scaling Considerations*. IEDM.

Wong, H.-S. (1996). Technology and device scaling considerations for CMOS imagers. *IEEE Transactions on* Electron Devices, *43*(12), 2131–2142. doi:10.1109/16.544384

Xu, R., Ng, W. C., Yuan, J., Yin, S., & Wei, S. (2014). A 1 / 2. 5 inch VGA 400 fps CMOS Image Sensor with High Sensitivity for Machine Vision. *IEEE Journal of Solid-State Circuits*, *49*(10), 2342–2351. doi:10.1109/JSSC.2014.2345018

Yao, Q. (2013). *The design of a 16*16 pixels CMOS image sensor with 0.5 e- RMS noise*. TUDelft.

Yoshida, K. (2005). *Japan Patent No. 3706385*. Japan.

Yoshida, K., Kamruzzaman, M., Jewel, F. A., & Sajal, R. F. (2007). Design and implementation of a machine vision based but low cost standalone system for real time counterfeit Bangladeshi bank notes detection. *Proceedings of the 10th international conference on Computer and information technology ICCIT '07.*

Zhang, M., & Bermak, A. (2010). Compressive Acquisition CMOS Image Sensor: From the Algorithm to Hardware Implementation. *IEEE Transactions on Very Large Scale Integration (VLSI) Systems*, *18*(3), 490–500. doi:10.1109/TVLSI.2008.2011489

Zhao, W., Wang, T., Pham, H., Hu-Guo, C., Dorokhov, A., & Hu, Y. (2014). Development of CMOS Pixel Sensors with digital pixel dedicated to future particle physics experiments. *Journal of Instrumentation*, *9*(02), C02004–C02004. doi:10.1088/1748-0221/9/02/C02004

Chapter 3
Automated Visual Inspection System for Printed Circuit Boards for Small Series Production:
A Multiagent Context Approach

Alexandre Reeberg de Mello
Federal University of Santa Catarina, Brazil

Marcelo Ricardo Stemmer
Federal University of Santa Catarina, Brazil

ABSTRACT

There is a crescent need to produce Printed Circuit Boards (PCB) in a customized and efficient way, therefore, there is an effort from the scientific and industrial community to improve image processing techniques for PCB inspection. The methods proposed at this chapter aim the formation of a system to inspect SMD (Surface Mounted Devices) components in a SSP (Small Series Production), ensuring a satisfactory production quality. This way, a 3-step inspection system is proposed, formed by image preprocessing, feature extraction and evaluation components, based on characteristics related to shape, positioning and histogram of the component. The inspection machine used in this project is inserted in a cooperation among machines context, in order to provide a fully autonomous factory, coordinated by a multi-agent system. Experimental obtained results show that the proposed inspection system is suitable for the case, reaching a success rate above 89% when using actual components.

DOI: 10.4018/978-1-5225-0632-4.ch003

INTRODUCTION

Printed circuit board production is growing economically and technologically in an expressive and constant rate, where technologically is evidenced by a consistent advance of electronic devices functionalities, and economically by a 60 billion USD share in the world market, considering only electronics manufacture and final equipment assembly (IPC, Association Connecting Electronics Industries, 2014). Concomitantly with technological evolution comes electronics devices personalization, bringing a raise of SSP in the production systems scenario, as a result of the possibility to produce in multiple perspective and manners (Hitomi, 1996). To produce a variety of products in a short time space production, a flexible manufacture process with high reliability is demanded, because each product has a high aggregated cost. To minimize the potential failures during production, inspections systems are used to identify these failures as soon as possible.

One approach of inspection is the automatic optical inspection (AOI), which is capable to detect deficiencies in a non-destructive and precise manner, in this case, detect failures in component placement, such as misalignment, shifting, wrong type or missing component.

This work is addressed to inspect SMD without considering any serigraphy marking. The inspection position in the production line is located after insertion of all components and before the welding process, hence we don't have the presence of solder paste in the PCB.

- **Image Processing Pipeline:** An image processing pipeline that permits an adequate visual automatic inspection of SMD components in a SSP context.
- **Image Recognition Pipeline:** An image recognition pipeline, based on artificial intelligence techniques, to identify the status of a component.
- **An Inspection System Agent:** An agent that represents the inspection machine in the multiagent environment, and the communication system between agent and machine.

Background

In order to improve efficiency in a flexible production line, the project "Cognitive Metrology for Flexible Small Series Production" (COGMET) was created to raise production quality with adaptive capability and high cognitive perception in relation to product and process (Pfeifer, et al., 2010). The concept of self-organizing system (Frank, et al., 2004) is inserted in the COGMET project, which is characterized by analysis of the current situation, new system goals determination and an adaptive system behavior to new environment conditions (Pfeifer, et al., 2010). In this context,

Figure 1. General topology of the diagnostics expert system
Stemmer, Costa, Vargas, & Roloff, 2014.

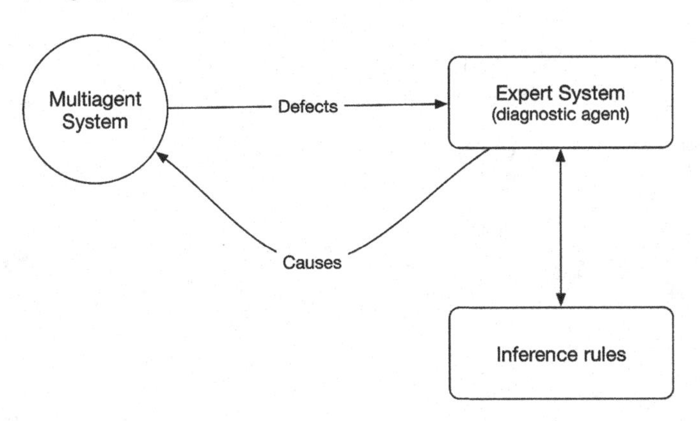

failure diagnostics in a production line is performed by an expert system (ES), and the communication between this system and the other devices in a manufacturing plant is executed by a multiagent system (MAS), this way the ES becomes a diagnostic agent, which seeks to achieve the previously assigned goals (Stemmer, Costa, Vargas, & Roloff, 2014), as presented in Figure 1.

A PCB assembly line is used to perform the experiments, and is composed – in assembly order – by a printer, a solder paste inspection machine, an automatic insertion machine, a manual inspection area, a reflow oven, an AOI machine, where and all machines are contextualized in a MAS.

A reference architecture to implement the configuration and supervision of a MAS in a SSP context was proposed by (Roloff, 2014), using the *JaCaMo* framework – developed by (Boissier, Bordini, Hubner, Ricci, & Santi, 2013) – which is a combination of: *Jason*, an agent programming language; *Cartago*, environment artifact programming tool and *Moise*, a MAS organization programming tool. The multidimensional context of proposal architecture applied to a line production of a SSP is presented in Figure 2.

Production line machines are monitored and controlled by a SCADA system, and a web service integrates MAS and manufacturing line. As these machines are non-autonomous, they behave like an artifact, providing information to agents. Among all agents, there is an inspector agent, whose objective is to receive diagnostics from automatic visual inspection machine (which is an artifact). A system diagnosis presents the inspection status, indicating existing issues. From this diagnosis, an inspection agent communicates with other agents to evaluate founded defects, and if necessary actions are performed. A transducer method was proposed to integrate vision system artifact and inspector agent to inspection software – working as a mediator between the agent – using an exclusive agent communication

Figure 2. Multidimensional context
Roloff, 2014.

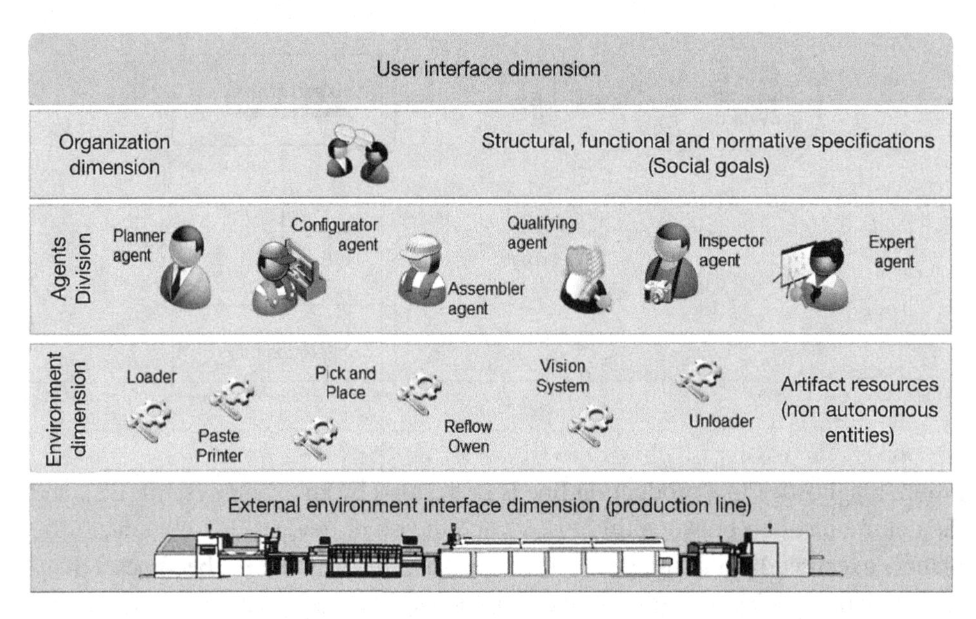

language (ACL). This communication occurs by message exchange; accordingly, the translator has to know the sentences from both sides. This way, turning the AOI system into an agent brings (Wooldridge, 1997) (Bordini, Hubner, & Wooldridge, 2007):

- **Autonomy:** Make decision based on encapsulated state, without intervention from human or other agent.
- **Reactivity:** Agents that are situated in an environment are able to perceive this environment, besides create deal answers by changes.
- **Proactivity:** Beside answer the environment, agents are capable of exhibit a directed behavior to a goal trough an initiative.
- **Social Skill:** Agent interact among others trough an ACL.

Concerning AOI systems, many methods were proposed to detect failure in component placement during a PCB production, as the template matching method using discrete wavelet transform presented by (Cho & Park, 2008), focusing on minimizing the need of memory storage and calculation time. (Wu, Zhang, & Hong, 2009) developed a visual inspection system for surface mounted components based on the features of these components under RGB structure light sources. The inspection is based on the electrodes found, by comparing the size, symmetry, shift and color. A

Bayesian classifier is used as well to identify the type of component. An automatic inspection system based on Support Vector Machines (SVM) and mean-shift was proposed by (Wang, Sun, & Zhang, 2009). The mean-shift method and an adaptive segmentation algorithm were performed to extract a binary form of the component marking. SVM classify it by diagnosing the result in right or wrong. (Xu & Liu, 2010) applied a fuzzy threshold segmentation algorithm to segment a gray-scale PCB image in silk-screen layer and wiring layer. (Luo, Gao, Sun, & Zhao, 2013) implemented an artificial neural network (ANN) to inspect the characters that exist on the components, reducing the false detection rate to approximately 0.1%, by not tolerating position and rotation nuances, matching the component in 0 and 180 degrees to double check and using a good template of character images. (Xie, Dau, Uitdenbogerd, & Song, 2013) presented a study on evolving an AOI program with genetic-programming. The input for training are images from a perfect board, and the output is either "defect" or "no defect", working as a binary classifier.

An extended study performed by (Krippner & Beer, 2004) compares AOI following printing, placement and reflow. The faults to be considered in the study are related to solder paste application and component placement. The test was made in 2404 PCBs during assembly process, and the AOI inspection found faults in 167 PCBs, so the First Pass Yield (FPY) was 93.1%. In the 167 defective circuit boards, 189 single or component faults were present. When the AOI was performed after paste print, the FPY was 97.9%, and when inspected after component assembly the FPY was 94.7%. After post reflow the FPY was 93.1%, and the percentage of defects detected was 99.5%, because the faults found after paste and component assembly was present in this stage as well. The test has shown that the inspection detected over 50% of process faults, and 48% of the optically recognizable faults could not be recognized electrically.

A visual inspection system for SMD (surface mounted devices) on PCB was developed by (Lin & Su, 2006) using a neural network sorter with five types of image indexes (white pixel count index, histogram index, correlation coefficient, regional index, and high contrast index), analyzing revered, missing and skewed samples. 558 training samples were used to train a 5-9-2 neural network applying the backpropagation technique, and 1949 samples collected from production line were tested, resulting in a false alarm rate of 2.12%, a missing rate of 2.44% and a wrong classification rate of 4.88%.

Industrial inspection systems effectiveness rate is evaluated by counting defects per million opportunities (DPMO) during a PCB production. (Oresjo, 2009) developed a study with 14 different companies evolving ICT (in-circuit test), AXI (automatic x-ray inspection), AOI, FT (functional test) and MVI (manual visual inspection), where the last three are grouped in the same category. 1138 board types were inspected, totaling 726091 boards and 3675298930 joints. There were a total of 865 defects, and the AOI, MVI and FT combined found 209 defects, that is, 24% of effectiveness. Post-reflow AOI is 25-75% effective, and on average AOI finds around 40-50% of defects, considering AOI systems from various vendors.

A first approach to inspect SMD components for small series production (SSP) was made by (Szymanski & Stemmer, 2015), using the Scale-Invariant Feature Transform (SIFT) technique to extract keypoints related to components serigraphy marking, then comparing good and test components. This approach was very efficient for components with serigraphy, but had problems to find good keypoints in components without any markings.

Thus, the inspection method proposed here is meant as a complement of (Szymanski & Stemmer, 2015) method, and is composed of two steps. The first is a preprocessing step consisting of a gray scale component image region of interest (ROI) extraction, angle, reference point position, ROI histogram and Fourier descriptors acquisition. The second step is to identify the existence of a component on the board using k-nearest neighbor (KNN) algorithm, with information taken from the histogram of the ROI as input. Knowing that a component exists in the tested region, compare angle and position gives us enough information to ensure that a component exists in the right place. In order to identify the component, a MLP (Multilayer Perceptron) neural network using back-propagation is adopted and trained using the contour Fourier descriptors from each component.

VISUAL SMD INSPECTION SYSTEM AND COMMUNICATION PROTOCOL BETWEEN AGENT AND MACHINE

Overview

The inspection machine is controlled by a PC trough a controller software, responsible to provide a graphical user interface (GUI), create new and follow inspection projects, get pictures, and insert conveyor movement commands to move the camera and the machine axes. The developed GUI was made in Java, while the conveyor movement is made by step-motors attached to sliders, and images are acquired by an industrial camera model *Basler scA1390-17gc*. Each image acquired contains a desired area of a board, thus the result is not a figure of all the board, but a small area with one or more components.

The inspection system is composed by an inspection algorithm and a UDP (user datagram protocol) communication protocol, that allows the machine provide information to agents regarding manufacturing process. Inspection algorithm uses a classical pipeline approach, starting with a preprocessing stage, followed by a features extraction phase and lastly a device evaluation moment. The objective of

Figure 3. Inspection system overview

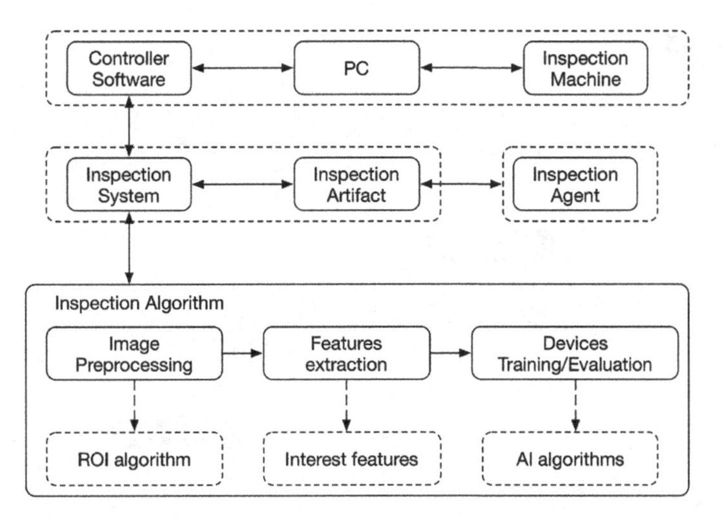

preprocessing techniques is to enable the features extraction related to shape and positioning trough a region of interest (ROI) algorithm. At features extractions phase, histogram related index and *Fourier* descriptors are calculated from the ROI, resulting in two vectors, that together represents each type of component in a unique way. Evaluation system verify devices existence during manufacturing trough a *K Nearest Neighbor* (KNN) classifier, and the positioning, orientation and device type are classified by an artificial neural network (ANN), over a multilayer backpropagation method. Figure 3 explicit proposal inspection system overview.

Proposal image processing architecture is composed by three steps: firstly, create the evaluation system, that is, extract features from reference (*gold*) devices, secondly extract features from PCB background and finally inspect test components. Figure 4 presents the inspection system general workflow.

Image Preprocessing

Image preprocessing architecture is a graphical pipeline to set the ROI of a frame, this is, define the best region that defines a component, as well as provide the region representation as a 2-D vector.

A SMD capacitor, represented by Figure 5, is used to exemplify the image preprocessing architecture. Before enter in the ROI pipeline, the operator selects a frame that contains the component, to reduce the area of processing, this way, Figure 6 describes the frame of a capacitor image.

ROI graphical pipeline is shown in Figure 7.

Figure 4. Inspection system general workflow

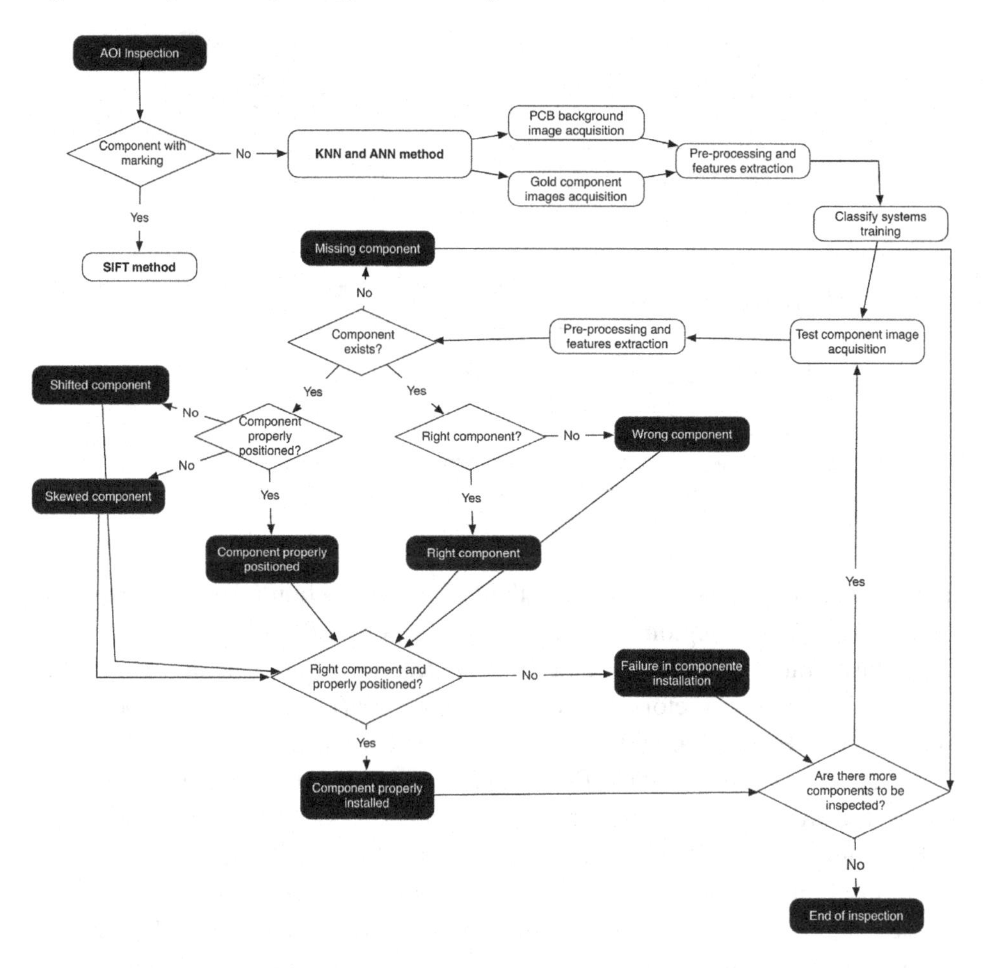

Figure 5. Reference (gold) component

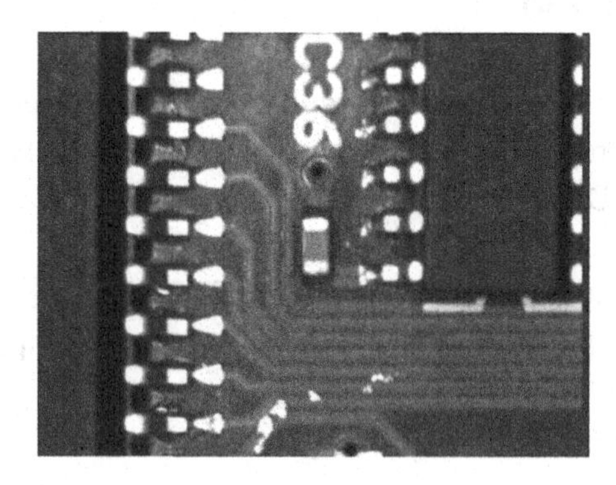

Figure 6. Cropped reference component

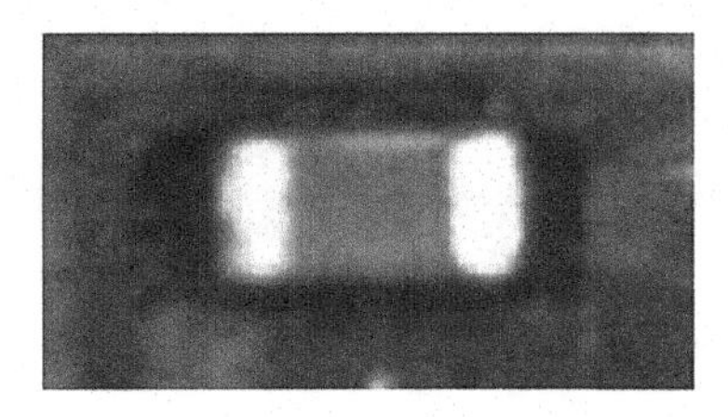

Figure 7. ROI graphical pipeline

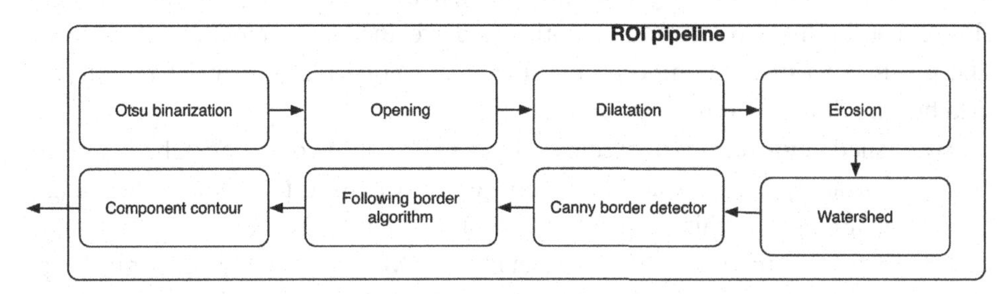

Effectiveness of entire system depends on the success of the ROI algorithm in extracting component contour, because all other features are derived from it. PCB illumination is constant, this way, the levels of exposition, brightness and contrast can be configured while extracting golden images. This way, the ROI pipeline first step is to turn the frame binary, to reduce computation cost of the following processes. As frame histogram have a bimodal characteristic, Otsu binarization method is applied (Otsu, 1979), separating coarsely object from background. The result is a frame with a large region, which it may contain the device and noises, as shown in Figure 8.

Opening morphological operation is applied to eliminate presented noises around device area. Using an 3×3 ellipsoidal shape elements structure, the result is a large area without noises. To eliminate non device regions or unwanted small intercon-

Figure 8. Otsu method outcome

Figure 9. Watershed algorithm outcome

nections between devices, dilation operation is applied for 3 times (it shows enough for this application), showing a potential device area. As consequence, border pixels are lost, in this way, erosion operation is performed the same amount of times. The result is an image composed by device and noises, however the region that matches the device is now more explicit.

Watershed algorithm, proposed by (Beucher & Lantuejoul, 1979), had only the principle of analyzing sample binary images, and is used to segment the image. The algorithm used in this chapter is a morphological approach to segment, using watershed transformation, made by (Beucher & Meyer, The Morphological Approach to Segmentation: The Watershed Transformation, 1993), and is responsible to determine which region belongs to the device and which should be discarded, resulting in the ROI (see Figure 9).

To ensure that only one region is found, Canny edge detector – proposed by (Canny, 1986) – is performed, which is an edge detector algorithm based on choosing a derivative estimation filter by using the continuous model to optimize a combination of signal to noise ratio, localization and low false positive (Forsyth & Ponce, 2002). The outcome may contain several contours, because it depends on selected values of upper and lower threshold. For this approach, upper threshold is fixed in 255, and lower threshold is iterated from 0 to 254. Each value of the lower threshold creates an image as output, with a potential contour of the component, and by the end of the iteration process, the 254 images are indexed forming a 3D matrix with dimensions of *[254 x Imagewidth x Imageheight]*.

The step now is to iterate trough all images of the 3D matrix to identify which one represents better the component contour, applying a border following algorithm, proposed by (Suzuki & Be, 1985), resulting in vectors composed by all closed and outermost borders pixels, following the rule of given two subsequent points *x1, y1* and *x2, y2* of the contour,

$$\max \sqrt{(x1 - x2)^2, (y2 - y1)^2} = 1 .$$

Figure 10. ROI of a SMD capacitor

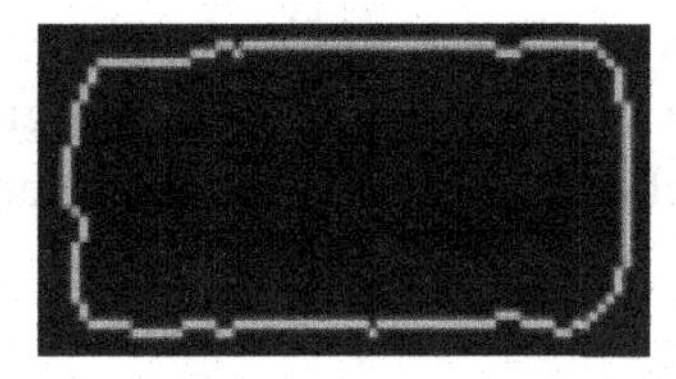

If more than one contour is found, a shape analysis is performed, searching for quadrilaterals shape. As devices are symmetrical, the analysis compare these shapes, and if they behavior symmetrically, they are connected by straight lines, resulting in a good ROI approximation. Figure 10 represents the final result of ROI pipeline for the SMD capacitor.

Features Extraction

Characteristics extraction objective is to obtain an image related metrics in a vector format, so that it represents an object in a unique manner. Consequently, a smaller memory space is used to store image information, increasing inspection software time efficiency. PCB background images are represented by a histogram related vector, while devices are represented by metrics related to ROI histogram, shape and position. Figure 11 presents the feature extraction process tree.

Figure 11. Features extraction process tree

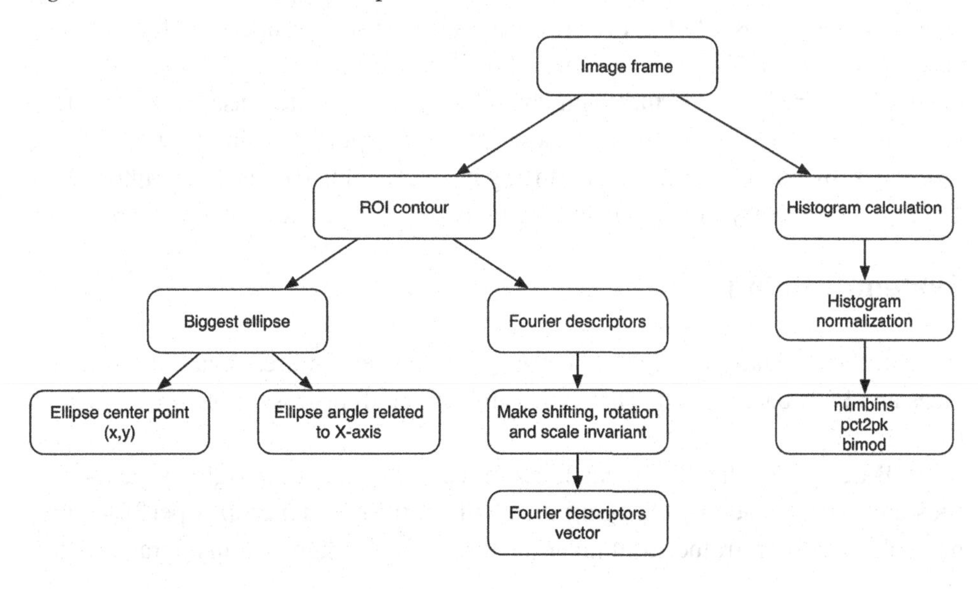

- **Reference Point and Angle:** As the contour has an irregular shape, in order to extract ROI reference point and angle, the biggest possible ellipse is drawn inside the component contour, therewith the angle is acquired considering longest axis of the ellipse and X-axis, and reference point is the ellipse center.
- **Histogram Related Features:** The selected frame has information besides the ROI, making a good area to extract information about the existence of a component. Frame extracted histogram is normalized, and extracted features are: number of histogram bins with > 3% of the pixels (*numBins*), percent of histogram range within the largest 2 peaks (*pct2pk*) and the average of sums of nearest neighbor pixel differences(*absdiff*).
- **Fourier Descriptors:** To ensure that a closed curve will correspond to any set of descriptors, a complex Fourier descriptors approach proposed by (Granlund, 1972) is performed, describing the set of *N* vertex $z(i) : i = 1,...N$ corresponding to *N*-outline points, this way the representation is made by Fourier coefficients of Z-transform, as shown by (1).

$$Z_i = \sum_{k=-\frac{N}{2}+1}^{\frac{N}{2}} c_k e^{2\pi j\frac{ki}{N}} \tag{1}$$

Lower frequency descriptors of the contour characterize an approximation of the shape, and high frequencies characterize details. This way, there is a quantity of descriptors that are enough to give singularity for a type of component. Also (Bertrand, Queval, & Maitre, 1982) describes some useful properties of the Fourier descriptors as position of the center of gravity of the shape, size of the shape and to check if there is any perturbation about rotation. To avoid classification system interference (that uses the Fourier descriptors as input), we should ignore these properties, by not using the $k = 0$ (shifting property), normalizing descriptors with $k = 1$ (scale property) and not using the phase of $c(k)$ (rotation of a perturbation).

System Training

To train the evaluation system, is necessary to gather features regarding gold devices and PCB background. The process to extract all necessary features is shown in Figure 12.

PCB device localization is made by moving the camera trough *x,y,z* axis, images adjusts are made by changing gain and contrast, and a crop is performed by manual selecting a frame from taken picture. PCB background histogram related

Figure 12. Features extraction process

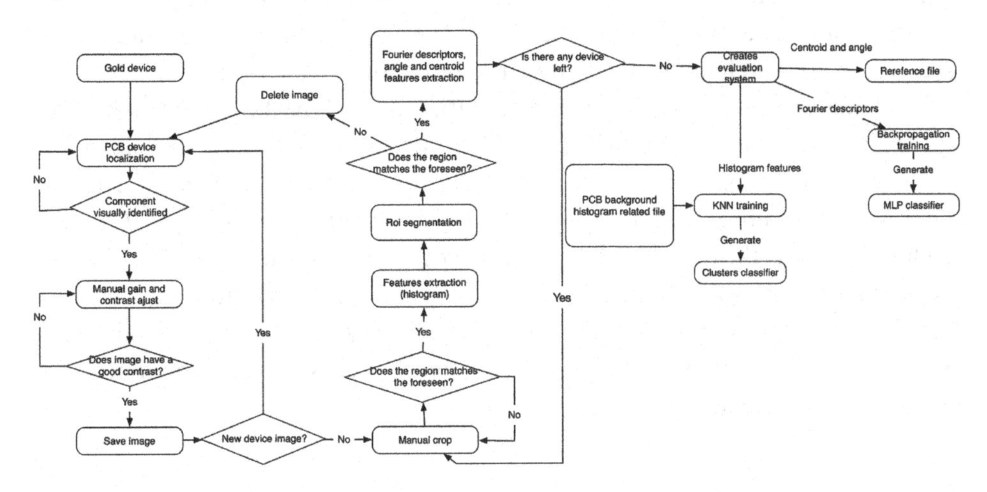

file contains all image information related to histogram, providing data to KNN training process, and the multilayer classifier request information only related to gold images features.

KNN Classifier

A KNN classification system used in this paper was proposed by (Aha, Kibler, & Albert, 1991), and the classifier set is made of golden and background images of a PCB, where each image is represented by a row of the classifier. The features used in the classifier are: histogram percent bins with > 0.5% of pixels (*numBins*), histogram percent range within the largest 2 peaks (*pct2pk*), average of various sums of nearest neighbor pixel difference (*absdiff*) and type of component (*component/ background*), forming a 4-instances vector v = [*numBins*, *pct2pk*, *absdiff*, *component/ background*]. Features set search space is initially empty, and for each iteration one feature is added. By the end of the process, a4-intances vector must be completed. The criteria function to classify the instance is the KNN classifier, and the result have to fit in one of the two available clusters, which are component or background. Table 1 shows the relationship between features and clusters.

Table 1. KNN clusters

Clusters	*numBins*	*pct2pk*	*absdiff*	**Type**
Component	39<x<65	829<x<12100	7<x<30	component
Background	12<x<65	4300<x<16000	4<x<8	background

Training instances of the KNN are dropped following a FIFO order, and the classifier is initialized with 1 nearest neighbor. The nearest neighbor search algorithm used is a linear search algorithm, which is a brute force search method.

To validate the use of the KNN as a classifier method for this application, a 10-fold cross validation based on (Kohavi, 1995) paper, is performed in golden and background images samples. The method consists in divide known information in two sets, one subset of training *T* and one of validation *V*. This way, *T* are the neighbors and *V* is the test subset, turning into new observations to be classified. To characterize a cross validation, different training and validation subsets are needed, and to identify it as a *K* cross validation, the initial set of data is divided into *K* subsets with same size. Accordingly, the validation occurs *K* times, and for the *i-th* iteration, cross validation is defined by (2).

$$\frac{1}{K}\sum_{i=1}^{K}\frac{1}{\#V_i}\sum_{xj\in V_i} l(tipo_j, h_k(nb_j, pk_j, bk_j \mid \{nb, pk, bd, tipo\} \in T_i \quad (2)$$

where l is the loss function defined by h_k, which represents nearest neighbor method, which in turn identify the *k* nearest neighbors from the inputs nb_i, pk_i, bd_i relative to training subset. The prediction expected error is approximate by a mean, represented by $\frac{1}{\#V}$. In this manner, is now possible to approximate the expected loss with the observed mean loss. Applying this method for *K=10* folders, and utilizing 60 instances, being 30 components and 30 backgrounds, the results of the validation method presents only 1 misclassified instance when identifying component presence or absence using histogram related features, as shown in Table 2.

Table 2. KNN accuracy

Total Number of Instances	60	
Correctly Classified Instances	59	98,3333%
Incorrectly Classified Instances	1	1,6667%
Kappa statistic	0,9667	
Mean absolute error	0,0263	
Root mean squared error	0,1289	
Relative absolute error	5,2666%	
Root relative squared error	25,7709%	

Multilayer Perceptron

In order to use the neural network as a classifier method, the input and output needs to be normalized. The input used in this paper is a vector formed by the 6 first Fourier descriptors and the component type of a golden image, this way both Fourier descriptors and component type needs to be normalized. The normalization formula used in the Fourier descriptors have to be the same to train and classify, this way, the biggest and smallest values found in the golden images are used. However, to ensure that when classifying (using images different from the golden) no bigger or smaller value are found, a safety margin of 1000 is added to the biggest value, and a half of the smallest value is used, as follows (3).

$$nv = \frac{value - \frac{sv}{2}}{\left(bv + 1000\right) - \frac{sv}{2}} \tag{3}$$

where nv is the normalized value, sv is the smallest value and bv is the biggest value, while the constant *1000* is used to avoid division by 0. The component type normalization is determined by the number of types of golden images registered, equally distributed along 0 and 1. For example, for 5 different component types, the output is 0 for the first type, 0.25 for the second, and so on, until 1 for the last type of component. Therewith the MLP ANN of this paper is structured by: a vector formed by the 6 first normalized Fourier descriptors with the component type of a golden image as input, 30-neurons hidden layer and the normalized component type as output. A sigmoid function is used as activator, proposed by (Saul, Jaakkola, & Jordan, 1996), with a learning rate of 0.4 and momentum of 0.9. The maximum error is 0.00001 and 5000 epochs are used for training.

K-fold cross validation is used as a MLP architecture selection strategy, aiming overfitting and underfitting avoidance. It is possible to identify the overfitting when validation error increase and training error decrease, resulting in a memorization of the training data. Under fitting occurs when validation and training error remains large, resulting in an incomplete learning. As is not possible to minimize the expected error using any type of generalization, the cross validation method is applied.

Assuming that the x input standards are taken from a P probability distribution, and all data sets are derived from f^* truth function, cross validation sets are defined by $\left\{\left(x^{(1)}, d^{(1)}\right), \ldots, \left(x^{(k)}, d^{(k)}\right)\right\}$, with $x^{(i)} \sim P$ and $d^{(i)} = f^*\left(x^i\right)$, resulting in a approximation between expected and training error, as represented by (4).

$$\frac{1}{K}\sum_{i=1}^{K}\left(\frac{1}{2}\left(f\left(x^{(i)}-d^{(i)}\right)^{2}\right)\right) \tag{4}$$

Therefore, when there is a good approximation, the hypothesis learned through data sets will have a good performance in all related data sets. However, the previous equation deals with only one set, and the learning function must have a satisfactory performance regardless of the test and validation sets. Therefore, Algorithm 1 is applied.

$$E\left(\vec{w};D;\lambda\right):=E\left(\vec{w};D\right)+\frac{1}{2}\lambda\left\|\vec{w}\right\|^{2} \tag{5}$$

According to literature, λvaries between 0 and 10, and if the value is small to application in question, generates overfitting, and if it is big, occurs underfitting. It was found that 30 hidden layers results in a good representation capability, and applying the same weight to all neurons, a smaller number of parameters need to be considered, therefore the process is less complex, creating an ideal mask to object identifying cases. To define MLP architecture, 30 components of 6 different device models were used (5 components of each model) in a backpropagation training method. After parameters adjust, and keeping ANN configuration, the biggest error was 2.6%, as shown in Table 3.

Algorithm 1. K-fold cross validation algorithm

Set D, $2 \le k \le K$

Desired answer = d(n)

For i=1 to K

ANN (re)- initialize

ANN training for subsets $D_1 \cup \ldots \cup D_{i-1} \cup D_{i+1} \cup \ldots \cup D_k$

Calculates mean test error e_i over D_i

End for

Return $\frac{1}{K}\sum_{i-1}^{K} e_i$

where D set is made of Fourier descriptors from gold images, and it is subdivided into K disjoints sets with same size. Due to MLP utilize severs adjusts parameters, weight decay technique is used, and It modifies the error when it is minimized over the learning, manipulating λ variable in (5).

Table 3. Epoch error

Error after 5000 Epochs	
Fold 0	0,0000011
Fold 1	0,0000012
Fold 2	0,0000001
Fold 3	0,0000012
Fold 4	0,0000009
Fold 5	0,0000012
Fold 6	0,0000001
Fold 7	0,0052916
Fold 8	0,0264577
Fold 9	0,0000001
Fold 10	0,0052921

The K=10 realized training result is one misclassified instanced, as presented by Table 4.

Only one device type was misclassified during all tests, and It happened because the used frame was not ideal, this way proposal architecture is sufficient to evaluate SMD devices using Fourier descriptors.

Machine Agent Interaction

To allow the agent to inspect the PCB process, is necessary that the inspection system be ready, this is, the machine must be calibrated, images acquired and evaluation system trained. Therefore, agent objective is machine calibrating, acquire images,

Table 4. Backpropagation accuracy

Total Number of Instances	30	
Correctly Classified Instances	29	96,667%
Incorrectly Classified Instances	1	3,334%
Kappa statistic	0.96	
Mean absolute error	0,0642	
Root mean squared error	0,1389	
Relative absolute error	26,8225%	
Root relative squared error	25,7709%	

Figure 13. Inspect agent objective diagram

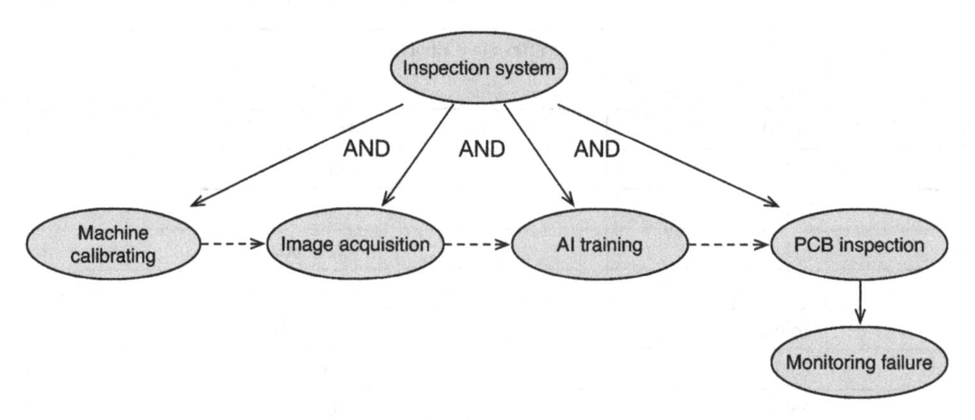

train evaluation system, inspect PCB and monitor the inspection process, as shown by Figure 13.

From information prevenient from the inspection machine, the agent may interrupt or reinitiate inspection process, whose decision is made through message exchanges between agents, by the use of current status from training, acquisition, calibration and inspection steps in the form of perceptions, as presented by Figure 14.

Interaction actions and perceptions between agent and artifact are explicit by the environment overview diagram, presented by Figure 15.

Inspection artifact obtain information from inspection machine trough operation *status*, whose representation method is made by a set of Boolean variables. Machine

Figure 14. System overview diagram

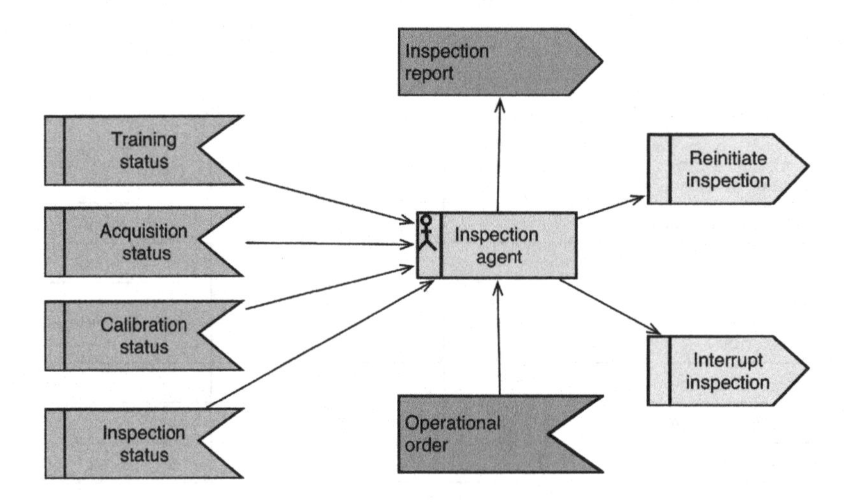

Figure 15. Environment overview diagram

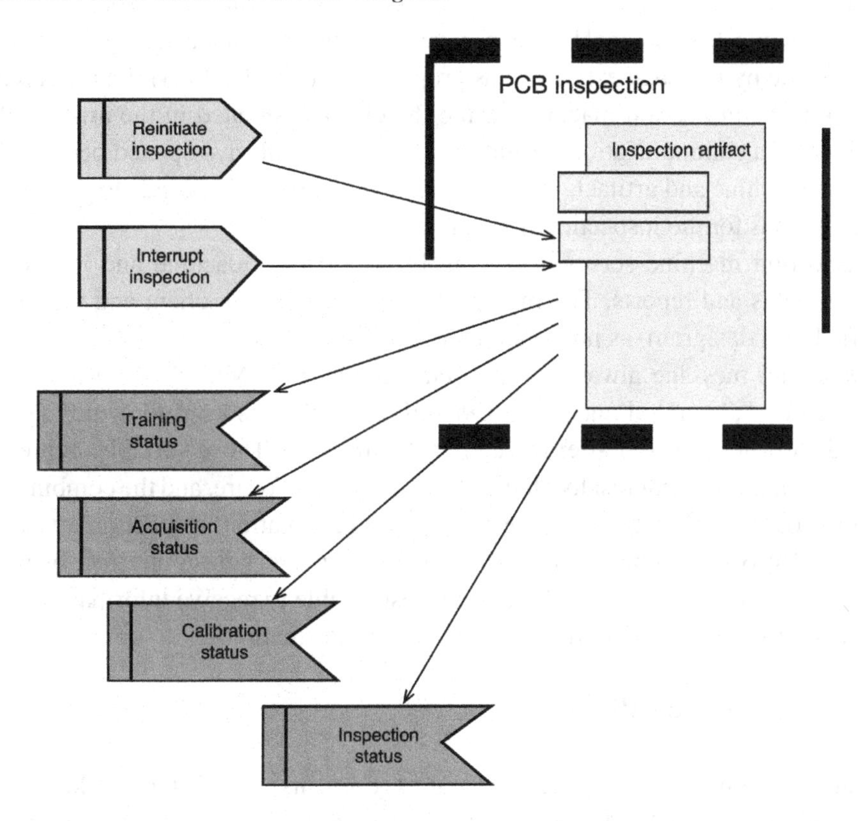

status parameters (also Booleans) are made of training, acquisition, calibration and inspection status, this way, observables properties are logical derivation from those parameters. Table 5 presents inspection artifact refinement.

Table 5. Inspection artifact description

Inspection Artifact	
Description	This artifact is used to obtain information from inspection machine e act as a repository to agent consultancy.
Operations	Status(Boolean true/false)
Observable properties	incmsg: incoming message; training: system training; systrained: system trained; getgold: acquiring reference images; goldok: references images acquired; sysrdy2insp: system ready to inspection
Observable events	

Inspection machine and agent interaction occurs trough an UDP protocol, which set a communication channel between inspection agent and artifact. Agent's interaction is made by message exchange, as proposed by (Roloff, 2014). UDP server is located in the inspection machine, while the client is situated in the artifact, this way there is a communication channel made by socket always opened between inspection machine and artifact, updated periodically (every 2 seconds for the agent, and 5 seconds for the inspection machine).

Inspection machine server is responsible to provide machine and inspection process status and reports. The messages exchange between client and server are codified by a datagram, as presented by Figure 16.

Datagram message always have a start identifier – "MSG;" –, variables are separated by ";" symbol and an end identifier – ";#" – is used to avoid any unwanted garbage localized after the end of the message. Those variables represent agent information of interest located in the inspection software, and the combination among those variables generates agent required information. Inspection artifact is composed by two main functions, a *perceive* and an *act*, serving as interface between *Java-Jason* variables. *Perceive* function is responsible to receive information from inspection machine, and *act* function send back information.

Experimental Results

To validate proposal AOI system, several experiments with different SMD types were made, using a real a PCB with issues. Images are acquired in grayscale, and the following SMD devices were used: 5 capacitors (Figure 17 top left), 5 resistors type

Figure 16. Message exchange using datagram format

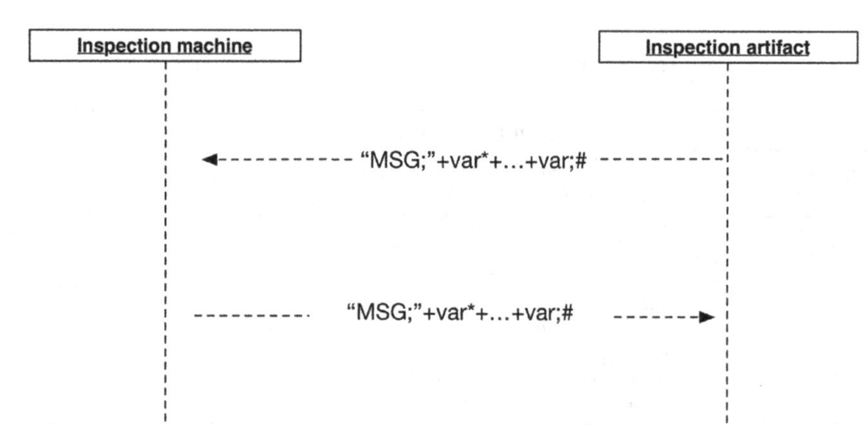

Figure 17. Example of component types and background images

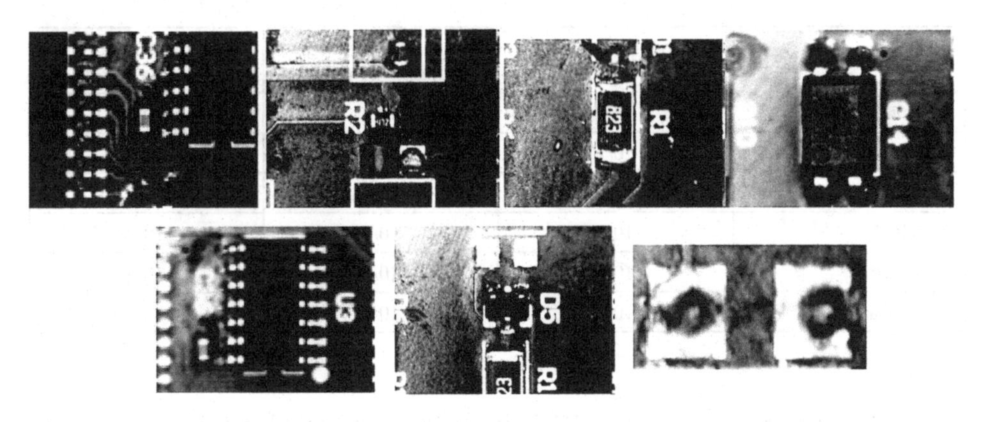

1 (figure top middle left), 5 resistors type 2 (Figure 17 top middle right), 5 power transistors (Figure 17 top right), 5 transistors (17 down left), 5 Schmitt trigger (17 down middle) and complemented with 6 background type images (17 down right), presented by Figure 17.

As already said, this inspection system aims to inspect the existence, positioning and type of components, in this manner, Figure 18 presents shifting (left), rotation (middle) and absence (right) of device, characterizing as failures.

Besides the approached failures, different levels of gain and contrast, undesired spots or other devices, and solder paste application failure were present during tests.

Figure 18. Component with installation failure

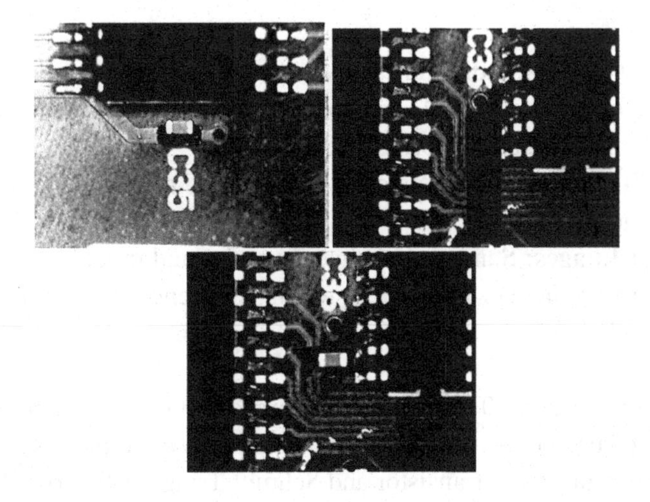

Table 6. ROI pipeline algorithm performance

	Low Brightness Quantity/Errors	Medium Brightness Quantity/Errors	High Brightness Quantity/Errors
Capacitor	10/0	10/0	10/0
Resistor type 1	10/1	10/0	10/0
Resistor type 2	10/1	10/0	10/0
Schmitt Trigger	10/1	10/0	10/0
Transistor	10/0	10/0	10/0
Power Transistor	10/0	10/0	10/0

The first experimental stage is ROI pipeline evaluation. All 6 device types were evaluated under different gain and contrast levels. Independently of the levels used, the capacitor, power transistor and transistor had the expected ROI, while for the Schmitt Trigger and resistors, two patterns of ROI occurred, as presented by Table 6.

Two ROI patterns possibility for the same device is not indicated for this approach, but when the acquired ROI do not match with the expected, it is possible to set new levels of brightness, contrast and gain and get a new image. As the failure frequency shows small, this approach seems to suit the propose. Features extraction process did not present any unexpected behavior.

Device evaluation regards only shifting, rotation, absence or wrong type, not considering solder paste, trails, board or device manufactory failure. Real installation failures were physically produced on boards, while virtual defects were created digitally, aiming simulate device shifting, rotation and absence. 5 experiments were created, in which each of them represents an inspection test.

Experiment 1:

- **Gold Images:** 5 different copies of each component type, made by: capacitor, resistor type 1, resistor type 2, power transistor, transistor and Schmitt Trigger, all correctly installed.
- **PCB Background:** For each gold image, a related PCB background image.
- **Test Images:** Same gold images quantity and models, yet for each device type there is 1 correct, 1 wrong, 1 absence, 1 rotated and 1 shifted model.

Experiment 2:

- **Gold Images:** 10 copies of each component type acquired different levels of brightness, made by: capacitor, resistor type 1, resistor type 2, power transistor, transistor and Schmitt Trigger, all correctly installed.

- **PCB Background:** For each gold image, a related PCB background image.
- **Test Images:** Same gold images quantity and models, yet for each device type there are 2 correct, 2 wrong, 2 absences, 2 rotated and 2 shifted model.

Experiment 3:

- **Gold Images:** 5 different copies of capacitors.
- **PCB Background:** For each gold image, a related PCB background image.
- **Test Images:** Same gold images quantity and models, yet for each device type there is 1 correct, 1 wrong, 1 absence, 1 rotated and 1 shifted model.

Experiment 4:

- **Gold Images:** 10 different copies of capacitors acquired with different levels of brightness.
- **PCB Background:** For each gold image, a related PCB background image.
- **Test Images:** Same gold images quantity and models, yet for each device type there are 2 correct, 2 wrong, 2 absences, 2 rotated and 2 shifted model.

Experiment 5:

- **Gold Images:** 10 different copies of capacitors and transistors acquired with different levels of brightness.
- **PCB Background:** For each gold image, a related PCB background image.
- **Test Images:** Same gold images quantity and models, yet for each device type there are 2 correct, 2 wrong, 2 absences, 2 rotated and 2 shifted model.

For each experiments 10 trials were made, and Table 7 presents the obtained results summary focusing a relation between existing (“*ex*”) and found (“*fd*”) defects.

Most of experiments resulted in a 100% performance, and this number is correlated to ROI pipeline algorithm success, as shown Table 8, whose focus is in the relation between inspection devices images acquired (“*ima*”) and ROI algorithm success (“*suc*”).

As mentioned, a graphical pipeline is used to perform the inspection, thus the preprocessing step output is the input of features extraction step, soon a failure in the first step propagate trough evaluation. This error propagation behavior presented by trial 1 extends to trial 2, where for both cases, wrong devices type issues found by evaluation stage are related to a non-desired ROI. Rotations and shifting

Table 7. Performance relation between existing and found defects

	Absence		Rotated		Shifted		Wrong		Correct		Total	
Experiment	*ex*	*fd*	*ex*	*fd*	*ex*	*fd*	*ex*	*fd*	*ex*	*fd*	*ex*	*fd*
1	6	6	6	5	6	5	6	5	6	5	30	26
2	12	12	12	11	12	11	12	10	12	10	60	54
3	1	1	1	1	1	1	1	0	1	1	5	4
4	2	2	2	2	2	2	2	0	2	2	10	8
5	4	4	4	4	4	4	4	4	4	4	20	20
Total	25	25	25	23	25	23	25	20	25	22	125	112
%	100%		92%		92%		80%		88%		89.6%	

Table 8. Dependency relation between preprocessing, features extraction and evaluation stages

	Preprocessing		Features extraction		Evaluation	
Experiment 1	*ima*	*suc*	*ima*	*suc*	*ima*	*suc*
Trial 1	30	30	30	30	30	30
Trial 2	30	30	30	30	30	30
Trial 3	30	28	30	28	30	28
Trial 4	30	30	30	30	30	30
Trial 5	30	30	30	30	30	30
Trial 6	30	30	30	30	30	30
Trial 7	30	30	30	30	30	30
Trial 8	30	26	30	26	30	26
Trial 9	30	30	30	30	30	30
Trial 10	30	30	30	30	30	30

errors occurs due to the 5% error margin used in the comparison criteria be too sensitive to ROI accuracy, this manner, it may not identify shift and rotation failure or identify inexistent errors. Trials 3 and 4 were made using only one type of device, therefore the found failures were expected, because during ANN training step only one type of device was used. Test 5 did not present any failure.

The communication bridge between inspection agent and machine was successfully established, keeping a dedicated channel in an uninterrupted manner.

FUTURE RESEARCH DIRECTIONS

The inspection approach proposed by this project is based exclusively on device shape and Cartesian position, not considering serigraphy marking or surface texture analysis. This way, further work is developing new evaluation methods based on new features, or integrate already developed method, as proposed by (Szymanski & Stemmer, 2015). Aiming robustness increase in preprocessing and features extraction stages, it is indicated use colorful images, because in each image channel there is a matrix to process, as proposed by (Wu, Zhang, & Hong, 2009). Also, use PCB project CAD (computer aided design) files to train the learning systems is a viable upgrade. Device angle and position inspection is made by comparing values, thus, develop a more reliable method to realize this verification is indicated.

CONCLUSION

This chapter proposes an AOI system for SMT devices, relying upon features extracted from a camera and applied to a backpropagation multilayer perceptron network and a K-Nearest Neighbor classifier. Results have shown that proposal algorithm identify rotation, shifting, missing or wrong device. As the system of this paper was projected for PCB inspection during a SSP, DPMO is not indicated to measure the effectiveness rate of the system. This way, the yield is the parameter used to measure the robustness of the inspection system. The yield obtained in the tests is 99%, which means that the system proposed is effective to inspect SMDs. This approach fails when different components have very similar shapes and sizes, or when the ROI algorithm extracts a wrong contour, because the features used are shape and size related. The error propagates trough the angle and relative position comparative criteria, as further in the classification system.

A highlight is no need to use bad-examples to train the learning systems, avoiding resources waste. Also, this approach does not specify what type of failure to inspect, acting in a more general manner with low computational costs. The communication protocol between inspection agent and machine shows efficient to present case, due to maintain an exclusive open channel to message exchange.

A system limitation is presented when different devices have very similar dimensions, generating similar characteristics for different devices, which in turn proposal ANN method is not capable to distinguish.

This way, proposal approach achieved a high successful rate, showing itself robust enough to inspect SMD devices in a SSP.

REFERENCES

Aha, D. W., Kibler, D., & Albert, M. K. (1991). Instance-based learning algorithms. *Machine Learning*, *6*(1), 37–66. doi:10.1007/BF00153759

Bertrand, O., Queval, R., & Maitre, H. (1982). Shape interpolation using Fourier descriptors with application to animation graphics. *Signal Processing*, *4*(1), 53–58. doi:10.1016/0165-1684(82)90039-1

Beucher, S., & Lantuejoul, C. (1979). Use of Watershed in contour detection. *Proceedings of theInternational Workshop on Image Processing: Real-time Edge and Motion Detection/Estimation*.

Beucher, S., & Meyer, F. (1993). *The Morphological Approach to Segmentation: The Watershed Transformation*. Palo Alto, CA: E. R. Dougherty.

Boissier, O., Bordini, R. H., Hubner, J. F., Ricci, A., & Santi, A. (2013). Multi-agent Oriented Programming with JaCaMo. *Science of Computer Programming*, *78*(6), 747–761. doi:10.1016/j.scico.2011.10.004

Bordini, R. H., Hubner, J. F., & Wooldridge, M. (2007). *Programming Multi-Agent Systems in AgentSpeak Using Jason*. Chichester, UK: John Wiley & Sons.

Canny, J. (1986). A Computational Approach to Edge Detection. *IEEE Transactions on Pattern Analysis and Machine Intelligence*, *PAMI-8*(6), 679–698. doi:10.1109/TPAMI.1986.4767851 PMID:21869365

Cho, H.-J., & Park, T.-H. (2008). Template matching method for SMD inspection using discrete wavelet transform. *Proceedings of the2008 SICE Annual Conference* (pp. 3198-3201). Tokyo: IEEE.

Forsyth, D. A., & Ponce, J. (2002). *Computer Vision: A modern Approach*. Prentice Hall Professional Technical Reference.

Frank, U., Giese, H., Klein, F., Oberschelp, O., Schmidt, A., Schulz, B., et al. (2004). *Selbstoptimierende Systeme des Maschinenbaus - Definitionen und Konzepte* (Vol. 155). Padeborn: W. V. Westfalia Druck.

Granlund, G. H. (1972). Fourier Preprocessing for Hand Print Character Recognition. *IEEE Transactions on Computers*, *C-21*(2), 195–201. doi:10.1109/TC.1972.5008926

Hitomi, K. (1996). *Manufacturing Systems Engineering: A Unified Approach to Manufacturing Technology, Production Management, and Industrial Economics* (2nd ed.). London: Taylor & Francis.

IPC, Association Connecting Electronics Industries. (2014, February 1). *Industry Data: Trends New Orders*. Retrieved from http://www.ipc.org/3.0_Industry/3.1_Industry_Data/2014/Trends-New-Orders-0214.pdf

Kohavi, R. (1995). A Study of Cross-Validation and Bootstrap for Accuracy Estimation and Model Selection.*Proceedings of the 14th international joint conference on Artificial intelligence*.

Krippner, P., & Beer, D. (2004, April 1). AOI testing positions in comparison. In *Viscom Vision Technology*. Hanover, Germany: Viscom AG.

Lin, S.-c., & Su, C.-h. (2006). A Visual Inspection System for Surface Mounted Devices on Printed Circuit Board.*IEEE Conference on Cybernetics and Intelligent Systems* (pp. 1-4). Bangkok: IEEE. doi:10.1109/ICCIS.2006.252237

Luo, B., Gao, Y., Sun, Z., & Zhao, S. (2013). SMT Components Model Inspection Based on Characters Image Matching and Verification.*IEEE International Conference on Green Computing and Communications and IEEE Internet of Things and IEEE Cyber, Physical and Social Computing*. doi:10.1109/GreenCom-iThings-CPSCom.2013.252

Mello, A. R., & Stemmer, M. R. (2015). Inspecting Surface Mounted Devices Using K Nearest Neighbor and Multilayer Perceptron.*International Symposium on Industrial Electronics*. doi:10.1109/ISIE.2015.7281599

Oresjo, S. (2009). *Results from 2007 Industry Defect Level and Test Effectiveness Studies*. APEX and Designers Summit.

Otsu, N. (1979). A Threshold Selection Method from Gray-Level Histograms. *IEEE Transactions on Systems, Man, and Cybernetics*, 62–66.

Pfeifer, T., Schmitt, R., Pavim, A., Stemmer, M., Roloff, M., & Schneider, C. (2010). Cognitive Production Metrology: A new concept for flexibly attending the inspection requirements of small series production.*Proceedings of the 36th International MATADOR Conference*. doi:10.1007/978-1-84996-432-6_81

Roloff, M. L. (2014, September 14). Uma nova Abordagem para a Implementação de um Sistema Multiagente para a Configuração e o Monitoramento da Produção de Pequenas Spéries. *Uma nova Abordagem para a Implementação de um Sistema Multiagente para a Configuração e o Monitoramento da Produção de Pequenas Spéries*. Florianópolis, SC, Brazil: Universidade Federal de Santa Catarina.

Saul, L. K., Jaakkola, T., & Jordan, M. I. (1996). Mean field theory for sigmoid belief networks. *Journal of Artificial Intelligence Research*, 61–76.

Stemmer, M. R., Costa, C. P., Vargas, J., & Roloff, M. L. (2014). Artificial Intelligent Systems for Quality Assurance in Small Series Production. *Key Engineering Materials*, *613*, 279–287. doi:10.4028/www.scientific.net/KEM.613.279

Suzuki, S., & Be, K. (1985). Topological structural analysis of digitized binary images by border following. *Computer Vision Graphics and Image Processing*, *30*(1), 32–46. doi:10.1016/0734-189X(85)90016-7

Szymanski, C., & Stemmer, M. R. (2015, June6). Automated PCB inspection in small series production based on SIFT algorithm.*International Symposium on Industrial Electronics*. doi:10.1109/ISIE.2015.7281535

Wang, Y., Sun, Y., & Zhang, W. (2009). Automatic Inspection of Small Component on Loaded PCB Based on Mean-Shift and Support Vector Machine.*Fifth International Conference on Natural Computation*. doi:10.1109/ICNC.2009.407

Wooldridge, M. (1997). Agent-based Software Engineering. *EE Proceedings on Software Engineering* (pp. 26-37). London: IEEE.

Wu, H.-H., Zhang, X.-M., & Hong, S.-L. (2009). A visual inspection system for surface mounted components based on color features.*2009 International Conference on Information and Automation* (pp. 571-576). Zhuhai: IEEE. doi:10.1109/ICINFA.2009.5204988

Xie, F., Dau, A. H., Uitdenbogerd, A. L., & Song, A. (2013). Evolving PCB visual inspection programs using genetic programming.*28th International Conference on Image and Vision Computing New Zealand*. doi:10.1109/IVCNZ.2013.6727049

Xu, W., & Liu, G. (2010). Multi-Layer Image Segmentation Based on Fuzzy Algorithm in PCB Inspection.*International Conference on Multimedia Technology*. doi:10.1109/ICMULT.2010.5630842

KEY TERMS AND DEFINITIONS

Artificial Intelligence: Computational system with characteristics associated to human behavior, as language comprehension, pattern recognition and learning process.

Artificial Neural Network: Computational methods that represents a mathematical model inspired by human neural network structure, which acquire knowledge over passed experiences.

Automatic Visual Inspection: To inspect in an automated and visual manner products during or after manufacturing, acting as a quality control.

Computational Vision: A science responsible to provide vision to a machine, extracting relevant features by camera, sensors or scanners, and allowing the recognition of objects that compose an image.

Multiagent System: A distributed system where computational components are autonomous, which is, they are able to control their own behavior in the furtherance of their own goals.

Printed Circuit Board: A board used to mechanically support and electrically connect electronic components, using conductive pathways.

Small Series Production: A manufacturing system which adopt several perspectives and ways to produce, generally with a high product variety in a short time space.

Chapter 4
Laser Scanners

Lars Lindner
Autonomous University Baja California, Mexico

ABSTRACT

The presented book chapter provides an overview and detailed description about actual used laser scanner systems. It explains and compares the mainly used coordinate measurement methods, like Time-of-flight, Phasing and Triangulation and Imaging. A Technical Vision System, developed by the engineering institute at the Autonomous University Baja California (UABC) is presented. The mostly used mechanical principles to position a laser beam in a field of view are described and which mechanical actuators are applied. The reflected laser beam gets measured by light sensors or image sensors, which are explained and some principle measuring circuits are provided. The received measuring data gets post-processed by different algorithm or principles, which close the chapter.

INTRODUCTION

Laser scanners are optical devices, which as the rule uses lasers to obtain certain information about surface topography, superficial coordinates or other characteristics, by physical sensing of light spot displacement across a surface relief. In opposition to stylus instruments, laser scanners' measuring is contactless and thereby have higher scanning speeds.

DOI: 10.4018/978-1-5225-0632-4.ch004

One main task of laser scanners is receiving 3D spatial coordinates of the scanned surface, which is mainly realized by optical methods in a very large amount of applications. For example, (Wendt, Franke, & Härtig, 2012) describes a various concepts of large 3D structure measurement using four portable high-accurate tracking laser interferometers. Each of these interferometers tracks a single moving retroreflector and sends their data to a central control, which calculates the 3D point coordinate using the Pythagorean Theorem in space. An optical scanning tomography for time-resolved measurements of kinematic fields in the volume of structures is presented in (Morandi, Breman, Doumalin, Germaneau, & Dupre, 2014). This tomography uses a plane laser beam, which illuminates a transparent probe in layers and the scattered light of each layer is recorded then with a single camera. Measuring spatial coordinates using a non-diffracting beam is presented in (Ma, Zhao, & Fan, 2013). It is used a combined system with a laser beam and an optical system to measure the attitude angle of a probe. Another application of spatial coordinate measurement can be found in (Colombo, Colosimo, & Previtali, 2013), where laser welding is used within the automotive industry. The main objective is to online-monitor the welding conditions, in order to guarantee a constant product quality.

Laser scanning systems, as remote sensing technology are also known as light detection and ranging (Lidar) systems, are widely used in many areas, as well as in mobile robot navigation. (Kumar, McElhinney, Lewis, & McCarthy, 2013) uses an algorithm and terrestrial mobile Lidar data, to compute the left and right road edge of a route corridor. In (Hiremath, van der Heijden, van Evert, Stein, & ter Braak, 2014) a mobile robot is equipped with a Lidar-system, operating with time-of-flight principle, and which navigates in a field of maize. Lidar systems are also used in airborne laser scanners to receive geo-referenced points from terrain. In (Alexander, Erenskjold Moeslund, Klith Bøcher, Arge, & Svenning, 2013) an airborne laser scanner is used to estimate understory light conditions of vegetation plots in semi-open habitants and forest ecosystems.

Another area, where laser scanners are used to measure 3D spatial coordinates can be found in confocal microscopy. (Hong-Seok & Chintal, 2015) are presenting a high speed 3D dental intra oral scanner, which uses a confocal laser scanning microscope (CLSM) to achieve a high accurate 3D image of the oral cavity. A CLSM uses a focused laser beam to point-wise scan the surface of an examined object. Due to the focused laser beam, a CLSM can only measure images one depth level at a time.

Laser scanners can be categorized by design principles, measurement methods and used measuring signals. Design principles describe the characteristics by which a laser scanner is composed and define its construction. Three different main design principles can be determined for laser scanners:

Figure 1. Motorola LI4278 barcode reader
Motorola, 2016.

- Hand-held laser scanners,
- Mobile laser scanners,
- Stationary laser scanners.

Hand-held laser scanners are small and compact devices, which increasingly find applications, where flexibility and mobility play a central role for coordinate determination. In opposite to stationary laser scanners, they have a shorter scanning range. Hand-held laser scanners are often used as barcode readers or hand-held coordinate measurement device. Figure 1 shows an image of a barcode reader brand Motorola® and Figure 2 an image of a hand-held scanner brand Faro®.

Figure 2. Faro hand-held scanner Freestyle 3D
Faro Freestyle, 2016.

Figure 3. Riegl VMX 450 mobile laser scanner
Riegl, 2016.

Mobile laser scanners are related to hand-held laser scanners, which are mainly used for autonomous mobile robot applications. Here they serve for navigation and orientation of the autonomous robot in a typically unknown environment and are part of the robot total system. Mobile laser scanners can also be installed on vehicles and even designed as portable measuring systems for man. Figure 3 shows a mobile laser scanner brand Riegl®.

Figure 4. Faro Focus 3D stationary laser scanner
Faro Focus, 2016.

Table 1. Coordinate measurement methods

Measurement Value	Measurement Method
Flight time of the laser beam	Time-of-flight (TOF)
Phase difference between the transmitted and received laser beam	Phasing
Incident angle of the reflected laser beam	Static and Dynamic Triangulation
Intensity of the reflected laser beam	Imaging

Table 2. Signals for coordinate measurement methods

Signal	Advantages	Disadvantages
Laser	Fast and for long distance.	Measurement readings depend of the scanning surface Cannot used underwater
Sonar	Cost efficient. Independent of the scanning surface and ambient conditions.	Wide beam angle Worse measurement results
Optic	Resembles the way of human vision.	Receive more data, then is necessary for distance measurement Depends on visible light existence

In stationary laser scanners (also called terrestrial laser scanner) the surface geometry of objects is recorded digitally. Thereby a set of discrete sample points is produced, referred to as a point cloud. This point cloud is then converted into user data by post-processing using mathematical methods. The stationary laser scanners, such as cameras, are operated on tripods and are also found permanently installed on buildings or in institutions, for control and supervision purpose. Figure 4 shows a stationary laser scanner brand Faro®.

Laser scanner determines 3D spatial coordinates from objects or 2D coordinates from surfaces, by determining the following measurement values, which corresponds to a specific measurement method shown in Table 1.

Different sensors are used, to determine this coordinates, which can be classified according to three different signals: Laser, Sonar and Optic. Table 2 gives an overview of the advantages of these used signals.

Measuring Methods

The introduced coordinate measurement methods from Table 1 get compared in following Table 3. Thereafter, these methods are described in terms of content and

Table 3. Advantages/disadvantages of coordinate measurement methods

Method	Advantages	Disadvantages	Range
Time-of-flight	Short reaction time. No optical aperture angle.	Expensive for high resolution.	1m – some kilometers
Phasing	Less technical effort for high resolution.	Less measuring distance. Ambiguous for distance more than half wavelength.	Depends of the frequency
Triangulation	Cost-efficient and robust.	Depends of the surface.	1μm – 100m

mathematically. In addition, current research references are given, which are using the respective methods.

Time-Of-Flight

By the TOF method, the laser scanner measures the absolute flight time Δt of an emitted and received laser pulse. Thereby the laser pulse must arrive and be reflected perpendicular on the measured surface, so that the transmitter and receiver can be combined in the same measuring unit. Figure 5 shows the principle of TOF with the emitted (red) and the reflected laser pulse (blue).

With the known speed c of the emitted laser pulse, the distance d between the measuring unit and the surface can be determined using:

$$D = c \cdot \frac{\Delta t}{2} \quad (1)$$

Figure 5. Time-of-flight principle

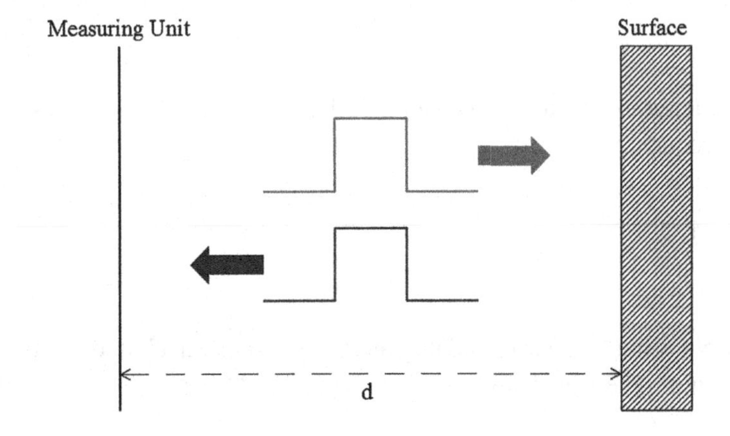

Figure 6. Reflection of a pulse train

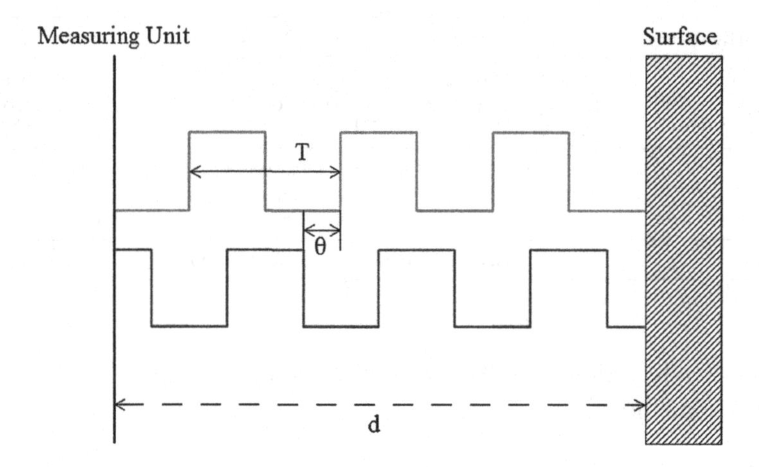

The speed c depends on the refractive index n of the medium and Δt denotes the measured time. Since the laser beam moves with approximately speed of light, short reaction times must be measured. This results either in expensive sensors for time measurement or in a low resolution of the measured time.

Phasing

A harmonic laser signal is sent and overlays with its reflected signal. The distance between the measuring unit and the measured surface is determined with the phase shift between the emitted (red) and the reflected laser pulse train (blue), see Figure 6.

With the period of the laser signal T and the measured phase delay time Δt, the phase difference angle is expressed by:

$$\theta = \frac{\Delta t}{T} 2\pi \tag{2}$$

The wavelength λ of a harmonic signal is defined by its propagation speed c and oscillation frequency f:

$$\lambda = \frac{c}{f} = c \cdot T \tag{3}$$

The double distance $2d$ is calculated by the sum of n wavelengths and the phase difference length, the wavelength λ is substituted with (3):

$$2d = n\lambda + \frac{\Delta t}{T}\lambda = ncT + \Delta tc \qquad (4)$$

Substitution of Δt from (2) and factor out cT, the final expression for the distance d is received:

$$d = \frac{cT}{4\pi}\left[\theta + 2\pi n\right] \qquad (5)$$

With accurate measurement of the phase angle θ the distance between the laser scanner and the examined object is determined. The most disadvantage of this method is the detection of the n signal periods. When the measurement object moves a distance of λ, the measured phase difference θ remains the same value and the measurement will not be unique (unambiguous). For this reason, this method is used for distance measurements of change in dimensions less than the signal wavelength.

Static and Dynamic Triangulation

Triangulation is the classic method for land surveying in geodesy and consists in division and measuring an area by triangles. From these triangles, the coordinates of a point (triangle vertex) are determined using the opposite side and the two adjacent angles of this side. The formulas for coordinate calculation are presented in section post-processing of measured values.

Laser scanners, which use triangulation, typically consist of a transmitter and receiver unit. The transmitter sends a constant or pulsed laser beam and positions it in the field of view (FOV) of the laser scanner. The laser beam gets reflected (totally or diffused) on the examined object surface and measured by the receiver unit. Figure 7 shows a scheme of laser triangulation, where the transmitter unit is indicated as the positioning laser (PL) and the receiver unit as the scanning aperture (SA). The use of the triangulation method in laser scanner systems can be of static or dynamic nature (Sergiyenko, et al., 2011). Static nature is referred to an operation mode without moving the positioning laser or the scanning aperture, which results in a statically behavior of the triangulation. The principle of dynamic triangulation varies the emitter and receiver angle, to eliminate this statically behavior and increase the laser scanner FOV.

Figure 7. Laser triangulation

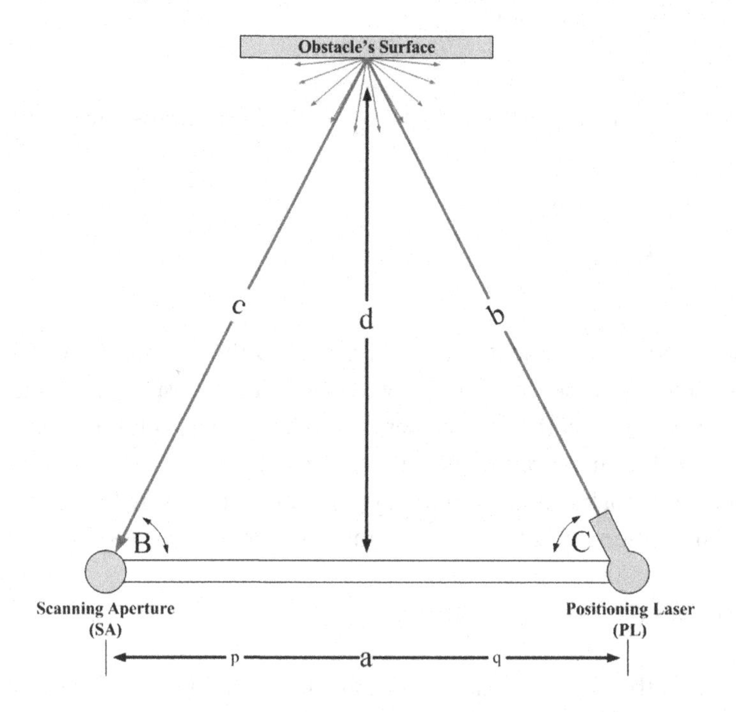

Imaging

A laser scanner, which detects the intensity of the reflected laser beam, is called a laser imaging scanner. These scanners measure the surface reflectance by processing the return signal energy from the laser beam. With this measuring method the

Figure 8. Grayscale panoramic image

true surface reflectance can be determined, as there are no distortions by shadows or darkening effects (Hug & Wehr, 1997). The result is similar to a grey-scale photo image of the surface. (Hinks, Carr, Gharibi, & Laefer, 2015). Figure 8 shows an example of a grayscale panoramic image captured by a laser scanner.

One measuring device which uses this method is the Scanning Laser Altitude and Reflectance Sensor (ScaLARS) developed by the Institute of Navigation of the University of Stuttgart in Germany in the mid-1990s. The ScaLARS uses the phasing measuring technique with a signal frequency of 1 MHz and 10 MHz, to obtain spatial data of a geographical area and to find the slant range from an airborne platform to the ground (Shan & Toth, 2008). Thereby this laser scanner belongs to the type of terrestrial laser scanners (TLS), which are used to survey and map a specific geographic area from an aircraft (Fang, Huang, Zhang, & Li, 2015). Another application of the imaging laser scanning can be found in paper (Garcia-Talegon, et al., 2015). Here, the intensity values of the reflected laser beam are used to assess pathologies of a complex building facade.

Technical Vision System

A laser scanning system must fulfill two main tasks: positioning a laser beam in the FOV and measuring the reflected laser beam on the scanned surface using sensors. The positioning laser is represented by PL and the measuring device by SA in Figure 7. The distance between PL and SA depends on the used measuring method. In the TOF method the PL and SA are combined in the same measuring unit, which can be found actually as laser rangefinders. Figure 9 shows a laser rangefinder from the company LightWare which can be used for open-source applications. It has included the positioning laser, detecting optics and amplifiers, to measure 3D coordinates in a range of 0.5 – 9m.

Figure 9. LightWare OSLRF-01

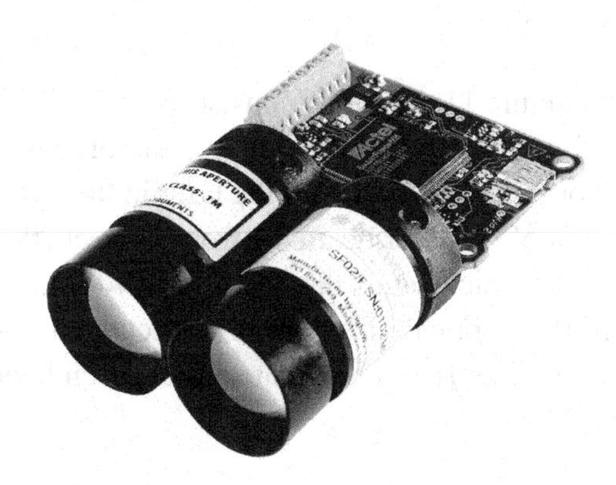

Figure 10. Technical Vision System (TVS)
Rodríguez-Quinoñez, et al., 2013.

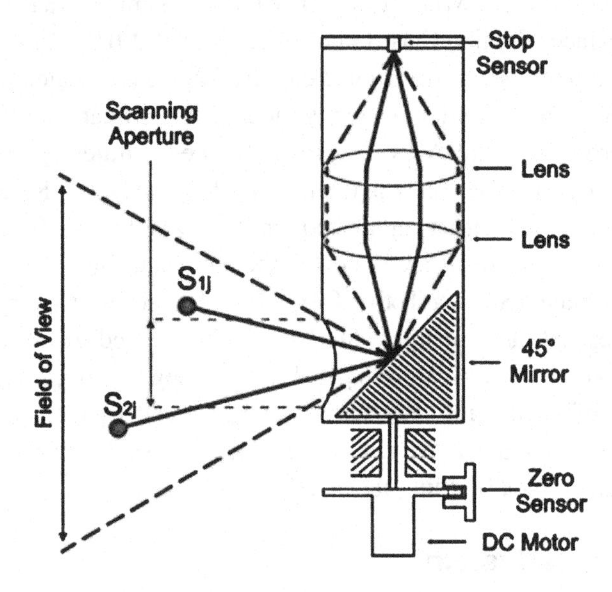

The distance between PL and SA in the triangulation method is greater, than in the TOF method, due to the forming triangle. A Technical Vision System (TVS, prototype No.2), developed by the engineering institute at the Autonomous University Baja California (UABC), is shown in Figure 10. This TVS uses the principle of Dynamic Triangulation for 3D spatial coordinate measurement.

With the measured angles and the formulas (13) – (15) the prototype determines the two-dimensional coordinates in space. The developed successor prototype No.3 (Lindner, et al., 2015) uses servomotors instead of stepping motors, to increase measurement speed and accuracy of these coordinates. The TVS is composed of two principle parts: the Positioning Laser and the Scanning Aperture.

Positioning Laser

The Positioning Laser (Figure 11) basically consists of two 45° staggered mirrors (I and II), a red laser (635 nm, 20 mW) and a stepping motor, which rotates mirror II over a transmission (worm drive) to position the laser in the FOV. Figure 11 shows the principal optical path of the laser beam. Mirror I deflects the laser beam vertically upward to mirror II, which sends the beam forward into the FOV. The used mirrors are made of N-BK7 optical glass with a protected aluminum coating. Due to the 45° construction of the reflecting mirror surface, each incident laser beam is deflected precisely by 90°.

Figure 11. Positioning Laser (PL)
Rodríguez-Quinoñez, et al., 2013.

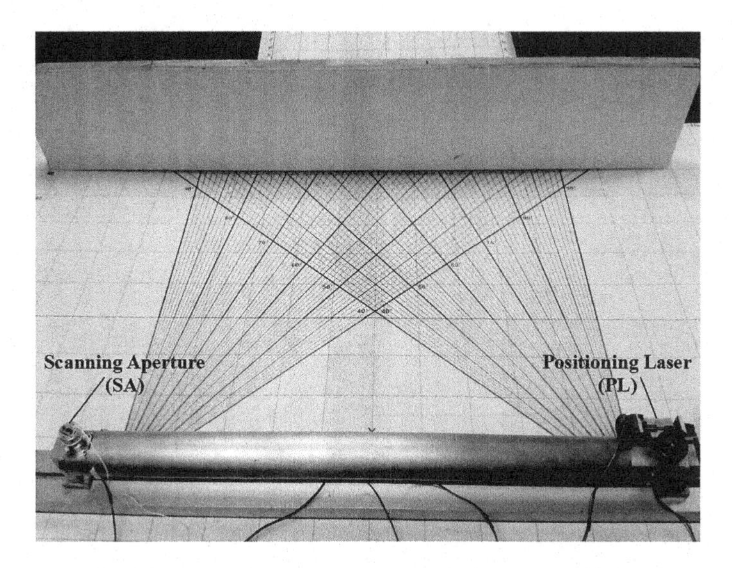

Due to the existing face clearance in every transmission, the successor prototype No.3 has mirror I directly connected the servomotor, without using gearings or transmissions.

Scanning Aperture

The reflected laser beam from the examined object surface is received by the Scanning Aperture (Figure 12), which mainly consists of one 45° mirror, two biconvex lenses, one stop and one zero sensor. The zero sensor produces a reference signal each full revolution and starts two counters to integrate pulses. The first counter integrates pulses $N_{2\pi}$ for each full revolution and the second counter integrates pulses N_A until the detection of an object. The mirror is rotated with (nearly) constant speed by a DC motor and reflects the received laser signal towards the lenses, which focus them for the stop sensor (high speed photodiode). When an object is encountered, that is the laser beam gets reflected, the photodiode is measuring the light incidence distribution of the reflected laser beam. By searching the power spectrum centroid of this distribution (Flores-Fuentes, et al., 2014), a reference value can be produced, which defines the moment when an object has been detected. At this moment, the second pulse counter is stopped.

Figure 12. Scanning Aperture (SA)
Rodríguez-Quinoñez, et al., 2013.

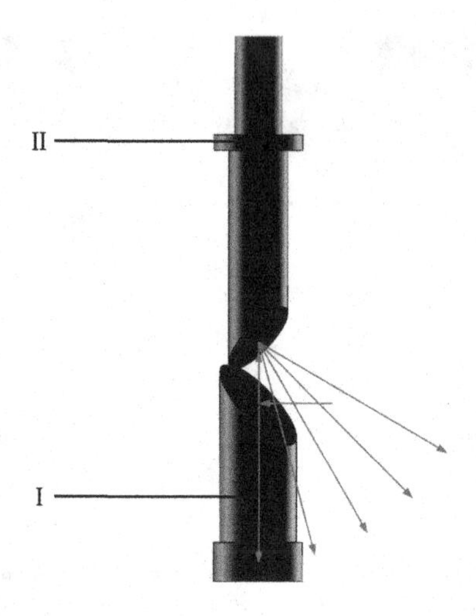

Then, the received laser beam angle is calculated using following equation:

$$B_{ij} = 2\pi \cdot \frac{N_A}{N_{2\pi}} \tag{6}$$

where $N_{2\pi}$ is the pulse number of the first pulse counter and N_A of the second pulse counter. Experimentations have shown, that in due to the nearly constant speed of the DC motor, the number of pulses N_A varies every revolution. This produces different measurement results for a same beam angle B_{ij}. Hence, in the developed successor prototype No.3, the DC motor is substituted by a servomotor in closed-loop speed control and a new design of the scanning aperture is used, depicted in Figure 13. This figure shows the optical path of the reflected laser beam (I), the servomotor (II), a 45° mirror (III), two biconvex lenses (IV), an optical filter (V) and a high-speed photodiode (VI). Like the servomotor has an integrated incremental encoder with index channel, there is no need for a zero sensor. The used coherent laser signal comes from a 10 mW laser module with 514 nm wavelength. The optical filter thereby has a center wavelength of 515 nm, full width-half wavelength of 10 nm and a transmission of ≥ 45%. The photodiode is used with a transimpedance amplifier (see section photodiode), to convert the small photocurrent of the diode to a greater voltage, which thereby can be measured using an analog-digital converter (ADC). The servomotor is digitally controlled by a microcontroller using a robust PI-algorithm, which stabilize the mirror III rotating speed around a reference value.

Figure 13. Scanning Aperture (SA)
Lindner, et al., 2015.

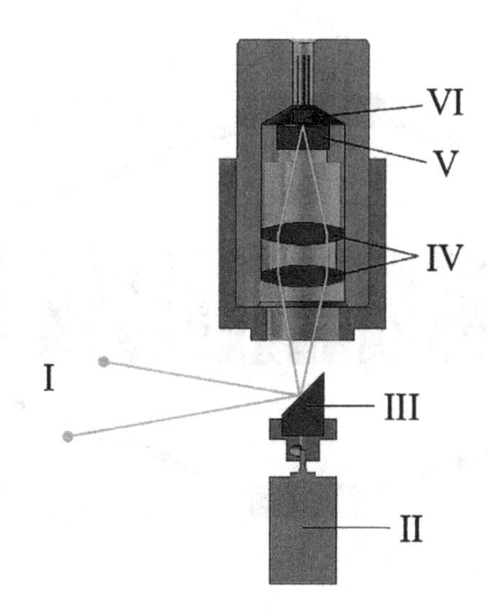

Laser Beam Positioning

Laser scanning systems, which position the laser beam using actuators, expand their FOV to a multidimensional scanning area (2D or 3D). This allows laser scanning systems to high dynamically capture data of their surroundings and in shortest time. For laser beam positioning in the FOV, four different mechanical principles are generally used:

- Scanning Mirrors (Figure 11),
- Galvoscanner (Figure 18 and Figure 19),
- Scanning Refractive Optics (Figure 20),
- Optical Modulators (Figure 21 and Figure 22).

These principles are explained in the following with examples from the present optical market and also current research references are given, which use the respective methods.

Scanning Mirrors

Scanning mirrors are optical devices with a planar surface, to deflect a laser beam in a certain direction. The previously introduced TVS use scanning mirrors (Figure 11), which reflect and position a laser beam in the laser scanner FOV. The principle of

Figure 14. Schematic of an amplified piezoelectric actuator

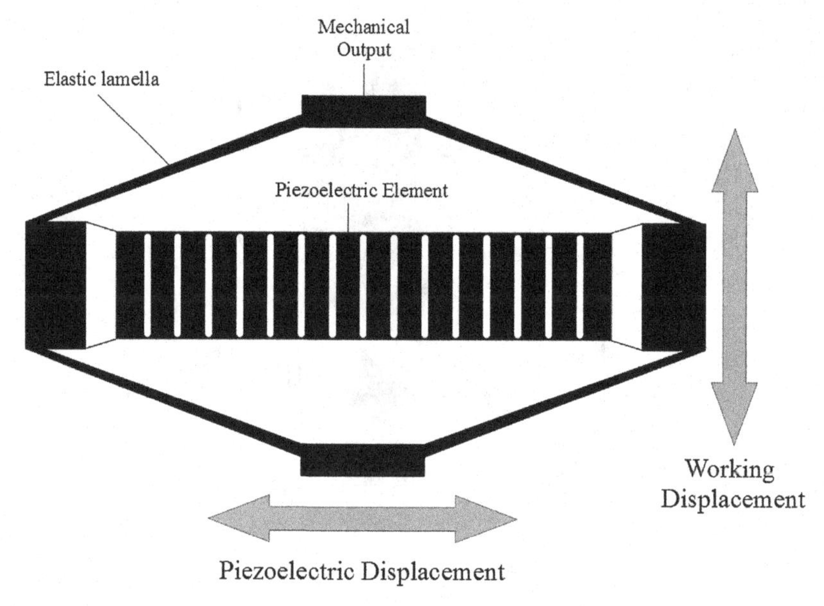

operation is explained previously. Also, the laser scanner Faro Focus 3D (Figure 4) uses a rotating mirror in order to position the laser beam in the FOV (Garcia-Talegon,

Figure 15. Technology schema of a stepping motor

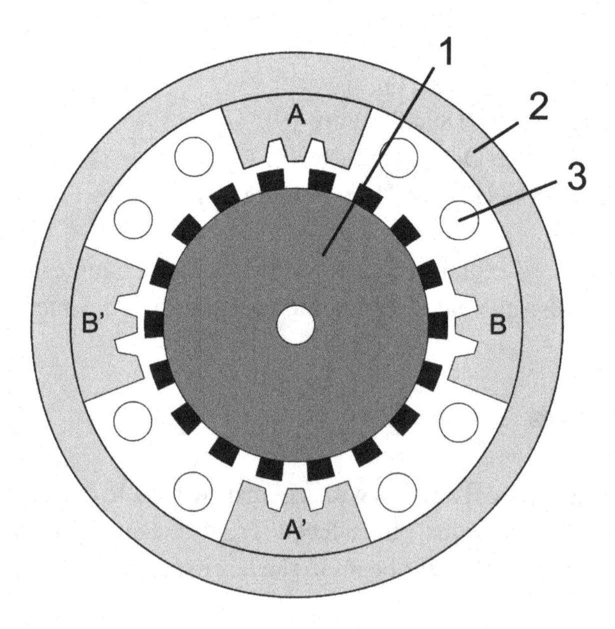

Table 4. Advantages/disadvantages of mechanical actuators for scanning mirrors

Method	Advantages	Disadvantages
Piezoelectric Actuators	Continuous working principle, no resolution. Offer a very fast response time. No interaction with magnetic fields.	Very small movement range. Still not widely used in the industry. Currently still high cost.
Stepping Motors	No need for position feedback, while $M_L < M_M$. With modern digital electronic, drive control is very easy. Availability in small sizes and with low cost.	Stalling of motor, when $M_L > M_M$, loss of steps. Torque drop-off at high speed. Due to their discontinuous behavior, a stepping motor can't reach any angle in his field of movement.
DC Motors	Can supply about twice their rated torque for short period (acceleration). Can theoretical reach any angle in his field of movement.	Require more complex drive circuits and positional feedback for accurate positioning.

et al., 2015). This laser scanner uses the continuous wave phase shift measuring method to determine 3D spatial coordinates. Another laser scanning device which uses mirrors to position the laser beam can be found described in paper (Hinks, Carr, Gharibi, & Laefer, 2015). Scanning mirrors in actual laser scanning systems are moved generally by three different mechanical actuators:

- Piezoelectric Actuators,
- Stepping Motors,
- DC Motors,
- Servomotors.

The advantages and disadvantages comparing piezoelectric actuators, stepping motors and dc motors are summarized in Table 4.

Piezoelectric Actuators

Piezoelectric actuators use the piezoelectric effect, which describes the property of certain materials to deform mechanically, when applying an electrical charge to it. Classical piezoelectric materials have a very low degree of strain which is too small for practical applications and thereby has to be amplified. A schematic image of an amplified piezoelectric actuator is shown in Figure 14 (Sixta, Linhart, & Nosek, 2013). It can be seen the piezoelectric element, which oscillations are amplified by the elliptic steel shell structure.

Another use and more description of these amplified actuators can be found in paper (Xiang, Chen, Wu, Xiao, & Zheng, 2010). Here an optical tip-tilt actuator based on piezoelectric actuators is used to deflect a laser beam.

Stepping Motors

A stepping motor is a special type of a synchronous motor (Figure 15), with a cogged permanent magnet rotor (1) and particular stator poles (2), whose windings (3) are driven by periodic current pulses (Fischer, 2003). An electronic circuit for stepping motors generates these pulses in the correct order to control the desired speed and direction. One pulse moves the stepping motor in one step-angle, which is a typical parameter of each motor. The pulse sequences generate a rotating magnetic field, that the rotor follows exactly while the load torque is less than the rated torque $M_L < M_M$. Under this condition, the driving of a stepping motor is simpler than that of a servomotor, and so no additional sensors are required for position feedback for example. Due to his discontinuous operation mode, the stepping motor is suitable as a switching mechanism (printer, quartz clock, etc.) and for positioning tasks. The previously introduced TVS (Figure 10) uses stepping motors, to position an emitted laser line-by-line in the FOV of the system (Sergiyenko O., 2010). Stepping motors have a discrete behavior, due to their stepwise rotation of the motor shaft and hence the laser light can only be positioned in discrete points, which represent an integer multiple of the step-angle. Despite the fact, that modern stepping motors in certain operating modes have step sizes less than one degree, they will always possess unreachable angles in their field of motion (FOM). Thus, by use of stepping motors in the developed TVS, the FOV is discretized and its resolution depends on the examined scanning object distance. Targets, which have dimensions less than the FOV resolution in a certain distance, cannot be scanned or captured.

DC Motors

A direct current motor (Figure 16) is a machine, which is energized by direct current and mainly consists of the following parts summarized in Table 5.

The stator produces the direct magnetic field by the main poles and the stator windings. The stator also can be made of permanent magnets in smaller machines to about 30 kW. The main pole ends are designed to pole shoes. The rotating part of the machine consists of a shaft, which carries the grooved armature. The grooves sustain the rotor windings, which consists of separate individual coils, which are electrically connected to the commutator. When rotating the shaft, this commutator realize the current reversal in the coils so, that a torque is produced on the shaft in actual rotating direction by the Lorentz force. Current is delivered to the commuta-

Figure 16. Technology scheme of a direct current motor

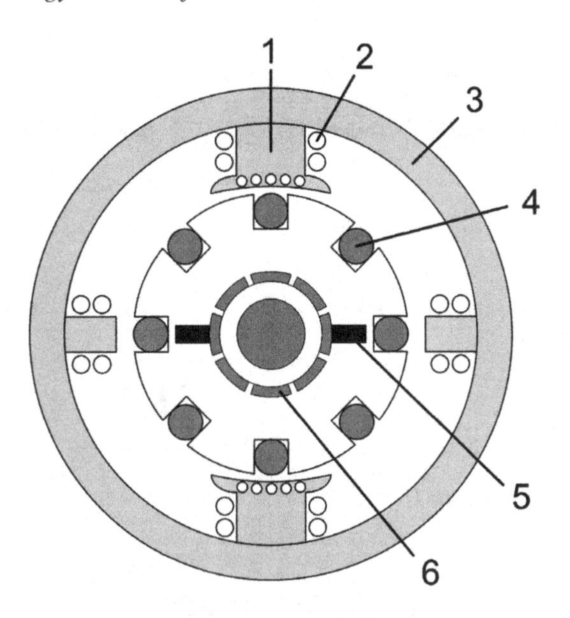

Table 5. Principal parts of a DC motor

1	Main pole	4	Rotor winding (armature)
2	Stator winding	5	Carbon brush
3	Stator (yoke ring)	6	Commutator

tor over the carbon brushes, which usually are made of self-lubricating graphite. These are being compressed with constant pressure to the commutator sections by a spring.

Servomotors

An electrical motor in a closed-loop leads to the servomotor, which has its original function as an auxiliary drive in the field of machine tool (Schulze, 2008). Like the term servomotor only refers to a servo system, where a servo actuator controls a motor in a closed-loop, this motor can be any type of. Often are used DC motors in a cascade structure, like is showing Figure 17.

This cascade structure contains three loops for current, speed and position control. The tachogenerator can be a resolver, an incremental or an absolute encoder. Each higher-level controller calculates the setpoint for the lower-level control loop, which must be faster, than his higher-level control. The actual values of current,

Figure 17. Typical cascade control of a servo system

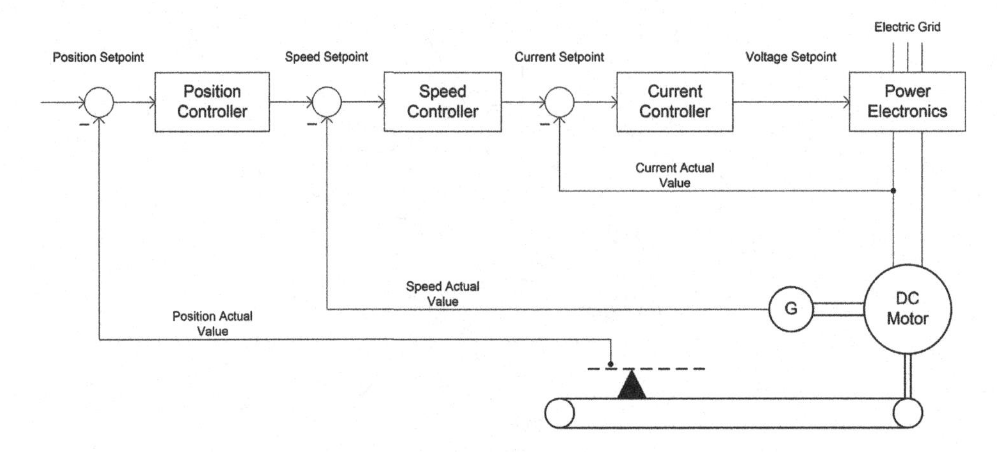

Figure 18. Raylase Miniscan II
Raylase, 2016.

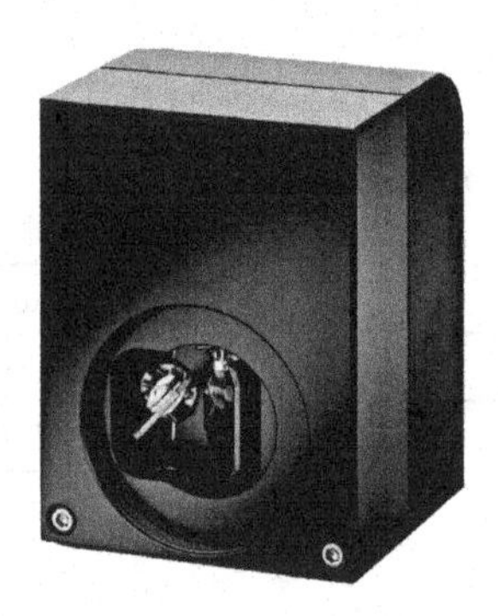

Figure 19. Scanlab dynAXIS S
Scanlab, 2016.

Figure 20. Rotating Risley prism

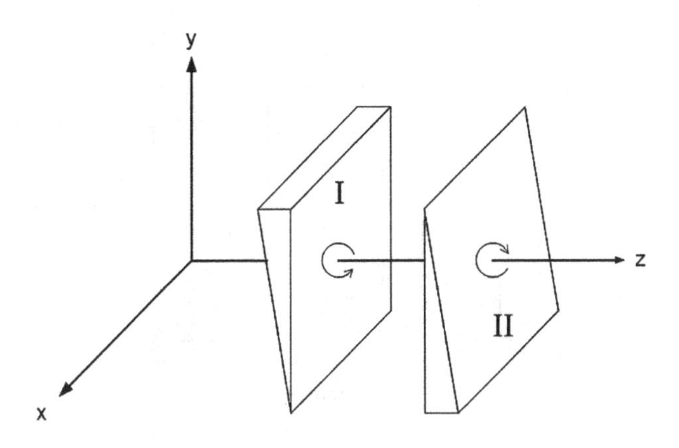

Figure 21. Acousto-optic deflector

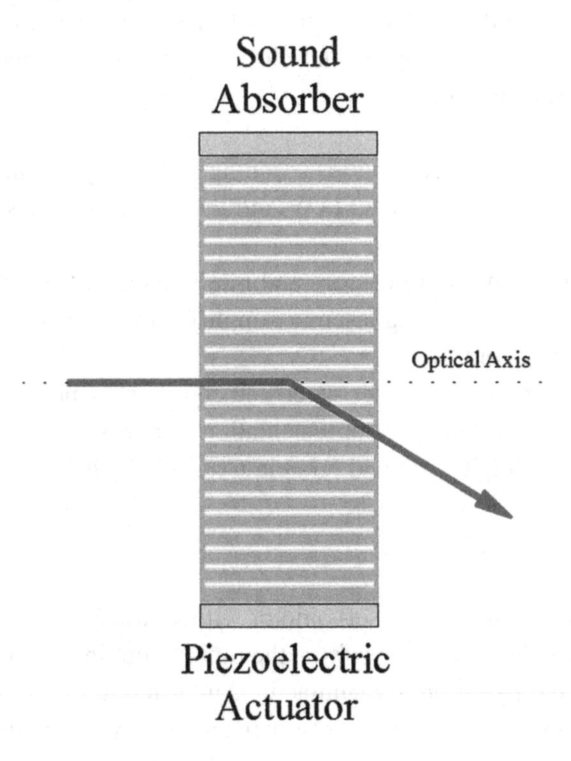

Figure 22. Electro-optic deflector

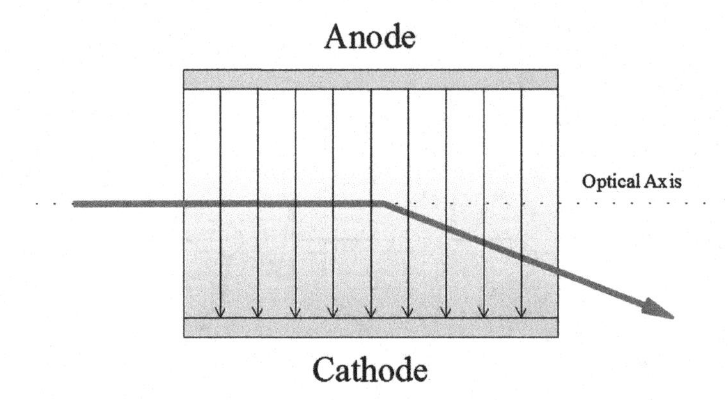

speed and position are detected by sensors and feed backed to their control loop. Paper (Lindner, et al., 2015) proposes dead zone elimination in optical laser dynamic triangulation by substitution of the used stepping motors with servomotors in the previously introduced TVS (Figure 10). By substitution of these stepping motors with servomotors in open- and closed-loop configuration, the following objectives shall be achieved:

1. Substitution of the discontinuous scanning method by a continuous one.
2. Elimination of the discretized FOV and thereby dead zones in the scanning plane.
3. Acceleration of the step response and minimization of the relative positioning error of the positioning laser by simplification of mechanical design and optimization of control.
4. Increasing of the laser scanner speed and coordinates measurement accuracy.
5. Opening the possibility for implementation of new scanning algorithms, e.g. variable step scanning method (Garcia-Cruz, et al., 2014).

Galvoscanner

Complete systems, consisting of servomotor with a built-in mirror are called Galvoscanner. These galvanometers position the rapidly rotating mirror in a very short time and are offered by different manufactures in different versions. Figure 18 and Figure 19 show modern galvoscanners from the manufactures Raylase and Scanlab.

Galvoscanners shall achieve high speeds and accelerations, thus there are optimized to have less friction in the bearings, commutator rings, etc. and less moment of inertia of all moving masses. An application of a 2D galvoscanner in combination with a fiber laser to provide spatial power modulation can be found for example in paper (Solchenbach & Plapper, 2013).

Scanning Refractive Optics

Scanning refractive optics is used with "Risley Prism", where the laser beam is positioned in the FOV only by rotation of two prisms independently about an axis parallel to the normal of their adjacent faces. Using these two prisms, a laser beam can be deflected in any direction within a ray cone (Wang, y otros, 2010). When these two prisms are oriented as in Figure 20, the laser beam is running parallel to the optical z-axis and got shifted in direction of y-axis. In this orientation, minimum deflection of the laser beam occurs. Maximum deflection is achieved through 180° rotation of the second prism II. In this orientation, both prism refract the laser beam in the same direction and function as a single prism with double prism angle.

An application of the Risley Prism for laser beam positioning can be found in paper (Song & Chang, 2012), where the optical system with MgF_2 ellipsoid dome and imaging system uses a pair of identical rotating Risley prisms. More research about refractive optics using Risley Prism can also be found in (Yang, Lee, Xu, Zhang, & Xu, 2001), (Yang Y., 2008) and (Tame & Stutzke, 2010).

Optical Modulators

Optical modulators are optical devices which produce a diffraction grating in a transparent solid to modulate incident light in frequency, propagation direction and intensity using sound waves or an electric field. An optical modulator, which uses the Acousto-Optic Effect based on sound waves, is shown in Figure 21. By compression and stretching of the solid, these sound waves periodically cause a change of material density and thus a varying refractive index n (Roemer & Bechthold, 2014).

An optical modulator, which uses the Electro-Optic Effect based on an electric field, is shown in Figure 22. Atom and molecular structures are disturbed in their position, orientation and shape by the electric field and thereby the refractive index n is changing periodically (Roemer & Bechthold, 2014).

Laser Beam Measuring

The reflected laser beam has to be detected and measured. Thereby, the intensity of the electromagnetic wave is converted into electrical signals by us of appropriate light sensors or image sensors. The laser intensity is measured in one point by light sensors and in a row or array by image sensors. The light intensity distribution in time is measured and other magnitudes obtained by application of mathematical calculations of this distribution. For example, the energy center of the distribution, the exceeding of a threshold value or the frequency of a periodic signal is determined. These magnitudes are post-processed then, to determine typically 3D coordinates of the examined object. The following describes physical principles and applications of the most commonly used light and image sensors in laser scanning systems.

Light Sensors

Light sensors convert light into a voltage, current or frequency, which can be further processed to evaluate the incident light. Here, the time depend signal (transient behavior) is being measured and / or static values such as maximum amplitude of the signal, threshold being exceeded, etc. determined. By application of algorithms and image processing methods to these measured signals, characteristics like 3D coordinates, surface reflectance, etc. are obtained. Light sensors can be distinguished by use of three electric components: Photoresistor, Photodiode and Phototransistor.

Photoresistor

The photoresistor (LDR, light dependent resistor) is an electrical component whose resistance decreases with higher light incidence. This process is based on the inner photoelectric effect, where electrons are emitted by the energy transfer of incident light. Advantages of photoresistors are simplicity, robustness and a quite large dynamic range from $k\Omega$ to $M\Omega$. The main disadvantage is the very low response time from a few *ms* to *min*. The measuring is performed via a simple voltage divider (Figure 23).

Figure 23. Photoresistor voltage divider

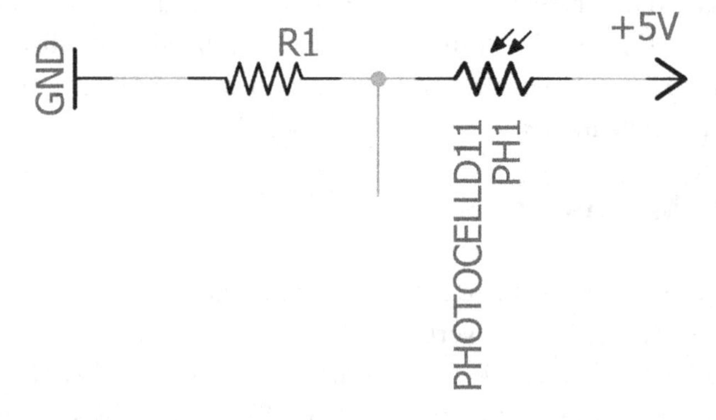

Figure 24. Schematic of a PIN photodiode

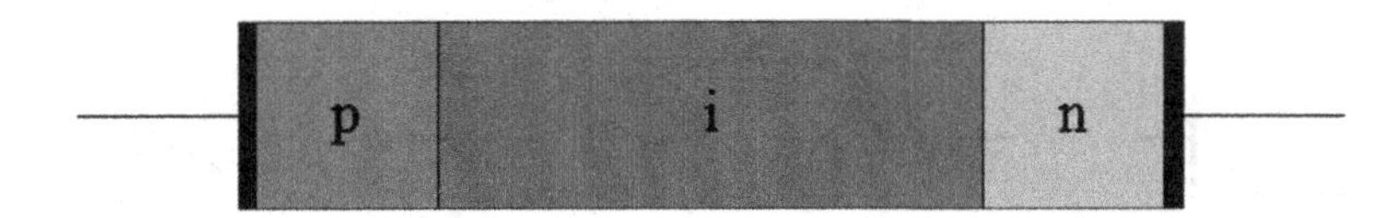

Photodiode

The photodiode is a semiconductor device in which free charge carriers are generated in a PN junction by light incidence. These charge carriers diffuse along the built-in electric field to the edges of the photodiode and form a voltage. Photodiodes have a response time in nanosecond range, but provide only a very small photocurrent from ηA to μA. Figure 24 shows the scheme of a PIN photodiode, which in addition to the PN junction (p-n) has an intrinsic layer (i), to increase the speed of response of a conventional photodiode. PIN photodiodes are more cost-effective and thermostable than conventional photodiodes and are mainly used in optoelectronics as light sensors.

Photodiodes can be used in the following three operating modes, with the three basic circuits:

Photo-Voltage Operation

In this mode, the photodiode is operated in forward direction as a photocell or solar cell. The photocell produces a very low power of the small exposure area ($< 1mm^2$). The output voltage depends logarithmically on the incident light power and reaches its maximum in the open-circuit voltage U_L. In this operation mode, the diode characteristic is running in the fourth quadrant of the current-voltage curve and is strongly temperature-sensitive. Also in this mode, the photodiode is relatively slow and is not suitable for detecting fast signals. This mode is mostly used to measure the brightness, e.g. in light meter or photometer. Unlike the photoresistor, in this mode, no external power source is needed (Figure 25).

Figure 25. Photo-voltage operation

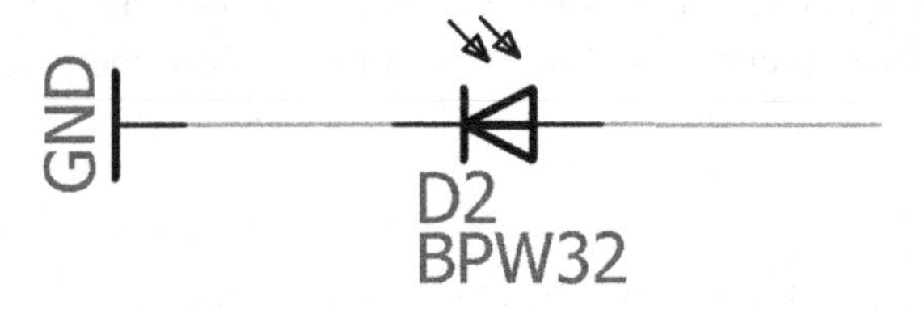

Figure 26. Current source with load resistor

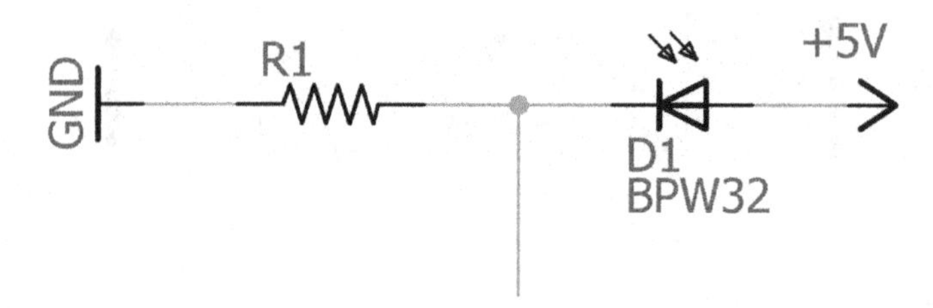

Current Source with Load Resistor

In this mode, the photodiode is operated in reverse direction and the applied voltage must not exceed the permissible photodiode reverse voltage. Over a voltage divider with a relatively high resistance, a light-dependent voltage across the load resistor can be measured. The output voltage is linearly proportional to the incident light power. This mode is faster than the photo-voltage operation mode, but generally slower than the transimpedance amplifier mode. The circuit is shown in Figure 26.

The cutoff frequency of this circuit can be calculated by the following approximation:

$$f_c = \frac{1}{2\pi \cdot R_1 \cdot C_D}, \tag{7}$$

with C_D representing the junction capacitance of the diode. This junction capacitance decreases with the applied voltage, so that the cut-off frequency increases with increased voltage. In this operation mode, the diode characteristic is running in the third quadrant of the current-voltage curve, why the differential resistance is very high and the generated photocurrent hardly depends on the operating voltage.

Current Source with Transimpedance Amplifier

In this mode, the photodiode is operated also in reverse direction and almost in short circuit ($U = 0V$) as a light-dependent current source. A transimpedance amplifier produces a virtual ground at the anode terminal of the photodiode and generates a linear voltage at the output from a very small photocurrent (Figure 27).

Since the voltage across the photodiode is zero, no capacity must be reloaded. Thus, high cut-off frequencies and short reaction times of the photodiode are possible:

$$f_c = \sqrt{\frac{GBP}{2\pi \cdot R_1 \cdot C_S}} \tag{8}$$

Figure 27. Current source with transimpedance amplifier

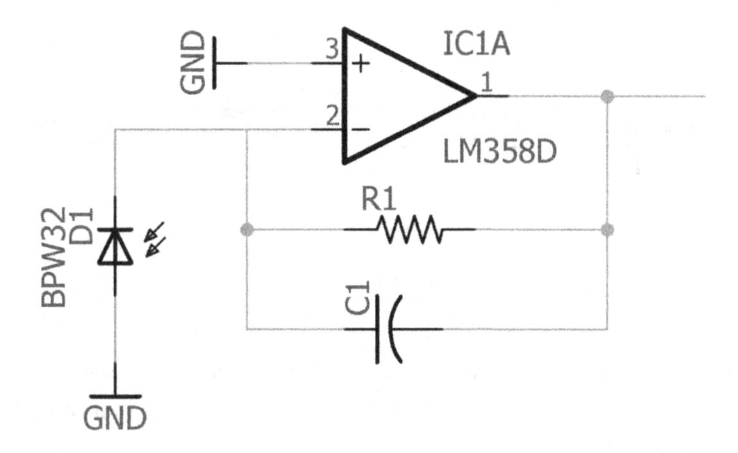

GBP represents the gain bandwidth product of the op-amp and:

$$C_S = C_D + C_I \tag{9}$$

the sum of the photodiode junction capacitance C_D and the op-amp input capacitance C_I. Important here is the use of the compensation capacitance C_I. If this capacitance is this too small or missing at all, the transimpedance amplifier oscillates. The calculation of this capacitance is made using following formula:

$$C_1 = \frac{R_D}{R_1} \cdot C_I, \tag{10}$$

with R_D the photodiode junction resistance.

Phototransistor

The phototransistor is a semiconductor device, which in principle represents the combination of a photodiode with a bipolar transistor. The p-n junction of the transistor base-collector is open to a light source. Thus, small photocurrents are amplified by the bipolar transistor to the collector current. The ratio of collector to photocurrent describes the current amplifier factor (small-signal short-circuit forward current transfer ratio). Due to this amplification, phototransistors are much more sensitive than photodiodes. Since the base of a phototransistor remains unconnected, the

Figure 28. Phototransistor with load resistor

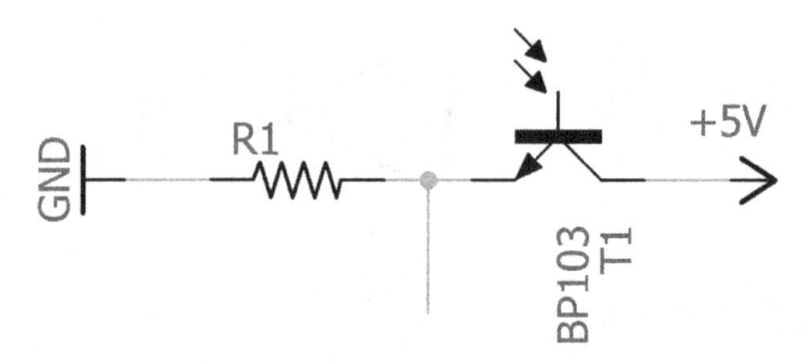

base-emitter zone is not actively discharged, whereby the phototransistor has a slow discharging behavior. The circuit of the phototransistor is like that of a photodiode with load resistor (Figure 28).

Image Sensors

Image sensors consist of a matrix-type array of light sensitive cells, which through individual pixel amplifiers and / or amplifiers in rows and columns convert the generated charges to voltages and which by use of attached hardware convert these voltages into usable image data. Typically, they apply a two-dimensional transformation of the incident image using semiconductor-based sensors that can record light to the mid-infrared range. Here are only described and compared the two main types of image sensors: CCD and CMOS sensors. Apart from these two sensors also exist image sensors for other spectral regions. As an example for these sensors are: Microbolometer, Pyroelectric Sensors and Focal Plane Arrays. The basic principle

Figure 29. Pixel cell schematic

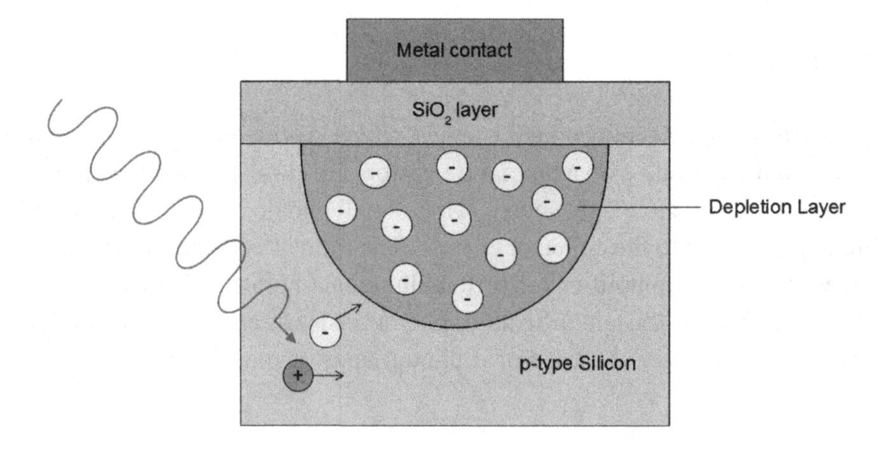

of charge generation in both CCD and CMOS sensors is the same and is shown in Figure 29. It shows the schematic representation of one-pixel cell. CCD and CMOS sensors are mostly used in cameras for private or industrial applications.

The fact that image sensors record a two-dimensional image of an object gives them the advantage over light sensors that they do not need to run a scanning process of the examined object. Thereby the opto-mechanical design is simplified in many cases, because no rotating mirrors are needed, for example. There are image sensors that read all charged cells at the same time and thus make a snapshot of the object. Due to time constraints or limited hardware, other image sensors read the cells in rows or columns, where may occur the rolling-shutter-effect.

CCD Sensor

An incident light photon creates an electron-hole pair. The free electrons are collected then in a depletion layer below the silicon dioxide layer. The depletion layer stores the charges similar to a capacitor, which is why these charges must be removed, before the open-circuit voltage of the pixel cell is reached. The amount of charge is proportional to the amount of incident light and therefore the charge is an analog magnitude. The individual pixel cells are not connected directly, but are separated from each other, depending on the sensor type, by straps or potential barriers. This on one hand prevents the free charge carriers (electrons) to overrun from one cell to his neighbor cell, on the other hand these straps are also used for reading the pixel cell content. Therefore, depending on the type of sensor, the light-sensitive cells are not filling the entire sensor, but a part of the sensor surface is used as a transport area or keep-out zone. The space occupied by the pixel cells on the sensor chip is called bulk factor, which in real sensor chips is always well under 100%. The charges from the pixel cells are read-out via vertical and horizontal shift registers. In a shift register, the content of a cell is transferred to an adjacent cell. This can be achieved through different circuits, which build up or down the potential barriers. During the read-out time, further charge is not allowed to be collected by the cells. Two generally used charge read-out principles are available: interline transfer and frame transfer. Due to the limited capacity of the pixel cell and charge transfers, blooming can occur, which is characterized by vertical or horizontal light streaks around a local overexposure of a recorded image.

CMOS (Active Pixel Sensor)

CMOS sensors use the same basic physical principle for light measurement as CCD sensors and are sensors manufactured with CMOS-technology, which describes semiconductor devices with two complementary MOSFETs (n- and p-channel) on a common substrate. These sensors are also called Active Pixel Sensors, as each pixel cell has an amplifier circuit in order to convert the stored charges into voltage values. Therefore, each pixel can be read separately and in any order, which gives advantages for time-critical or region of interest (ROI) applications. In ROI applications only some of the pixel cells are read out. Due to the direct reading of each pixel cell no charges are moved from one cell to another, which drastically reduces the blooming effect. Disadvantage of CMOS sensors is the higher space requirement through active pixel cell charge amplification and thus a smaller bulk factor compared to CCD sensors, which leads to a smaller overall pixel sensitivity. Also the small bulk factor allows little or no noise filtering, so CMOS sensors in general have a poorer signal to noise ratio than CCD sensors.

Post-Processing of Measured Values

The measured magnitudes of the laser beam by light or image sensors are getting post-processed, to improve the measurement results, to reduce the measurement error and to calculate typically 3D spatial coordinates of the examined object. With Artificial Neural Network, measurement errors from the laser scanner TVS (Figure 10) can be decreased and thereby this post-processing method can be used to calibrate the scanner. After improve the measurement results, Trigonometric Functions are used in laser triangulation (Figure 7), to calculate 3D spatial coordinates. Another post-processing method to determine 3D coordinates is represented by the Iterative Closest Points (ICP) algorithm, where a set of measured coordinates (point cloud) is adapted to a reference model of the examined object. The following describes these three post-processing methods and gives references to recent research.

Artificial Neural Network

While problems, for whose exact solution exist short-time algorithm, can be solved much faster by a computer than a human, the human brain needs less time for tasks like face recognition. Artificial neural networks (ANN) try to apply the human brain operation to machines using networks of artificial neurons (Samarasinghe, 2006). Information's are processed by the activation of neurons each other by means of directed connections. Neural networks are characterized by their ability to learn. They can learn a task based on training examples, without being programmed explicitly. Further advantages are the high parallelism in information processing, the high fault tolerance and the distributed knowledge representation, whereby a damaged neuron leads only to a relatively small knowledge loss (Valluru B. & Hayagriva, 1995).

Figure 30. ANN with two neurons

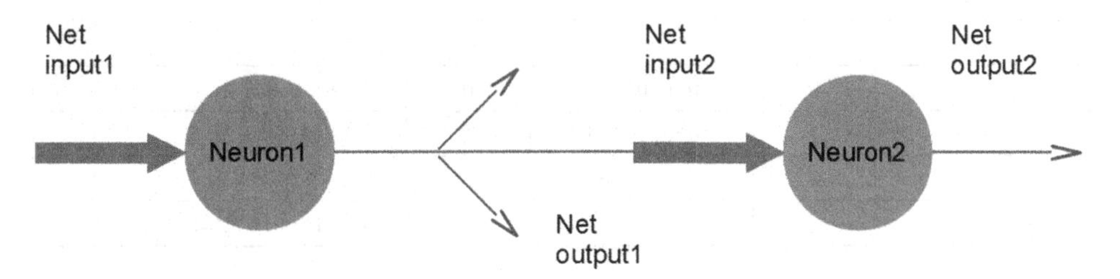

ANN´s therefore have a biological model; each neuron is characterized by the neuron body, the net input and the net output of this neuron. Figure 30 shows the basic principle of ANN with models of two neurons.

Each neuron body has an activation state Z determined by an activation function f_a. The net input of each neuron defines the allowed input values for the activation function f_a. It is used (quasi-) continuous and discrete ranges of values. The output function f_o of each neuron describes the net output generated by the activation state Z. Each connection between all neurons is a directed and weighted connection. The matrix W of all connections is called weight matrix. The propagation function f_p indicates how to calculate the net input of a neuron from the outputs of other neurons and the connection weights. It is the weighted sum of the outputs of the previous neuron.

As an example for an artificial neural network with 4 neurons, the most widely known XOR network shall serve. Figure 31 shows the network with the neurons and the weighted connections.

Figure 31. XOR network with 4 neurons

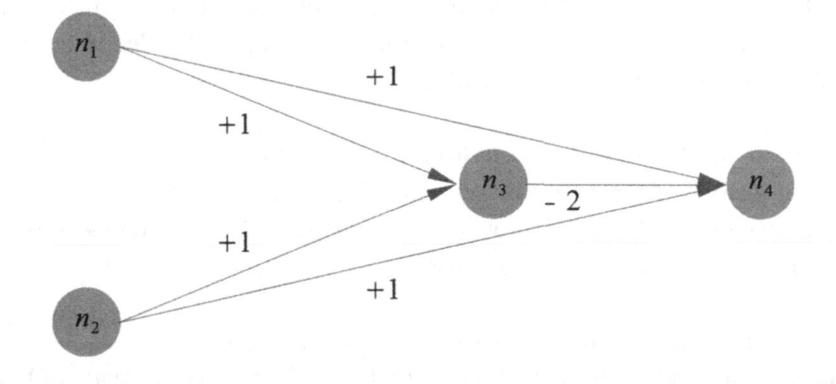

Table 6. XOR network with 4 neurons

o_1	o_2	n_3	o_3	n_4	o_4
0	0	0*1+0*1=0	0	0*1+0*1+0*(-2)=0	0
0	1	0*1+1*1=1	0	0*1+1*1+0*(-2)=1	1
1	0	1*1+0*1=1	0	1*1+0*1+0*(-2)=1	1
1	1	1*1+1*1=2	1	1*1+1*1+1*(-2)=0	0

Neurons 1 and 2 represent the net input, containing the only allowed binary values 0 and 1. The propagation function of this network is:

$$n_j(t) = \sum_i o_i \cdot w_{ij} \tag{11}$$

As activation function, a threshold function (12) with a threshold value S_j for each neuron is used. The output function here is the identity.

$$o_j(t) = \begin{cases} 1 : n_j(t) \geq S_j \\ 0 : otherwise \end{cases} \tag{12}$$

A table can now be generated, with $S_3 = 1.5$ and $S_4 = 0.5$, that calculates the output of neuron 4 for all possible inputs, which corresponds to a binary XOR combination between the neuron 1 and the neuron 2 output (see Table 6).

An artificial neural network can modify itself by using special learning algorithms. These learning algorithms aim to produce a desired output using a predetermined input into the neural network. By re-entering of training samples, the connections strength between the neurons is modified. It is aimed to minimize the error between the expected and the actual network output. The learning process basically consists of the following methods:

- Development or deleting of connections and neurons,
- Modification of the activation or output function,
- Modification of the connection strength (change of connection weights),
- Modification of the threshold values.

The last two methods represent parameter optimizations problems, for whose solution exist a variety of optimization methods of numerical mathematics. Optimization methods are commonly implemented in electronic computers and produce an approximation solution by iteration, starting from an initial value. One optimization method for example is defined by the Levenberg-Marquardt-Algorithm (LM), which represents a better variant of the Gauss-Newton-Algorithm (GN) and describes a numerical optimization algorithm for solving nonlinear curve fitting problems using the least squares method. Here, a model function is fitted to a series of measurements of function values such, that the mean square root of the distances is minimized. The LM algorithm is significantly more robust then the GN algorithm, which means that it converges with high but not guaranteed probability, even under poor start conditions.

The LM algorithm is described in further detail in paper (Rodríguez-Quinoñez, Sergiyenko, Gonzalez-Navarro, Basaca-Preciado, & Tyrsa, 2013) and used to train an artificial neural network, that transform the measured coordinates from the TVS (Figure 10) to the actual 3D coordinates. The once under certain boundary conditions trained ANN corrects continuously the measured data then, to minimize the measurement error.

Trigonometric Functions

Trigonometric functions are used for calculation of spatial coordinates by the triangulation measuring method. To measure distances using triangulation be considered a triangle in the two-dimensional plane (Figure 32), where the distance d is searched.

The distance a and the angles β and γ are known (measured) and by using the law of sines, the distance d, p and q can be determined: (Sergiyenko, et al., 2011)

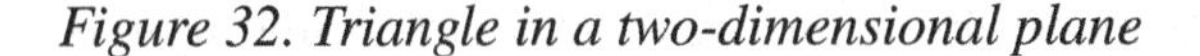

Figure 32. Triangle in a two-dimensional plane

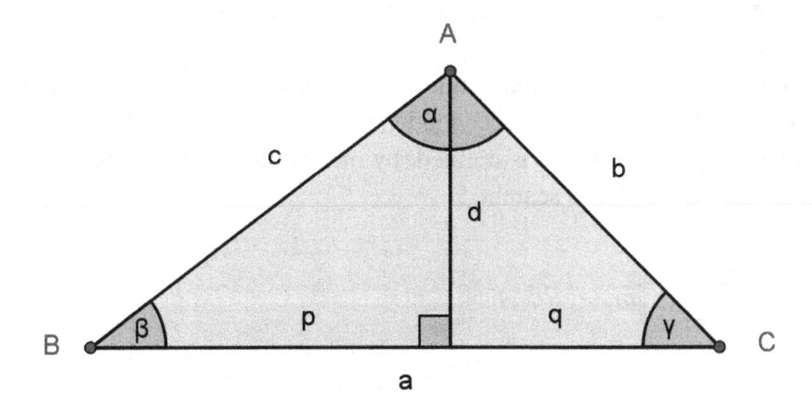

$$d = a \cdot \frac{\sin\beta \cdot \sin\gamma}{\sin\left(\beta + \gamma\right)} \tag{13}$$

$$p = \frac{d}{\tan\beta} = a \cdot \frac{\cos\beta \cdot \sin\gamma}{\sin\left(\beta + \gamma\right)} \tag{14}$$

$$q = \frac{d}{\tan\gamma} = a \cdot \frac{\sin\beta \cdot \cos\gamma}{\sin\left(\beta + \gamma\right)} \tag{15}$$

More research and use of Dynamic Triangulation can be found in paper (Sergiyenko, et al., 2011), (Rodríguez-Quinoñez, Sergiyenko, Gonzalez-Navarro, Basaca-Preciado, & Tyrsa, 2013), (Basaca-Preciado, et al., 2014), (Lindner, Sergiyenko, Tyrsa, & Mercorelli, 2014) and (Lindner, et al., 2015).

Iterative Closest Points (ICP) Algorithm

The ICP algorithm can calculate 3D spatial coordinates from an object using point clouds, which are taken from different perspectives. The point clouds overlap each other and are adjusted to each other by the ICP algorithm. Thereby, a point from one point cloud is assigned to the nearest point from another point cloud. The point clouds are adjusted by rigid coordinate transformation so (rotation matrix $\boldsymbol{R}$ and translation vector $\boldsymbol{t}$), that the root mean square of the distances of all points belonging to each other is minimized (Besl & McKay, 1992). The algorithm is mainly used in image registration, as various captured images are combined into a panorama image, as 3D coordinates are determined also in photogrammetry. The algorithm is also used in reconstruction of 2D surface profiles using 3D point clouds and using various programming languages.

There is a record of measured points $\boldsymbol{d}_j$ with $|D|=N_d$ that are transformed to the points $\boldsymbol{m}_i$ with $|M|=N_m$ of a reference model, by means of rotation $\boldsymbol{R}$ and translation $\boldsymbol{t}$. The following cost function is defined, by means of which the minimum of the sum of squared distances is searched:

$$E\left(R,t\right) = \sum_{i=1}^{N_m}\sum_{j=1}^{N_d} \omega_{ij} \left\| m_i - \left(Rd_j + t\right)\right\|^2 \tag{16}$$

with $\omega_{ij} = 1$, if $m_i = d_j$, otherwise $\omega_i = 0$. Now two things have to be calculated: the mutually associated points of the two point clouds and the transformations R and t which minimize the cost function (16). The ICP algorithm is iteratively executed as long, until all measured data points are uniquely associated with the points in the reference model (Surmann, Nuechter, & Hertzberg, 2003).

REFERENCES

Alexander, C., Erenskjold Moeslund, J., Klith Bøcher, P., Arge, L., & Svenning, J.-C. (2013, July). Airborne laser scanner (LiDAR) proxies for understory light conditions. *Remote Sensing of Environment*, *134*, 152–161. doi:10.1016/j.rse.2013.02.028

Basaca-Preciado, L., Sergiyenko, O., Rodríguez-Quinoñez, J., Garcia, X., Tyrsa, V., Rivas-Lopez, M., & Starostenko, O. (2014, March). Optical 3D laser measurement system for navigation of autonomous mobile robot. *Optics and Lasers in Engineering*, *54*, 159–169. doi:10.1016/j.optlaseng.2013.08.005

Besl, P., & McKay, N. (1992, February). A method for registration of 3-D shapes. In *Pattern Analysis and Machine Intelligence* (pp. 239-256).

Colombo, D., Colosimo, B., & Previtali, B. (2013, January). Comparison of methods for data analysis in the remote monitoring of remote laser welding. *Optics and Lasers in Engineering*, *51*(1), 34–46. doi:10.1016/j.optlaseng.2012.07.022

Fang, W., Huang, X., Zhang, F., & Li, D. (2015, February). Intensity Correction of Terrestrial Laser Scanning Data by Estimating Laser Transmission Function. *IEEE Transactions on Geoscience and Remote Sensing*, *53*(2), 942–951. doi:10.1109/TGRS.2014.2330852

Faro Focus. (2016). *Faro Focus 3D stationary laser scanner* [Online Image]. Retrieved from http://www.faro.com

Faro Freestyle. (2016). *Faro Hand-held Scanner Freestyle* [Online Image]. Retrieved from http://www.faro.com

Fischer, R. (2003). Elektrische Maschinen (12th ed.). Carl Hanser Verlag GmbH & Co. KG.

Flores-Fuentes, W., Rivas-Lopez, M., Sergiyenko, O., Gonzalez-Navarro, F., Rivera-Castillo, J., Hernandez-Balbuena, D., & Rodríguez-Quinoñez, J. (2014, May). Combined application of Power Spectrum Centroid and Support Vector Machines for measurement improvement in Optical Scanning Systems. *Signal Processing*, *98*, 37–51. doi:10.1016/j.sigpro.2013.11.008

Garcia-Cruz, X., Sergiyenko, O., Tyrsa, V., Rivas-Lopez, M., Hernandez-Balbuena, D., Rodríguez-Quinoñez, J., & Mercorelli, P. (2014, March). Optimization of 3D laser scanning speed by use of combined variable step. *Optics and Lasers in Engineering, 54*, 141–151. doi:10.1016/j.optlaseng.2013.08.011

Garcia-Talegon, J., Calabres, S., Fernandez-Lozano, J., Inigo, A., Herrero-Fernandez, H., Arias-Perez, B., & Gonzalez-Aguilera, D. (2015). Assessing Pathologies On Villamayor Stone (Salamanca, Spain) By Terrestrial Laser Scanner Intensity Data. *Remote Sensing and Spatial Information Sciences, XL-5/W4*, 445-451.

Grayscale Panoramic Image. (n.d.). Retrieved from http://www.iff.fraunhofer.de

Hinks, T., Carr, H., Gharibi, H., & Laefer, D. (2015, June). Visualisation of urban airborne laser scanning data with occlusion images. *ISPRS Journal of Photogrammetry and Remote Sensing, 104*, 77–87. doi:10.1016/j.isprsjprs.2015.01.014

Hiremath, S., van der Heijden, G., van Evert, F., Stein, A., & ter Braak, C. (2014, January). Laser range finder model for autonomous navigation of a robot in a maize field using a particle filter. *Computers and Electronics in Agriculture, 100*, 41–50. doi:10.1016/j.compag.2013.10.005

Hong-Seok, P., & Chintal, S. (2015). Development of High Speed and High Accuracy 3D Dental Intra Oral Scanner. *Proceedings of the25th DAAAM International Symposium on Intelligent Manufacturing and Automation*. Elsevier. doi:10.1016/j.proeng.2015.01.481

Hug, C., & Wehr, A. (1997). Detecting And Identifying Topographic Objects In Imaging Laser Altimeter Data. *IAPRS, 32*, 19-26.

Kumar, P., McElhinney, C., Lewis, P., & McCarthy, T. (2013, November). An automated algorithm for extracting road edges from terrestrial mobile LiDAR data. *ISPRS Journal of Photogrammetry and Remote Sensing, 85*, 44–55. doi:10.1016/j.isprsjprs.2013.08.003

Lindner, L., Sergiyenko, O., Rodríguez-Quinoñez, J., Tyrsa, V., Mercorelli, P., Fuentes-Flores, W., . . . Nieto-Hipolito, J. (2015). Continuous 3D scanning mode using servomotors instead of stepping motors in dynamic laser triangulation. *Industrial Electronics (ISIE), 2015 IEEE 24th International Symposium on* (pp. 944-949). Buzios: IEEE.

Lindner, L., Sergiyenko, O., Tyrsa, V., & Mercorelli, P. (2014, June 01-04). An approach for dynamic triangulation using servomotors. *Proceedings of the 2014 IEEE 23rd International Symposium on Industrial Electronics (ISIE)* (pp. 1926-1931). Istanbul: IEEE.

Ma, G., Zhao, B., & Fan, Y. (2013, May). Non-diffracting beam based probe technology for measuring coordinates of hidden parts. *Optics and Lasers in Engineering*, *51*(5), 585–591. doi:10.1016/j.optlaseng.2012.12.011

Morandi, P., Breman, F., Doumalin, P., Germaneau, A., & Dupre, J. (2014, July). New Optical Scanning Tomography using a rotating slicing for time-resolved measurements of 3D full field displacements in structures. *Optics and Lasers in Engineering*, *58*, 85–92. doi:10.1016/j.optlaseng.2014.02.007

Motorola. (2016). *Motorola LI4278 Barcode Reader* [Online Image]. Retrieved from https://portal.motorolasolutions.com

Raylase. (2016). *Raylase Miniscan II* [Online Image]. Retrieved from http://www.raylase.de

Riegl. (2016). *Riegl VMX 450 mobile laser scanner* [Online Image]. Retrieved from http://www.riegl.com

Rodríguez-Quinoñez, J., Sergiyenko, O., Gonzalez-Navarro, F., Basaca-Preciado, L., & Tyrsa, V. (2013, February). Surface recognition improvement in 3D medical laser scanner using Levenberg–Marquardt method. *Signal Processing*, *93*(2), 378–386. doi:10.1016/j.sigpro.2012.07.001

Roemer, G., & Bechthold, P. (2014). Electro-optic and Acousto-optic Laser Beam Scanners. *Proceedings of the8th International Conference on Laser Assisted Net Shape Engineering*. Elsevier.

Samarasinghe, S. (2006). *Neural Networks for Applied Sciences and Engineering: From Fundamentals to Complex Pattern Recognition* (1st ed.). Auerbach Publications.

Scanlab. (2016). *Scanlab dynAXIS S* [Online Image]. Retrieved from http://www.scanlab.de

Schulze, M. (2008). Elektrische Servoantriebe: Baugruppen mechatronischer Systeme. Carl Hanser Verlag GmbH & Co. KG.

Sergiyenko, O. (2010). Optoelectronic System for Mobile Robot Navigation. *Optoelectronics, Instrumentation and Data Processing*, *46*(5), 414–428. doi:10.3103/S8756699011050037

Sergiyenko, O., Tyrsa, V., Basaca-Preciado, L., Rodríguez-Quinoñez, J., Hernandez, W., Nieto-Hipolito, J., & Starostenko, O. (2011). Electromechanical 3D Optoelectronic Scanners: Resolution Constraints and Possible Ways of Improvement. In Optoelectronic Devices and Properties. InTech.

Shan, J., & Toth, C. (2008). *Topographic Laser Ranging And Scanning*. Boca Raton, FL: CRC Press. doi:10.1201/9781420051438

Sixta, Z., Linhart, J., & Nosek, J. (2013). Experimental Investigation of Electromechanical Properties of Amplified Piezoelectric Actuator. *Proceedings of the 2013 IEEE 11th International Workshop of Electronics, Control, Measurement, Signals and their application to Mechatronics (ECMSM)* (pp. 1-5). Toulouse: IEEE.

Solchenbach, T., & Plapper, P. (2013, December30). Mechanical characteristics of laser braze-welded aluminium–copper connections. *Optics & Laser Technology*, *54*, 249–256. doi:10.1016/j.optlastec.2013.06.003

Song, D., & Chang, J. (2012, February29). Super wide field-of-regard conformal optical imaging system using liquid crystal spatial light modulator. *Optik (Stuttgart)*, 2455–2458.

Surmann, H., Nuechter, A., & Hertzberg, J. (2003, December). An autonomous mobile robot with a 3D laser range finder for 3D exploration and digitalization of indoor environments. *Robotics and Autonomous Systems*, *45*(3-4), 181–198. doi:10.1016/j.robot.2003.09.004

Tame, B., & Stutzke, N. (2010). Steerable Risley Prism antennas with low side lobes in the Ka band. *Proceedings of the 2010 IEEE International Conference on Wireless Information Technology and Systems (ICWITS)* (pp. 1-4). IEEE.

Valluru, B. R., & Hayagriva, R. (1995). *C++ Neural Networks and Fuzzy Logic* (2nd ed.). M & T Books.

Wang, L., Liu, L., Sun, J., Zhou, Y., Luan, Z., & Liu, D. (2010, September). *Large-aperture double-focus laser collimator for PAT performance testing of inter-satellite laser communication terminal*. Academic Press.

Wendt, K., Franke, M., & Härtig, F. (2012, December). Measuring large 3D structures using four portable tracking laser interferometers. *Measurement*, *45*(10), 2339–2345. doi:10.1016/j.measurement.2011.09.020

Xiang, S., Chen, S., Wu, X., Xiao, D., & Zheng, X. (2010, February). Study on fast linear scanning for a new laser scanner. *Optics & Laser Technology*, *42*(1), 42–46. doi:10.1016/j.optlastec.2009.04.019

Yang, S., Lee, K., Xu, Z., Zhang, X., & Xu, X. (2001, October). An accurate method to calculate the negative dispersion generated by prism pairs. *Optics and Lasers in Engineering*, *36*(4), 381–387. doi:10.1016/S0143-8166(01)00055-0

Yang, Y. (2008, November21). Analytic Solution of Free Space Optical Beam Steering Using Risley Prisms. *Journal of Lightwave Technology*, *26*(21), 3576–3583. doi:10.1109/JLT.2008.917323

Chapter 5
Machine Vision Application on Science and Industry:
Real-Time Face Sensing and Recognition in Machine Vision - Trends and New Advances

Oleg Starostenko
Universidad de las Americas Puebla, Mexico

Vicente Alarcon-Aquino
Universidad de las Americas Puebla, Mexico

Claudia Cruz-Perez
Universidad de las Americas Puebla, Mexico

Viktor I. Melnik
Kharkiv National Technical University of Agriculture, Ukraine

Vera Tyrsa
Kharkiv National Technical University of Agriculture, Ukraine

DOI: 10.4018/978-1-5225-0632-4.ch005

ABSTRACT

Face detection, tracking and recognition is still actual field of human centered technologies used for developing more natural communication between computing artefacts and users. Analyzing modern trends and advances in this field, two approaches for face sensing and recognition have been proposed. The first color/shape-based approach uses sets of fuzzy saturated color regions providing face detection by Fourier descriptors and recognition by SVM. The second approach provides fast face detection by adaptive boosting algorithm, and recognition based on SIFT key point extraction into eye-nose-mouth regions has been improved using Bayesian approach. Designed systems have been tested in order to evaluate capability of the proposed approaches to detect, trace and interpret faces of known individuals registered into facial standard databases providing correct recognition rate in range of 94.5-99.0% with recall up to 46%. The conducted tests ensure that both approaches have satisfactory performance achieving less than 3 seconds for human face detection and recognition in live video streams.

INTRODUCTION

Face sensing and recognition is still open problem despite the fact that there are many publications and scientific reports in this research field. Currently, automatic face recognition is used in affective computing and usability testing, robot guidance, distance education, security monitoring, automatic environment inspection, marketing studies and consumption, control and surveillance systems, training people for more effective interpersonal communication, and others (Zhang, 2010; Garcia-Amaro, 2012; Singla, 2014; Granger, 2014; Nappi, 2015). In the mobile autonomous systems particularly, if this process is carried out in real time, the ability of a robot to recognize a human in video stream and track continuously his face is the most important feature of human-machine interaction (Parmar, 2013; Unar, 2014; Vezzetti, 2014; Huang, 2015).

After analysis of known relevant systems for face detection, recognition and tracking, the standard set of used approaches may be classified into a generalized architecture as it is shown in Figure 1.

As it is depicted in Figure 1 after image acquisition, noise reduction, scaling and enhancement the global or local feature extraction is applied. These processes as well as image segmentation and features representation in particular structures define the precision of the detection of regions of interest and simplify the face recognition. Each process is controlled by knowledge base (KB) that a priory con-

Figure 1. Generalized architecture of a systems for face detection, recognition, and tracking

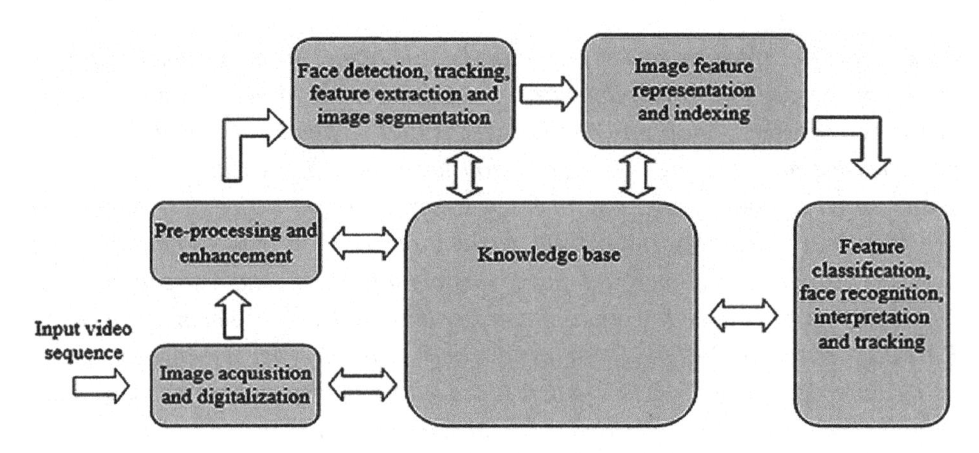

tains all the necessary data and rules. On the preprocessing step, for example, the knowledge base operates with appropriate parameters for used filters under specific conditions of input images (noise, contrast, color distribution, etc.) or in case of feature extraction and segmentation, KB provides appropriate method or transform and then defines how and which region of interest must be segmented. Particular importance the KB has for classification and interpretation of features providing recognition of regions of interest according to used classification approaches. Therefore, the generalized architecture provides standard reference for developing new systems ensuring their high performance.

The particular requirements for designing systems for face detection and recognition may be subdivided in the following way:

- A system must have a capacity to extract relevant features from images as well as to represent and encode them in suitable manner for fast and precise processing;
- Used classifier must realize inferences providing accurate recognition on base of incomplete information;
- A software for decision making must interpret the obtained data and learn on examples to be used in the future circumstances (Gonzalez, 2009; Meva, 2014; Starostenko, 2015b).

The particular attention of the researchers during last decades is attracted by performance improvement of pattern classification methods because they are not directly fitted to face recognition due to many specific conditions presented in complex machine vision applications (Kamencay, 2014; Singla, 2014; Moeini, 2015). The well-known models for pattern classification and particularly, for face recognition are subdivides into two groups: generative and discriminative models. Such approaches as Principal Component Analysis (PCA), Non-negative Matrix Factorization (NMF) or Independent Component Analysis (ICA) belong to generative models. They provide suitable representation of image data by model-based approaches (object approximation is done by sets of 3D geometrical primitives: boxes, cones, cylinders, spheres, etc.), shape-based approaches (extraction of object edges, contours or shapes is applied) and appearance-based approaches (object representation is provided by its 2D views). The classifiers based on these approaches learns the probability of a pattern or particularly, face of interest f_i using retrieved information from low-level features *s* of the image, denoted as $P\left(f_i \mid s\right)$, for each face *i* stored in memory.

The discriminative classifiers based on such approaches as Support Vector Machines (SVM), Decision-generalization Boosting or Linear Discriminant Analysis (LDA) provide face classification by assigning pattern of interest f_i directly from the image feature vector $\vec{S}$ as $f_i = function(\vec{S})$ based on estimated decision boundary (Roth, 2008; Cruz, 2008; Hidayat, 2011; Zhang, 2013).

Although the potential performance of classification methods has been widely discussed in scientific literature, recent approaches adhere to well-known generalized architecture and methodology that, as usual, involves the following steps. Initially, face region detection is provided, when for each individual frame in the input video one or many regions of interest with face are searched and their positions are detected. Then the filtering and enhancement of detected regions with face are applied by processing each consecutive frame. In the next step the obtained regions are tracked and whole face or face particular features are extracted using appropriate methods. The extracted face features are encoded according to particular machine learning algorithms suitable for classifiers that must be trained with basic patterns to ensure their high precision of recognition. Finally, face classification and validation of the correct interpretation is provided (Raducanu, 2010; Granger, 2014; Meva, 2014; Huang, 2015).

Despite of the presence of several currently available face detection and recognition techniques with improved performance, most of these algorithms still use relatively heavy computing resources to process such low-level as semantic features of images especially, when the image size is large and under different illumination conditions, as well as face features representation, indexing and recognition is not fast and precise enough when multiple objects in video streams must be interpreted and real-time processing is required.

In this chapter a particular attention is given to analysis of emergent pattern recognition approaches and as consequence the development methodology for design of systems for human face sensing and recognition is discussed. Particularly, following this methodology, two examples of the proposed face detection and interpretation approaches that belongs to both groups of generative and discriminative methods are introduced: first is based on analysis of global kernel face features and second one provides face recognition using local key points of the particular face such as eye-mouth-nose during its detection and interpretation. The first proposed color/shape-based system for face sensing uses sets of fuzzy saturated color regions providing non-parametric color image segmentation. The face detection is carried out by computing Fourier descriptors, which then are used as feature vector on the recognition step provided by SVM. For improvement of non-parametric color region detection, the face shape analysis in tangent space is used providing very simple way for shape matching and face recognition. Although these proposals satisfactory detect and recognize faces under controlled conditions, the designed system has some disadvantages derived from principles of the methods that use global face features sensible to rotation, occlusion and scaling.

The second proposed SIFT (Scale Invariant Feature Transform) key local point technique is developed and evaluated as an improved alternative to color/shape-based approach. It has been proposed for face recognition by a mobile robot that can learn new faces and recognize them in real time within realistic indoor environments. For fast face detection and tracking the preprocessing step is applied to reduce effects of different illumination and then adaptive boosting algorithm has been used to provide the selection of Haar features for the representation of objects. Finally, a novel face recognition algorithm based on SIFT key point extraction into three distinctive regions in face such as left eye, right eye and nose–mouth has been proposed. The matching strategy is able to discard correctly unknown faces and the accuracy of the recognition process has been improved using Bayesian approach for video streams. This approach is able to learn in real time a new face and then recognize it even under a different viewpoint and environmental conditions.

Experimental results with more than hundreds images within indoor environments as well as images of 28 different individuals from Yale B and CAS-PEAL standard face databases have been tested confirming that the proposed approaches are able to learn new faces and recognize them during less than three seconds. Both proposals demonstrate very competitive results comparing them to recent related works.

The chapter is structured as follows. First, an overview of background, recent trends and new advances for face detection, processing and recognition are presented. Then, following to the introduced methodology for development of systems that integrate existing and novel methods for face sensing and recognition, two color/shape and SIFT-based approaches have been proposed, implemented and discussed. Then, the evaluation of the designed systems used the proposed approaches is provided analyzing their performance, functionality and ability to recognize face in real-time invariant to changing ambient conditions. Finally, the critical discussion of obtained results, contribution and future works are presented in conclusion.

Emergent Pattern Recognition Approaches

The quality of modern systems for pattern recognition and particularly, for face sensing depend on the precision and speed of methods, approaches and techniques used for image preprocessing, feature extraction and indexing, as well as recognition and interpretation. The most widely used approaches may be subdivided in some such groups as: knowledge based, feature invariant, template matching and appearance-based methods. The knowledge based approaches include several techniques based on dimensionality reduction, which consider a region of interest as whole entity and try to localize its core components that make it distinct to others. The use of formalisms for knowledge representation and reasoning for solution of complex problems improves the accuracy, recall and precision of these methods. Under controlled conditions, they provide good results but often they are sensible to rotation and scaling as well as require a structured environment (Lu, 2012; Gil 2012; Granger, 2014; Cruz-Perez, 2015).

The group of feature invariant techniques uses local features of single pixel or small region (color, gradient, gray value, dimensionality, texture, etc.) invariant to illumination variation, noise, scale, relative position, orientations and changes in viewing direction providing robust pattern recognition under different conditions such as Scale Invariant Feature Transform (SIFT), Difference of Gaussian Points (DoG), Entropy Based Salient Region detector (EBSR), Harris and SUSAN corner detectors, Intensity Based Regions and Edge Based Regions, Gradient based algorithms and others. However, the computational complexity of these approaches frequently is too high, as well as speed of processing is low to use them in real-time applications. Additionally, ideally distinctive properties of local features cannot be reached due to the simplicity of the features themselves (Quaglia, 2012; Zhang, 2013, Deshmukh, 2015; Starostenko, 2015a).

The template matching techniques involve the evaluation of similarity of predefined template with image regions during translation of template to every possible position in that image. If the correlation between test region and template is large enough, then detector returns an exact position of region of interest within an image with description of shape and orientation similar to template (Mahmood, 2012; Starostenko, 2015b). Various formal metrics are frequently used to find the location of region of interest into image, and the most popular quantitative similarity measures are the sum of absolute differences (SAD), the sum of squared difference (SSD) and the normalized cross correlation (NCC). The SAD and SSD do not require a complex computation providing fast and simple template searching process. However, for these measures some factors like tolerance to deformation, robustness against noise, feasibility of template matching during image distortions and other factors must be taken into account because they are sensitive to occlusions and variations within template. On the other hand, the NCC measure is more accurate and robust under uniform illumination changes, but it is much slower than previous approaches for computing similarity and it requires high computational efforts if exhaustive search is applied (Sharma, 2013; Founda, 2015).

Into the group of appearance-based object recognition methods only the pattern appearance is used, which is commonly captured by different two-dimensional views of that tested pattern. In some research reports the local and global appearance-based approaches are distinguished. Local approaches are used to process regions of interest characterized by corners, edges, shapes while global approaches use color, entropy, gradient moments and sometimes semantics, which in formal manner may be described by proper feature descriptor or vector (Roth, 2008; Starostenko, 2008; Rivera-Rubio, 2015).

During recognition and classification process the obtained after face detection, feature extraction and indexing feature vectors are compared to the descriptors of previously learned and used for training images. In contrast to local, the global approaches transform the whole input image onto a suitable lower dimensional data set that does not require complex time consuming processing. Among the global appearance-based approaches the well-known and widely used techniques vary from simple histogram-based statistical measures to more sophisticated dimensionality reduction approaches, such as principle component analysis (PCA) or Karhunen-Lo'eve transformation (KLT), independent component analysis (ICA) and non-negative matrix factorization (NMF). Although in the global methods image structure information can be modeled and processed in more suitable and simple way, the encoding size of dimensional data sets are enlarged considerably as well as used iterative training and the evaluation processes limit their use in real-time applications (Liu, 2013; Rivera-Rubio, 2015; Cruz-Perez, 2015).

After analysis of relevant techniques and processes for face recognition, the generic methodology used as a guide for development of new systems may be established on definition of necessary designing requirements, analysis of applicability of face sensing and recognition methods and expectation of relevant conceptual and practical contributions.

Among the main requirements and challenges of methods for face recognition in video sequences, some aspects are very important. First of all, a face detection problem can be defined in the following way: for given image the task is to know whether there are faces in an image and what the locations of these faces are. There are several technical problems that make this solution quite difficult. They are complexity of human pose estimation and analysis of face orientation, presence of structural components like glasses, beard, hat, and possible occlusions, changes of facial expressions, face translation in dynamic scenes, etc. The quality of an image obtained by capture device is not always the best. The quality is affected by factors like low resolution and variations in color depth. This implies processing low quality images, where objects of interest are not well defined or important details are lost for the recognition process. Frequently, a size of region with sought face may be too small and this requires other specific methods that differ of those used for face recognition in still images, where a size of face is predefined. There are many applications where detected face tracking is required. In this case a frame by frame monitoring is applied computing location of moving object in current position and providing the prediction of an object in the following frames. It is common, that in face detection and feature extraction processes the obtained data in many cases cannot be directly applied to recognition step due to differences in feature description, indexing, matching and classification. In case of real time applications, a time interval for face sensing and recognition in live video streams is quite restricted. For example, in experiments presented in Viola (2004), Apostoloff (2007) and Cruz (2008) five - ten subjects can be successfully tracked at 15 - 30 fps on multi-core machine at 1.86 - 3.3 GHz, while feature localization and interpretation can only be computed at least at 2 fps obtaining 97±2% of precision with 20% of recall.

Another important component of systems for face sensing and recognition is database (corpus of faces) that contains previously processed and stored images with patterns (faces) to be used for comparison with input ones. Frequently, the ad-hoc face databases are generated for evaluation of novel systems for face recognition. That makes difficult formal quantitative performance validation of those systems. Nowadays, there are many standard face databases, which may be used to compare recognition accuracy and speed of new proposals in identical homogeneous conditions. Among standard face databases some commonly used by researchers are: Yale extended B (16128 640×480 gray scale images of 28 persons with 9 poses and 64 different illumination conditions); CAS-PEAL face database (99,594 gray-scale images of 1040 individuals with 15 different illumination conditions); FERET data base (with gallery of 1196 de images of 1196 persons with variable facial expressions and 194 images with variation in illumination); ORL database (gray-scale 10 images for each of 40 distinct subjects, 92×112 resolution and with varying lighting and facial expressions); PIE Database, CMU (41,368 images of 68 subjects, with 13 different poses, 43 different illumination conditions and with 4 different expressions), and others: NLPR Face Database, the Extended M2VTS Database of University of

Surrey UK, Georgia Tech face database, VidTIMIT Database, FRAV3D Database, UMB database of 3D occluded faces, Project - Face In Action (FIA), Face Video Dat, etc. (Gao, 2008, Yale, 2015; Phillips, 2015; ORL, 2015; Databases, 2015).

DESIGN OF FACE SENSING AND RECOGNITION SYSTEMS

According to generalized architecture presented in Figure 1 two practical examples of designing face sensing and recognition systems are presented in this section. The first designed system takes advantages of the methods, where global features of image are used to detect a face, describe its properties by specific feature vector that later will be used for face recognition by comparison corresponding feature vectors of input and previously indexed images in face database. In the second approach the methods based on local feature analysis are used that permits to overcome some disadvantages of the global methods. Both proposed approaches and designed systems based on introduced concepts have been tested in order to evaluate their capability to detect, trace and interpret faces of known individuals registered into face standard databases.

Practical Solution for Face Recognition: Color/Shape-Based Approach

The color/shape-based approach has been developed taking advantage of two principal features of a face: skin color and face silhouette. A skin color is used as filter for image segmentation and generation of the regions of interest with faces, neck, arms, hands or parts of body. Then a shape analysis is used for selection only regions with faces and for recognition of the individuals with the similar face silhouette. On the image preprocessing and enhancement step, image size normalization and filtering is provided. Therefore, image with homogenized pixels is generated specifying regions with human skin. Skin color varies depending on race or among people of the same race as well as the differences in illumination and presence of noise, shadows and camera distortions may also have a considerable effect on skin detection process. Additionally, input images may be color, pseudo-color or grayscale images, where regions with skin must be also detected. For solution of these problems an adaptive skin region filter has been proposed. The algorithm for skin color detection and saturation is presented in Algorithm 1.

For detection of regions of interest with faces, the hue-saturation-lightness HSL representation of skin color has been used discarding those pixels that don't correspond to skin tones or have low hue or low/high lightness for dark/light regions. The values of low and high limits of hue, saturation and lightness for skin have been

Algorithm 1. Pseudocode for skin color detection

```
FOR y=0 TO HeightImage DO
    FOR x=0 TO WidthImage DO
            r = getRed (x,Y)
            g = getGreen (x,y)
            b = getBlue (x,y)
            lightness = 0.3*r+0.59*g+0.11*b
            IF saturation (x,y) > LOW_SATURATION(ls) AND
               lightness > LOW_LIGHTNESS(ll) AND
               lightness < HIGH_LIGHTNESS(hl) AND
               (hue(x,y) <= LOW_HUE(lh) OR hue (x,y) >=
HIGH_HUE(hh) THEN
                    setHSL(x, y, hue(x,y), 1, 1)
            ELSE
                    setHSL(x, y, 0, 0, 1)
```

defined experimentally: *lh=5%* from maximum value of hue, *ls=60, hs=336, ll=60* and *hl=250*, respectively (Sanchez-Garcia, 2004). To make the filter adaptive to changes of skin color in images, their histograms are computed to select automatically the value of HSL limits. During face tracking process the skin color and its limits are updated iteratively according to histogram of previous and actual image frame. After image preprocessing and segmentation step, the image I_s that contains region of pixels with hue corresponding to skin color and other regions in black or white are obtained. In Figure 2a filtered skin pixels detected in color image are shown as well as in case of images with pseudo coloring or gray-scale images the regions corresponding to skin are also detected as it is depicted in Figure 2b and c.

Unfortunately, obtained set of pixels that probably are part of a face cannot be accepted as well. Since the segmentation process joins adjacent homogeneous pixels, it is very probable that different objects are grouped in the same region for example, when two faces are placed together or when in hugging scenes the arms or hands are mixed with the face due to similar colors. To solve this problem, the border detection first derivative operator is applied to produce gradient image I_g using variable thresholds for selection of relevant values of pixels on the border. The threshold is computed as inversely proportional value to quantity of pixels, whose hue lies in the range of the red component of the image I_s. Multiplying then I_s by I_g the resulting image with detected regions of interest separated by borders is obtained as it is shown in Figure 3c.

Figure 2. a) Filtered skin pixels detected in color image; b) skin regions detected in pseudo-colored image; c) skin detection in gray-scale image

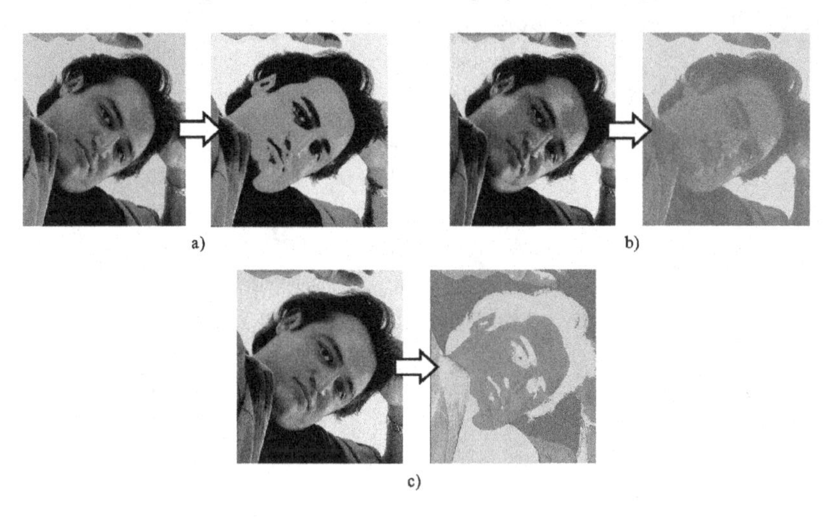

Figure 3. a) Saturated skin filtered image I_s, b) image I_g with border detection using gradient, c) result of multiplying I_s by I_g, where distinct regions (face, neck and dress) are separated

Due to only hue representation is used as principal feature, black and white pixels of $I_s * I_g$ are removed generating the image, where the range of skin color is substituted by one representative color and contiguous pixels that belongs to the same set are grouped (see Figure 4b). Finally, the obtained region with skin colors is contoured by representative rectangle as it is shown in Figure 4c.

Frequently, regions detected during segmentation have occlusions and consequently, the rectangles that represent the detected regions are intersected. As it depicted in Figure 5 each object in range of skin color (face, neck, arm, dress, etc.)

Figure 4. a) Image I_s with detected skin color region, b) face region presented by representative skin color, c) face region detection contoured by rectangle

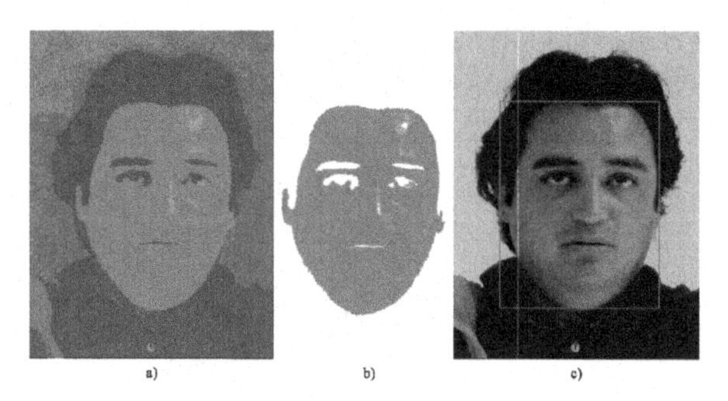

Figure 5. a) Occlusion of the rectangular regions that represent objects of interest with skin color, b) separated objects of interest

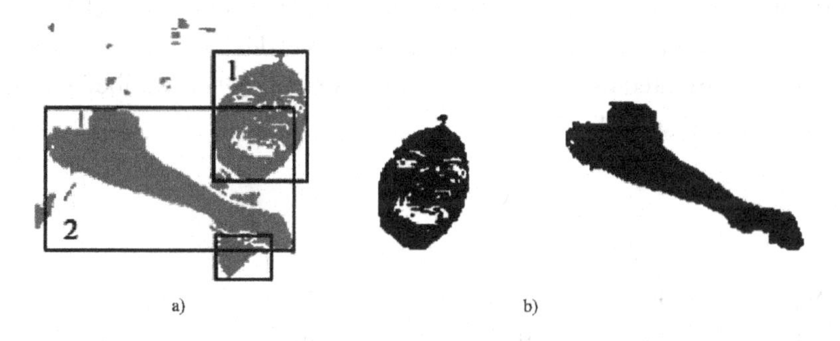

has its contoured rectangle that are intersected and must be "cleaned". The cleaning process consists in separating object of interest analyzing whether a pixel belongs to an actual region or not. The separation of regions is applied using Algorithm 2.

The result of applying this algorithm is a binary image with separated objects of interest as it is depicted in Figure 5b.

Unfortunately, a color is not trustworthy feature due to its variability. Additionally, to color, low-level image features such as texture or shape frequently are used for face detection and segmentation. Due to a shape has significant cognitive value, it is appropriate feature to use for representing each region. There are a lot of approaches invariant to scaling and rotation that may be used for shape representation and comparison. They are subdivided into contour-cased approaches (chain code, B-spline, two-segment turning function, Hausdoff descriptors, scale space, elastic matching, star field, wavelet descriptors, etc.) and region-based approaches (structural methods, convex hull, media axis, Euler number, geometric, Zemike and Legendre moments, generic Fourier descriptors, grid method, etc.), where similarity between

Algorithm 2. Pseudocode for separating objects of interest

```
For each region of image
   FOR y = 0 TO height of the region DO
      FOR x = 0 TO width of the region DO
           IF pixel (x,y) is associated with the actual region
   THEN
      Assign to pixel (x,y) value "0" (black)
         ELSE
            Assign to pixel (x,y) value "1" (white)
```

shapes may be found using Jacobs equation, Q.Tian metrics, Euclidian distance and others (Zhang D., 2004; Flores-Pulido, 2008).

To represent the shape of object the best alternative is a method of Fourier descriptors due to its properties of invariance to scale, rotation translation, initial contour tracing point and reduced complexity (Gonzalez, 2009). The shape of an object may be presented as a sequence of complex numbers described by function b_t presented in Equation 1

$$b_t = x_t + jy_t \tag{1}$$

where *N* is a total number of points and $t = 0,...,N-1$

For given function b_t the DFT and inverse DFT are computed according to Equation 2:

$$a(u) = \frac{1}{N}\sum_{t=0}^{N-1} b_t \exp\left(-j\frac{2\pi ut}{N}\right);\ b_t = \sum_{u=0}^{N-1} a(u)\exp\left(j\frac{2\pi ut}{N}\right) \tag{2}$$

For $u = 0,...,N-1$ the complex coefficient $a(u)$ is called the Fourier descriptor for a shape. Using inverse DFT a shape b_t may be reconstructed according to Equation 2. This shape b_t is approximated without taking into account all the coefficients $a(u)$, for example, using only *M* coefficients, where $M < N$ and $u = 0,...,M-1$. In Figure 6 the results of shape reconstruction using different number of descriptors $a(u)$ are presented.

The shape tracing is provided using 4-connected neighborhood. The coefficients of higher frequencies represent details of a shape, while the coefficients of low frequencies provide general description (structure) of a shape. According to invari-

Figure 6. a) Original image, b) face detection bounded by its contour, c) shape reconstruction with 159 descriptors, d) shape reconstruction with 100 descriptors and e) shape reconstruction with 10 descriptors

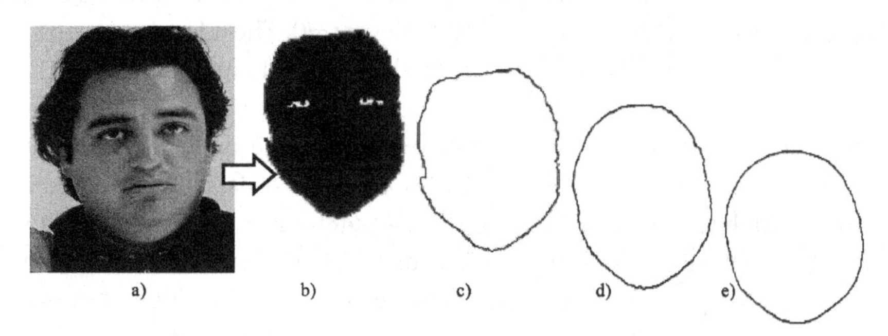

ance properties of Fourier descriptors, the module of coefficients $|a(u)|$ is invariant to rotation and translation as well as argument of $a(u)$ is invariant to scaling. The mean squared error *MSE* obtained by Equation 3 may be used as acceptable metrics for comparing shapes and computing matching of patterns, particularly, during retrieval of images from face database with indexed previously faces.

$$MSE = \frac{1}{N}\sum_{k=0}^{N-1}\left(\left|a_1(u)\right| - \left|a_2(u)\right|\right)^2 \tag{3}$$

In order to evaluate the effectiveness of the proposed face recognition approach, two metrics such as precision and recall are used. The recall (sensitivity) specifies the proportion of relevant images in the face database considered for comparison with a face into input image. The precision is proportion of the relevant found images with recognized faces from face database. Formally, recall and precision are defined by Equation 4 and Equation 5:

$$recall = \frac{a}{a+b} \tag{4}$$

$$precision = \frac{a}{a+c} \tag{5}$$

where *a* represents the number of instances correctly recognized, *b* is the number of instances incorrectly rejected and *c* is the number of instances incorrectly recognized.

The variables *a, b*, and *c* have been used for establishment of recognition relevance (similarity of input image with face to each image in face database). This is done by assigning three values to each image retrieved from a database for a single query: relevant =1, partially relevant =0.5, and not relevant =0. The relevance corresponds to the probability that the item is relevant to the analyzed input image.

For analyzing performance of the proposed approach, two sets of experiments have been conducted. The first test evaluates the system precision/recall during recognition of faces that are registered in non-standard database. This database consists of 800 digitized image sequences (640×480 pixel arrays with 8-bit gray-scale or 24-bit color values) from 42 adults of ages between of 18 and 50 years. In the second test the precision/recall are computed during recognition of the known faces from standard databases, particularly, the extended Yale B and CAS-PEAL face databases. They consist of thousand images with males and females (from 28 to 1040 individuals) with different poses and 15-64 illumination conditions (Yale, 2015; Gao, 2008). These databases have been selected to show the invariance of the proposed approach to particular images. The face recognition has been provided on a computer with core duo processor of 3.3 GHz with 2GB of RAM.

In Table 1 the average precision and recall of recognizing 28 subjects from non-standard and Yale B/CAS-PEAL face databases are shown for different sets of images. Each set have different number of images to be compared with input known face. For non-standard database Set 1 has 10 images while for standard databases Set 1 has 5 images from Yale B and 5 images from CAS-PEAL databases. Each next set has twice more images than previous set from each database. Therefore, in Set 7 comparison of input image is provided with 640 images from non-standard or both standard databases.

Table 1 shows that the recognition accuracy of relevant images is small (average recall lies between 0.18 and 0.4). It means that the method is rigorous and does not accept wide range of images with similar faces. The precision metric shows that

Table 1. Precision and recall for color and shape-based approach for face detection and recognition

	Set 1	Set 2	Set 3	Set 4	Set 5	Set 6	Set 7
Non-Standard DB							
Recall	0.40	0.35	0.33	0.28	0.27	0.21	0.18
Precision	0.94	0.92	0.88	0.87	0.88	0.86	0.86
Standard DBs							
Recall	0.38	0.39	0.35	0.33	0.33	0.25	0.24
Precision	0.94	0.94	0.90	0.89	0.91	0.88	0.88

sometimes the average recognition accuracy is high (up to 0.94) confirming that the majority of images with similar faces are relevant.

After numerous conducted experiments with different face databases, some advantages of color/shape-based approach for face recognition have been detected. The system performance is better when the image is processed in well-separated sub-regions. The analysis of the color and shapes represented by Fourier descriptors shows that approach is faster than contour-cased and some region-based approaches. Therefore, the proposed approach may be used in real-time applications such for face detection as for recognition. The disadvantages of the system are the errors in spatial sampling during generation of the image feature vector, inefficient application of color thresholds, reduced number of used shape Fourier descriptors as well as precision that still is not high enough. However, the proposed approach is invariant to rotation, translation, scale and partial occlusion of a face. The obtained satisfactory experimental results show that the introduced color/shape-based approach may be considered as alternative way for the development of applications for real-time face sensing and recognition.

Practical Solution for Face Recognition: SIFT-Based Approach

A second approach has been developed in order to evaluate the performance of the methods where face local feature analysis is applied. According to previously introduced methodology for development of face sensing and recognition systems, this approach presents another way to solve the problem and partially improve the color/shape-based technique. The proposed approach for face recognition consists of three stages. The first stage was developed for fast face detection into each video frame without any previous enhancement. The second stage consists in face tracking, when face searching is reduced to processing only space, where location of face is predicted using its locations in previous frames. In the third stage, face recognition is provided using Scale Invariant Feature Transform (SIFT) for face representation. Additionally, in order to determine the identity of recognized person, a Bayesian approach is used in live videos.

In image preprocessing step the fast frontal detection of face with variations about $\pm 20°$ is obtained using fast Viola and Jones algorithm (Viola, 2001). The adaptive boosting algorithm provides selection of Haar features for the representation of objects. In digital imaging, the Haar features are defined by difference between sums of all the pixels under two regions of Haar patterns. For face recognition the extended rotated Haar feature set of Lienhart and Maydt is more appropriate (Lienhart, 2002). In this method only three or four arithmetic operations for calculation of Haar features are used reducing significantly the computational complexity and time. Finally, adaptive boosting algorithm seeks to improve the classification results by boosted cascade of simple feature classifiers (weak classifiers) providing cascade decision tree, where the first classifier is very simple but fast and the latter is more complex but accurate. Particularly, in the proposed approach for face detection, 20 cascade classifiers in binary decision trees have been exploited, as well as a training set of 4916 images with presence of faces and 10000 images without faces has been used.

During face tracing in live video the main difficulty consists in association of the location of target in consecutive frames especially, when objects move relatively fast (above 5 km/h corresponding to the average walking speed of a human). After detecting a face, its size defined by rectangular region is used to generate a searching window in the next frame, reducing a space to be processed. This searching window is defined by increasing each rectangular side by 2/3 of its previous length. This increment is enough to define new possible position of a face in the next frame taking into account the common speed of human translation (no more than 5 Km/h) and considering limited speed of used face detection algorithm. Despite its simplic-

Figure 7. a) Detection of region with face contoured by black rectangle, b) searching a face within white window and detected face is contoured by black box, c) black box defines face position found in the white windows used as searched region, d) face region defined by white window has no face, d) correct face detection invariant to scaling, translation and illumination variations

ity, this method is extremely robust to variation of lighting conditions and even in presence of some partial occlusions. In Figure 7, some examples of face detection and tracking with face scaling and different illumination are presented.

The face tracing process starts by seeking a face in the whole image shown in Figure 7a and if a face is detected, the region of interest is highlighted by a black box. For the next frame shown in Figure 7b the search is done within the window (white box) increased by a factor of 2/3 with respect to black rectangle that contains a detected face from the previous frame colored now as gray box. If the face is located in this white window, the region with detected face is highlighted with a new black box. The same process is repeated for the next frames. Figure 7c highlights by gray box the region with face detected in the previous frame, by white window the region where face is searched in the current frame and by black box the detected face in the current frame. Figure 7d depicts a particular case, when region defined by the white window has no face. Therefore, a face tracker starts to search a face again in the whole image. Finally, Figure 7e shows the correct tracking of a face despite of variation in size, translation and illumination.

Using just introduced face searching process it can be reaffirmed that adaptive boosting approach provides a very satisfactory face tracking in limited space contoured by the white box as well as reduces significantly the time for face detection in consecutive frames.

After face detection, the preprocessing step is applied to input images in order to reduce the influence of different illumination conditions in face recognition process. The histogram equalization is used for enhancing images and adjusting their contrast. Additionally, to contrast enhancement process, the compensation of illumination is recommended. One simple and fast method applied for correction of non-uniform illumination consists in subdivision of image in some regions and then, the average illumination in each region is used as illumination compensation of that region (Gonzalez, 2009). The original image is divided, for example, into 16 regular regions (see Figure 8a) and the average of each region is obtained generating an image with 4×4 regions and assigned corresponding values as it is shown in Figure 8b. Then the obtained image is smoothed using bilinear interpolation (see Figure 8c). Finally, the complement of the smoothed image shown in (Figure 8d) is added to original image to get the enhanced image with uniform illumination (Figure 8e).

The next stage of feature extraction starts from eye region detection using again the fast adaptive boosting algorithm but this time trained to find eyes (Viola, 2001). When the position of at least one eye is found, three characteristic regions are obtained. Particularly, region for the right eye, for the left eye and for the mouth and nose are detected. The positions and sizes of eye or nose-mouth regions are estimated taking into account standard face measures. The detection of three regions for eyes and mouth/nose is provided applying the following proposed algorithm,

Figure 8. a) Original image with non-uniform illumination, b) 16 regions with corresponding average value of each region, c) smoothed image using bilinear interpolation, d) complement of the smoothed image used as illumination compensation, e) resulting image as the sum of complement of smoothed and original images

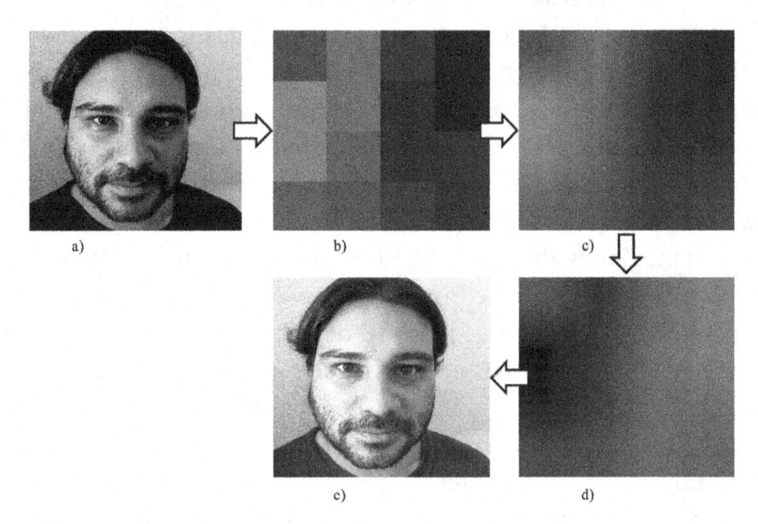

where input data are: coordinates (x, y) of top left corner of rectangle that contoured left eye (*eye_left.x* and *eye_left.y*) and right eye (*eye_right.x* and *eye_right.y*) as well as height and width of both eye region rectangles (*eye_left.height, eye_left.width, eye_right.height* and *eye_right.width*). The output of the procedure is a the feature vectors *pt[]* that specifies regions of eyes and mouth/noise separation (Algorithm 3).

The obtained feature vectors *pt[]* is used to describe eye and mouth/nose regions as it is shown in Figure 9a. In Figure 9b the combinations of vectors *pt[i]* returned by algorithm are used for computing of the coordinates of three regions of interest.

In order to provide face recognition using local feature methods, the SIFT approach has been implemented modifying it to specific operations of face analysis in live videos. Each feature point of SIFT is composed of four parts: location (the coordinates (x, y) in image), scale, dominant position and feature descriptor vector of *128* integers (Lowe, 2004; Cruz, 2008). Apart from SIFT invariance to scale and rotation, some following advantages can be distinguished:

- The locations of feature points are independent of each other, and also the robustness is provided to partial occlusions and conglomerate environments without previous segmentation;

Algorithm 3. Pseudocode for generation of region coordinates for eyes and mouth/ nose separation

```
Local variables:
   width1 = eye_left.x + eye_left.width
   bridge = eye_right.x - width1
   average = (eye_right.height + eye_left.height)/5
   width2 = eye_right.x + eye_right.width
      IF eye_left.height < eye_right.height THEN
           third = eye_left.height/3
      ELSE
         third = eye_right.height/3
      END IF
      IF eye_right.y < eye_left.y THEN
         major = eye_right.y
      ELSE
         major = eye_left.y
      END IF
      IF bridge ≥ 0 THEN
         half = bridge/2
         bridge = width1 + half
      ELSE
         half = (bridge × -1)/2
         bridge = eye_right.x + half
      END IF
      pt[0] = eye_left.y - average
      pt[1] = eye_right.y - average
      pt[2] = eye_left.y + eye_left.height
      pt[3] = eye_right.y + eye_right.height
      pt[4] = major + width2 - eye_left.x
      pt[5] = eye_left.x - average
      pt[6] = eye_left.x
      pt[7] = bridge
      pt[8] = width2
      pt[9] = width2 + average
   RETURN pt[]
```

Figure 9. a) Eye and mouth/nose regions detected by adaptive boosting algorithm with their descriptions by feature vectors pt[ ], b) location of three regions of interest, where SIFT key points will be found

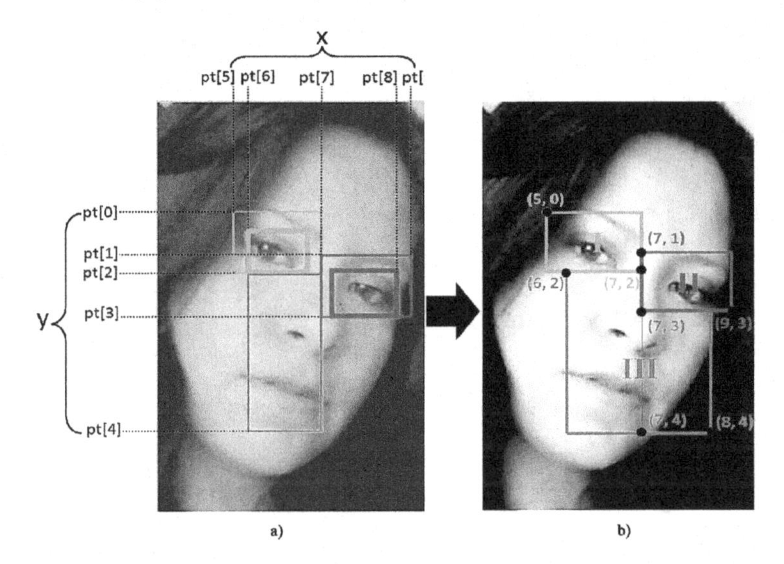

- The particularity of features defined by descriptor composed of 128 elements can be easily compared with other similar homogeneous SIFT features in large databases;
- High efficiency and low computing time are the principal characteristics of this technique.

The process to extract the SIFT descriptors consists of four stages:

1. **Detection of Scale-Space Extrema:** This step corresponds to the construction of a pyramid of scaled images, which then are smoothed with Gaussian filter to find maxima and minima in the different scales of the image. These points are candidates to be invariant and will be analyzed in the following stages. Initially, the scale-space representation is obtained as a set of images defined by a function shown in Equation 6:

$$L(x, y, \sigma) = G(x, y, \sigma) * I(x, y) \tag{6}$$

where convolution of original image $I(x, y)$ with Gaussian $G(x, y, \sigma)$ is computed with different variance σ. Scales are subtracted according to Equation 7 to form a pyramid of Gaussians as:

$$D(x,y,\sigma) = (G(x,y,k\sigma) - G(x,y,\sigma) * I(x,y) = L(x,y,k\sigma) - L(x,y,\sigma) \quad (7)$$

where k is a constant factor between two scales. A set of scales forms an octave, after which the image is subsampled by a factor of 2 to serve as input to the next octave. This allows to find candidate SIFT key points at different scales. Finally, maximum and minimum of the DoG (Difference of Gaussians) are detected by comparing each pixel with its 26 neighbors in a region of 3×3 pixels in the current scale and the two adjacent scales (Gonzales, 2009).

2. **Localization of Feature Points:** All the points generated in the previous step are evaluated to measure their stability according to their scale, location and radius, eliminating feature points with low contrast (sensitive to noise in the image or which are misplaced at the edges of objects). Therefore, for each key point position selected as candidate of interest, the detailed stability model is generated according to its scale and location, where stability is determined by Equation 8:

$$D(\hat{x}) = D + \frac{1}{2}\frac{\delta D^T}{\delta x}\hat{x} \quad (8)$$

where D represents a level of the pyramid of Gaussians in which the point has been found, and $\hat{x} = (x,y,\sigma)$ define this point. If the obtained value of $\left|D(\hat{x})\right|$ for point exceeds the threshold equal 0.03, it is considered as candidate for the next stage (Lowe, 2004). This operation is used to reject feature points that are sensible to noise or misplaced at the border of region of interest.

3. **Assigning Orientation to Feature Points:** For all not rejected points of stage 2 their gradient magnitude $m(x,y)$ and orientation (angle of gradient vector $\theta(x,y)$) are calculated according to Equation 9 and Equation 10:

$$m(x,y) = \sqrt{\left((L(x+1,y) - L(x-1,y)\right)^2 + \left((L(x,y+1) - L(x,y-1)\right)^2} \quad (9)$$

$$\theta(x,y) = \tan^{-1}\left(\frac{L(x,y+1) - L(x,y-1)}{L(x+1,y) - L(x-1,y)}\right) \quad (10)$$

where $L(x,y)$ represents a value of point (x,y) for each scale.

4. **Feature Description:** Finally, the descriptor of 128 elements is generated for every key point taking into account the information from neighboring pixels, ensuring for selected points invariance to changes in the lighting and changes in 3D perspective. Particularly, a region of interest is divided into 16 sub-regions (4×4), where each sub-region contains a grid (4×4) of 16 gradient vectors. Then, an orientation histogram is generated from gradient orientations of points within a sub-region around the feature point. Each sub-region is converted to a histogram of eight directions and then each gradient vector is mapped to one of these directions. The maximum values in the orientation histogram represent dominant directions of local gradients into each sub-region. Now, the descriptor of SIFT key point is formed by the union of the 8 directions of each of the 16 regions resulting 128 element vector describing point location (x, y) in original image, magnitude and scale.

In the presented approach for face recognition, once the eye and mouth/nose regions have been detected, the SIFT key points are extracted only into three defined regions. Figure 10a shows the phase of eye region detection. Figure 10b depicts extracted SIFT key points and 10c presents grouped points in regions of interest.

In is important to mention that the proposed approach is quite simple and flexible when new faces must to be added to database during their processing. Once new face is processed and SIFT points within three regions are obtained the user may store new face descriptor providing for it a particular name for later identification. The total number n of recognized SIFT key points for each face is stored in the feature vector $\vec{PT} = \{pt_1, pt_2, ..., pt_n\}$. For each face the different number of

Figure 10. a) Detection of eye regions, b) extraction of SIFT key points, c) SIFT points in the eye and mouth/nose regions used for recognition

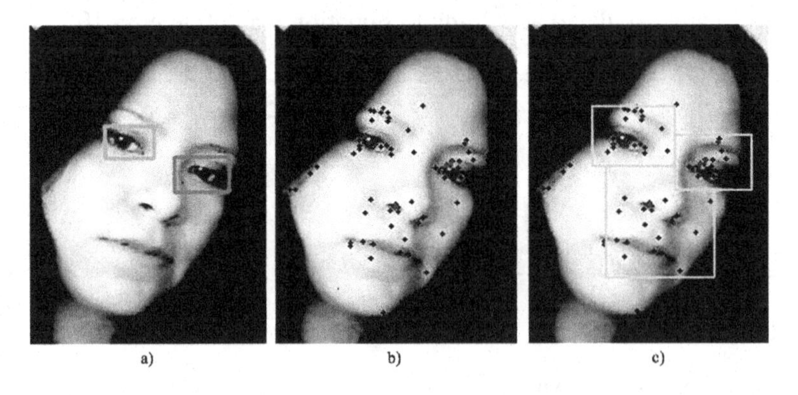

points can be extracted, so additionally a threshold vector $\vec{T} = \{t_1, t_2, ..., t_n\}$ is computed and stored, where each entry is the percentage of the total SIFT points in a face (in the conducted experiments this value lies in range from 5% to 20%).

For face recognition process the matching strategy consists in computing Euclidean distance between SIFT points restricted to three eye and mouth/noise regions of input image I_{new} and images in face database I_{DB}. The quantity of the similar points between two images is defined by Equation 11:

$$similarity\left(I_{DB}, I_{new} = \sum_{\forall reg \in face} match\left(I_{DB}^{reg}, I_{new}^{reg}\right)\right) \quad (11)$$

After obtaining total number of matching points, the similarity vector $\vec{S} = \{s_1, s_2, ..., s_n\}$ is generated for *n* faces registered in database.

For successful face recognition the fulfilment of at least one of three different criteria of matching should happen.

Criterion 1: There exists at least one element s_i of similarity vector $\vec{S}$, where $s_i \in \vec{S}$, which value exceeds percentage of number of points t_i in the vector of thresholds $\vec{T}$, where $t_i \in \vec{T}$, i.e. $s_i > t_i$ for any $i = 1, ..., n$

Criterion 2: If the criterion 1 is satisfied and additionally the difference between the maximum value in $\vec{S}$ and 2nd maximum value in $\vec{S}$ exceeds twice the 2nd maximum value in $\vec{S}$, i.e. $\max(\vec{S}) - 2nd(\vec{S}) \geq 2 \times 2nd(\vec{S})$

Criterion 3: If the criterion 1 is satisfied and additionally the difference between the maximum value in $\vec{S}$ and average $avrg(\vec{S})$ of all the $s_i \in \vec{S}$ except the maximum exceeds twice that average, i.e. $\max(\vec{S}) - avrg(\vec{S}) \geq 2 \times avrg(\vec{S})$

The criterion 1 is more relaxed and allows that some unknown faces will be accepted for next stage of recognition, while criteria 2 and 3 are more stringent and eliminate most of the vectors of unregistered faces.

If a face is not discarded applying presented matching strategy, the final step of face recognition is carried out using Bayesian approach. Equation 12 describes probability $P(f_i \mid s)$ of a face f_i using features *s* from several image frames in video stream:

$$P(f_i|s) = \frac{P(s|f_i)P(f_i)}{P(s)} = \frac{P(s|f_i)P(f_i)}{\sum_{k=1}^{n} P(s|f_k)P(f_k)} \tag{12}$$

where $P\left(f_i\right)$ is probability a priory that at initial step is taken as *1/n* for *n* registered faces. The relationship between contiguous frames is considered by taking $P\left(f\right)_i^t$ at instance *t* equal to $P\left(f_i \mid s\right)^{t-1}$ at the previous instance *t–1*. $P\left(s \mid f_i\right)$ is obtained from vector $\vec{S}$ of similarities s_i applying absolute and relative approaches. The absolute approach provides computing probability of similarity as the percentage of similarities for each face with respect to similarities of all the faces as it described by Equation 13

$$P(s|f_i)_{abs} = \frac{s_i}{\sum_{k=1}^{n} s_k} \tag{13}$$

where s_i is the number of similar points between the current image and the image with *i*-th face in database. In the relative approach the probability $P(s|f_i)_{rel}$ is estimated as the percentage of similarities relative to the total number of feature points recognized in each face to relative similarities of all the faces defined by Equation 14:

$$P(s|f_i)_{rel} = \frac{\frac{s_i}{pt_i}}{\sum_{k=1}^{n} \frac{s_i}{pt_i}} \tag{14}$$

where $pt_i \in P\vec{T}$ is a total number of SIFT key points of *i*-th face in database.

All the probabilities $P(f_i|s)$ are arranged into a global probability vector $\vec{P}$ and the recognition is carried out if the vector $\vec{P}$ of the current face is clearly higher than probability of other faces. Equations 15 describe global probability vector and the condition used to recognize a particular face:

$$\vec{P} = \left\{P(f_1|s), P(f_2|s), \ldots, P(f_n|s)\right\},\ \max(\vec{P}) - 2nd(\vec{P}) \geq 2 \times 2nd(\vec{P}) \tag{15}$$

where the *max* function calculates the maximum value of probability for the vector $\vec{P}$ and $2nd(\vec{P})$ is second the highest elements of $\vec{P}$. Once a face is recognized, the probabilities $P(f_i|s)$ are reinitialized to *1/n* for *n* registered faces.

For evaluation of SIFT-based approach three sets of tests have been conducted using designed for this proposal system for face recognition. The first test detects how well the system rejects new faces that are not in database; the second one evaluates the precision of known face recognition that are registered in database and the third test shows the precision of the known face recognition in standard data bases, particularly, the extended Yale B and CAS-PEAL face databases described in tests of the first proposed approach have been used.

For generation of input video stream in experiments with non-standard face database, the Canon VC-C4 CCD camera with resolution 640×480 has been used. A system for face recognition designed according to SIFT-based approach has been tested on a computer with core duo processor of 3.3 GHz with 2 GB of RAM.

Table 2 resumes the obtained in the first test results (precision of correct rejection of unknown faces), when different values of vector of thresholds $\vec{T}$ (in range from 5% to 20%) are used and three criteria of matching *C1*, *C2* and *C3* with absolute *A* and relative *R* approaches to compute probability of similarity are applied. Particularly, the precision has been computed according to Equation 5.

The accuracy of correct detection that new face has not been found in database varies from 86.7% to 99.4%. It means that from 100 images with unknown individuals, about 1 image has been detected incorrectly as a similar to some subject, who really is not registered in face database.

Table 2. Precision of correct rejection of unknown faces that are not registered in database

Threshold $\vec{T}$	C1-A	C1-R	C2-A	C2-R	C3-A	C3-R
5%	0.928	0.867	0.982	0.988	0.969	0.958
7%	0.934	0.898	0.982	0.988	0.976	0.976
9%	0.976	0.928	0.988	0.994	0.988	0.982
11%	0.988	0.939	0.988	0.994	0.988	0.982
13%	0.982	0.958	0.988	0.994	0.988	0.988
15%	0.988	0.969	0.994	0.994	0.994	0.994
20%	0.994	0.994	0.994	0.994	0.994	0.994

Table 3. Precision of correct recognition of known individuals, whose faces are registered in database

Threshold $\vec{T}$	C1-A	C1-R	C2-A	C2-R	C3-A	C3-R
5%	0.966	0.983	0.974	0.994	0.978	0.985
7%	0.969	0.986	0.978	0.994	0.981	0.989
9%	0.980	0.995	0.982	1.000	0.984	0.994
11%	0.989	1.000	0.993	1.000	0.994	0.994
13%	0.988	0.994	0.993	1.000	0.994	0.994
15%	0.988	0.994	0.993	1.000	0.994	0.993
20%	1.000	1.000	1.000	1.000	1.000	1.000

In the similar manner the second set of experiments has been carried out with the individuals that have been registered in database providing the precision computed by Equation 5. The obtained results lie in the range between 100% of precision (with 42% of recall computed according to Equation 4) and 97% of precision (with 57% of recall). In Table 3, the precision of recognition of known individual with faces registered in database is presented.

In order to show the invariance of the proposed approach to particular images, the images of 28 subjects from standard Yale B and CAS-PEAL databases have been used in the third set of experiments. Table 4 shows the obtained results for average precision and recall of face recognition in each set, when Set 1 contains 20 images of four subjects taken from Yale B database, Set 2 has images with 4 subjects more than Set 1, and so on, finishing with Set 7 containing images of 28 individuals. The similar results have been obtained for CAS-PEAL standard face database.

As it can be seen, the accuracy with increasing the number of individuals in face database is stable with correct recognition higher than 94.5%; however, the recall is affected because the greater number of frames is analyzed and the number of

Table 4. Average precision and recall for recognition of known subjects in sets from standard face databases

	Set 1	Set 2	Set 3	Set 4	Set 5	Set 6	Set 7
Precision	0.990	0.970	0.970	0.975	0.963	0.957	0.945
Recall	0.460	0.292	0.240	0.202	0.198	0.203	0.198

incorrectly rejected frames is increased. The time spent in the detection, tracking and face recognition per frame is in average is about one second. For used in tests equipment with previously described characteristics the average number of frames required to take a decision on the identity of a subject is about 2.5. That means that recognition of any individual requires less than 3 seconds.

In this proposal for face recognition only three regions have been used with set of some particularly selected fiducial points defined by vector $\vec{PT} = \{pt_1, pt_2, ..., pt_n\}$. However, for future research it is important to define subsets of face SIFT features that truly relevant and precisely describe a face. This will decrement the number false positives during matching process and reduce the number of iteration and nonsensical images during image retrieval from face databases.

FUTURE RESEARCH DIRECTIONS

Despite of recent advances in automatic unobtrusive face sensing from still images and video sequences, there are many issues that need to be improved in order to provide effective and high speed sensors and environments as well as precise algorithms for face detection, tracking, feature extraction, indexing, recognition and interpretation. Currently, researchers are beginning to demonstrate that unobtrusive audio-and-video based person identification systems can achieve high recognition rates without requiring that user must be in highly controlled environments. Therefore, the modern face recognition systems are quite efficient as usual under controlled or constrained conditions, when frontal view of static face in still images are processed without changes in lighting or presence of face scaling, translation, rotation or occlusion. However, all the face recognition algorithms developed for particular single modality or scene fail under significant varying conditions. Thus, novel recognition systems of next generation must be able to recognize people in natural real-time scenes, with a presence of noise and variable illumination (Garcia-Amaro, 2013; Zhang, 2013; Granger, 2014; Singla, 2014; Deshmukh, 2015; Rivera-Rubio, 2015).

Additionally, among emerging research directions related to face analysis and recognition the following trends may be mentioned:

- Develop new methods and algorithms providing enhanced environment before face detection;
- Provide continuous face detection and feature extraction invariant to pose, rotation, scaling, translation, changing facial expressions or age of tested people;

- Process the images with presence of cloth that may occlude part of face as well as design better techniques for face modeling, feature vector generation, improvement of matching process and similarity score estimation during face classification and recognition.

Taking into account easy implementation, wide portability and simple reproduction of applications for face detection, tracking and recognition, the existence of several non-compatible software development platforms and variety of standard facial databases represent a certain difficulty for performance validation of the reported approaches. The architectural solutions for development of efficient environments, which include physical infrastructure, related applications and humans, also remain still very challenging tasks.

Nowadays, there are many efforts of researches for extension of smart sensors integrating more face sensing systems that may process both behavioral and physiological features such as facial expressions, speech, gesture, vital signs and others. Improvement of facial action processing models and development of high efficient algorithms are required to ensure the precision and high speed of face recognition and classification (Bergman, 2014; Meva, 2014; Vezzetti, 2014; Fouda, 2015; Moeini, 2015; Starostenko, 2015a).

It is important to mention that appearance of efficient facial recognition systems may generate serious privacy concerns. For instance, face recognizers can be used to find out private information. That is why, the advances in facial recognition technologies must be used carefully taking in consideration the human fundamental right to control his private information.

CONCLUSION

This chapter presents some advances for expanding traditional approaches for face sensing and recognition in real time applications. The conceptual contribution of this research consists in development of two approaches and fast algorithms for face detection, tracking and recognition using extraction as global as local feature techniques for face description, indexing and interpretation. Particularly, color/shape-based and SIFT-based approaches have been extended and tested. The color/shape-based approach provides fast global feature extraction, face detection and tracking, with acceptable precision. In order to use the proposed approach in real-time applications, the speed of feature extraction and processing is provided by Fourier face descriptors that ensure a satisfactory compromise between low time of recognition process in complex conditions and high precision. Because only low-level image features such as color and shape have been used in the first approach, the average precision of recognition was not high enough (it lies in range from 86% to 94% with recall between 0.4 and 0.18, respectively for images in non-standard and standard face databases). For evaluating performance of face recognition using local feature approaches, the fast SIFT has been implemented, while fast face detection and tracking has been provided using the adaptive boosting algorithm. Despite of low complexity of image processing this approach achieves higher precision that lies in range from 95% to 99% with recall between 0.46 and 0.2, respectively. The number of frames in video stream required for face recognition is about 2.5 with overall time for face sensing and interpretation less than 3 seconds per person. It is important to mention that the proposed approaches and designed systems have some disadvantages that constrain their performance. Still low level of recognition is because of the limited number of features used for face representation and the simplicity of techniques that have been used for computing similarity of input image with the indexing images in face databases.

The discussed extended architecture as well as the proposed approaches and designed systems presented in this chapter have sufficient merit to be used as a reference for development of real-time face sensing and recognition applications.

ACKNOWLEDGMENT

This research is partially sponsored by European Grant: Cloud computing-based Advisory Services towards Energy-efficient manufacturing Systems and by Mexican National Council of Science and Technology, CONACyT, project #154438.

REFERENCES

Apostoloff, N., & Zisserman, A. (2007). Who are you? - real time person identification. *Proceedings of British Machine Vision Conference* (pp. 48.1-48.10). BMVA Press. doi:10.5244/C.21.48

Bergman, M. (2014). *Knowledge-based Artificial Intelligence.* AI3 Web page. Retrieved from http://www.mkbergman.com/1816/knowledge-based-artificial-intelligence

Cruz, C., Sucar, L. E., & Morales, E. F. (2008). Real–time face recognition for human–robot interaction.*Proceedings of 8th IEEE International Conference on Automatic Face & Gesture Recognition* (pp.1-6). IEEE Xplore Digital Library.

Cruz-Perez, C., Starostenko, O., Alarcon-Aquino, V., & Rodriguez-Asomoza, J. (2015). Automatic image annotation for description of urban and outdoor scenes. In T. Sobh & K. Elleithy (Eds.), *Innovations and advances in computing, informatics, systems sciences, networking and engineering* (vol. 313, pp.139-147). Retrieved from http://link.springer.com/book/10.1007%2F978-3-319-06773-5

Databases. (2015). *Face recognition home page*. Retrieved from http://www.face-rec.org/databases/

Deshmukh, A. C., Gaikwad, P. R., & Patil, M. P. (2015). A comparative study of feature detection methods. *International Journal of Electrical and Electronic Engineering & Telecommunications.*, *4*(2), 1–7. Retrieved from http://ijeetc.com/ijeetcadmin/upload/IJEETC_5524b6e0c6527.pdf

Flores-Pulido, L., Starostenko, O., Contreras-Gómez, R., & Alvarez-Ochoa, L. (2008). Wavelets families and similarity metrics analysis in VIR system design. *WSEAS Transactions on Information Science and Applications Journal.*, *5*(4), 436–448.

Fouda, Y. M. (2015). A robust template matching algorithm based on reducing dimensions. *Journal of Signal and Information Processing.*, *6*(02), 109–122. doi:10.4236/jsip.2015.62011

Gao, W., Cao, B., Shan, S., & Chen, X. (2008). The CAS-PEAL large-scale Chinese face database and baseline evaluations. *IEEE Transactions on Systems, Man, and Cybernetics. Part A, Systems and Humans*, *38*(1), 149–161. doi:10.1109/TSMCA.2007.909557

Garcia-Amaro, E., Nuño-Maganda, M., & Morales-Sandoval, M. (2012). Evaluation of machine learning techniques for face detection and recognition. *Proceedings of 22nd International Conference on Electrical Communications and Computers* (pp. 213–218). IEEE Xplore Digital Library. doi:10.1109/CONIELECOMP.2012.6189911

Gil, R. J., & Martin.Bautista, M. J. (2012). A novel integrated knowledge support system based on ontology learning: Model specification and a case study. *Knowledge-Based Systems*, *36*, 340–352. doi:10.1016/j.knosys.2012.07.007

Gonzalez, R. C., Woods, R. E., & Eddins, S. L. (2009). *Digital Image processing using MATLAB* (3rd ed.). Gatesmark Pub.

Granger, E., Radtke, P., & Gorodnichy, D. (2014). *Survey of academic research and prototypes for face recognition in video*. CBSA Science and Engineering Directorate, Division Report, 25. Retrieved from http://pubs.drdc-rddc.gc.ca/BASIS/pcandid/www/engpub/DDW?W%3DSYSNUM=800522

Hidayat, E., Fajrian, N. A., Muda, N. A., Huoy, C. Y., & Ahmad, S. (2011). A comparative study of feature extraction using PCA and LDA for face recognition. *Proceedings of 7th International Conference on Information Assurance and Security* (pp. 354-359). doi:10.1109/ISIAS.2011.6122779

Huang, Z., Li, W., Wang, J., & Zhang, T. (2015). Face recognition based on pixel-level and feature-level fusion of the top-level's wavelet sub-bands. *Information Fusion*, *22*, 95–104. doi:10.1016/j.inffus.2014.06.001

Kamencay, P., Hudec, R., Benco, M., & Zachariasova, M. (2014). 2D-3D face recognition method based on a modified cca-pca algorithm. *International Journal of Advanced Robotic Systems*, *11*(36), 1–8. doi:10.5772/58251

Nappi, M., Riccio, D., & Wechsler, H. (2015). Robust face recognition after plastic surgery using region-based approaches. *Pattern Recognition*, *48*(4), 1261–1276. doi:10.1016/j.patcog.2014.10.004

Lienhart, R., & Maydt, J. (2002). An extended set of Haar-like features for rapid object detection. *Proceedings of IEEE International Conference on Image Processing* (Vol. 1, pp. 900–903). Retrieved from http://www.lienhart.de/Prof._Dr._Rainer_Lienhart/Source_Code_files/ICIP2002.pdf

Liu, X., Tan, X., & Chen, S. (2012). Eyes Closeness detection using appearance based methods. In Z. Shi, D. Leake, & S. Vadera (Eds.), *IFIP advances in information and communication technology* (Vol. 385, pp. 398–408). Springer Berlin Heidelberg. doi:10.1007/978-3-642-32891-6_49

Lowe, D. (2004). Distinctive image features from scale-invariant keypoints. *International Journal of Computer Vision*, *60*(2), 91–110. Retrieved rom http://link.springer.com/article/10.1023/B:VISI.0000029664.99615.94

Lu, G.-F., Zou, J., & Wang, Y. (2012). Incremental learning of complete linear discriminant analysis for face recognition. *Knowledge-Based Systems*, *31*, 19–27. doi:10.1016/j.knosys.2012.01.016

Mahmood, A., & Khan, S. (2012). Correlation coefficient based fast template matching through partial elimination. *IEEE Transactions on Image Processing*, *21*(4), 2099–2108. doi:10.1109/TIP.2011.2171696 PMID:21997266

Meva, D. T., & Kumbharana, C. K. (2014). Study of different trends and techniques in face recognition. *International Journal of Computers and Applications*, *96*(8), 1–4. doi:10.5120/16811-6548

Moeini, A., Moeini, H., & Faez, K. (2015). Unrestricted pose-invariant face recognition by sparse dictionary matrix. *Image and Vision Computing*, *36*, 9–22. doi:10.1016/j.imavis.2015.01.007

ORL database. (2015). *Four face databases in Matlab format: Algorithms*. Retrieved from http://www.cad.zju.edu.cn/home/dengcai/Data/FaceData.html

Parmar, D., & Mehta, B. (2013). Face recognition methods & applications. *International Journal on Computer Technology & Applications*, *4*(1), 84–86.

Phillips, P. J. (2015). *The FERET database and evaluation procedure for face recognition algorithms, FERET documents*. Retrieved from http://www.nist.gov/itl/iad/ig/feret-docs.cfm

Quaglia, A., & Epifano, C. M. (2012). *Face recognition: methods, applications and technology*. Commack, NY: Nova Science Publishers, Inc.

Raducanu, B., Vitria, J., & Leonardis, A. (2010). Online pattern recognition and machine learning techniques for computer-vision: Theory and applications. *Image and Vision Computing*, *28*(7), 1063–1064. doi:10.1016/j.imavis.2010.03.007

Rivera-Rubio, J., Alexiou, I. I., & Bharath, A. A. (2015). Appearance-based indoor localization: A comparison of patch descriptor performance. *Pattern Recognition Letters*. Retrieved from http://www.sciencedirect.com/science/article/pii/S0167865515000744

Roth, P. M., & Winter, M. (2008). *Survey of appearance-based methods for object recognition*. Technical Report ICG–TR–01/08 Graz. Retrieved from http://machine-learning.wustl.edu/uploads/Main/appearance_based_methods.pdf

Sanchez-Garcia, I. (2004). *Model for automatic recognition of shapes based on Fourier descriptors* [Bachelor Thesis]. University de las Americas Puebla, Mexico.

Sharma, S. (2013). Template matching approach for face recognition system. *International Journal of Signal Processing Systems*, *1*(2), 284–289. doi:10.12720/ijsps.1.2.284-289

Singla, N., & Sharma, S. (2014). Advanced survey on face detection techniques in image processing. *International Journal of Advanced Research in Computer Science & Technology*, *2*(1), 22–24.

Starostenko, O., Cortés, X., Sánchez, J. A., & Alarcon-Aquino, V. (2015a). Unobtrusive emotion sensing and interpretation in smart environment. *Journal of Ambient Intelligence and Smart Environments*, *7*(1), 59–83.

Starostenko, O., Cruz-Perez, C., Uceda-Ponga, F., & Alarcon-Aquino, V. (2015b). Breaking textual-based CAPTCHAs with variable word and character orientation. *Journal Pattern Recognition*, *48*(4), 1101–1112. doi:10.1016/j.patcog.2014.09.006

Starostenko, O., Rodríguez-Asomoza, J., Sánchez-López, S., & Chávez-Aragón, J. (2008). Shape indexing and retrieval: a hybrid approach using ontological descriptions, In K. Elleithy (Ed.), Innovations and advanced techniques in systems, computing sciences and software engineering (pp. 381-386). Springer Science+Business Media B.V.

Unar, J. A., Seng, W. Ch., & Abbasi, A. (2014). A review of biometric technology along with trends and prospects. *Journal Pattern Recognition*, *47*(8), 2673–2688. doi:10.1016/j.patcog.2014.01.016

Vezzetti, E., Marcolin, F., & Fracastoro, G. (2014). 3D face recognition: An automatic strategy based on geometrical descriptors and landmarks. *Robotics and Autonomous Systems*, *62*(12), 1768–1776. doi:10.1016/j.robot.2014.07.009

Viola, P., & Jones, M. (2001). Rapid object detection using a boosted cascade of simple features, *Proceedings of IEEE Conference On Computer Vision And Pattern Recognition* (Vol. 1, pp. I-511 - I-518). Retrieved from https://www.cs.cmu.edu/~efros/courses/LBMV07/Papers/viola-cvpr-01.pdf

Viola, P., & Jones, M. (2004). Robust real-time face detection. *International Journal of Computer Vision*, *57*(2), 137–154. doi:10.1023/B:VISI.0000013087.49260.fb

Yale Extended Face Database B. (2015). Retrieved from http://vision.ucsd.edu/content/extended-yale-face-database-b-b

Zhang, C., & Zhang, Z. (2010). *A survey of recent advances in face detection.* Technical Report MSR-TR-2010-66. Retrieved from http://research.microsoft.com/pubs/132077/facedetsurvey.pdf

Zhang, D., & Lu, G. (2004). Review of shape representation and description techniques. *Pattern Recognition*, *37*(1), 1–19. doi:10.1016/j.patcog.2003.07.008

Zhang, Y., Hornfeck, K., & Lee, K. (2013). Adaptive face recognition for low-cost, embedded human-robot interaction. In *Intelligent Autonomous Systems* (Vol. 193, pp. 863-872). Springer. Retrieved from http://www.case.edu/mae/robotics/pdf/ZhangIAS2012.pdf

Chapter 6
Theoretical Methods of Images Processing in Optoelectronic Systems

Tatyana Strelkova
Kharkiv National University of Radio Electronics, Ukraine

Alexander P. Lytyuga
Kharkiv National University of Radio Electronics, Ukraine

Vladimir Kartashov
Kharkiv National University of Radio Electronics, Ukraine

Alexander I. Strelkov
Kharkiv National University of Radio Electronics, Ukraine

ABSTRACT

The chapter covers development of mathematical model of signals in output plane of optoelectronic system with registration of optical signals from objects. Analytical forms for mean values and dispersion of signal and interference components of photo receiver response are given. The mathematical model can be used as a base with detection algorithm development for optical signal from objects. An algorithm of signals' detection in output plane of optoelectronic system for the control is offered. The algorithm is synthesized taking into account corpuscular and statistical properties of optical signals. Analytical expressions for mean values and signal and noise components dispersion are cited. These expressions can be used for estimating efficiency of the offered algorithm by the criterion of detection probabilistic characteristics and criterion of signal/noise relation value. The possibility of signal detection characteristics improvement with low signal-to-noise ratio is shown.

DOI: 10.4018/978-1-5225-0632-4.ch006

1. PURPOSE AND METHODS OF IMPROVING OPTOELECTRONIC SYSTEMS

In recent decades the dynamic development of equipment and technologies has provided a way for technical realization of the potentially high possibilities of optoelectronic systems. The main advantages of optoelectronic systems are as follows: a great accuracy of objects coordinates determination, a high resolution in range, angular resolution. All the above allows a widespread use of television systems in many fields of science and technology, for example, in astronomy, vision systems, biology and medicine.

The development of modern optoelectronic systems provides an opportunity to deepen knowledge about the surrounding world and allows you to make discoveries in the field of natural sciences. A deep understanding of physical processes of emergence, propagation of optical radiation and also the theory of receiving and processing of optical signals with consideration of the peculiarities of their spatial-temporal structure, wave, corpuscular and statistical properties lies in creating systems with improving.

There are possible ways to improve and expand the capabilities of optoelectronic systems:

1. Reducing of sensitivity thresholds in the primary information processing devices through the improvement of individual system components, for example, development of new technologies of photodetector elements having higher energy sensitivity.
2. Development of new technologies for creation of elements of an optical link, and also the use of spectral and neutral density filters for matching the dynamic range of the systems during registration of optical signals.
3. Design and development of methods for describing information converted in optoelectronic systems.
4. Improvement of algorithms for data processing, taking into account the statistical characteristics of the received signals in the information processing systems.

The basis of each direction is a rigorous physical and mathematical description of the model of optical signals reception and image processing

Reasoning from the concepts of geometrical optics, image formation in optoelectronic systems is due to employing light rays, which are independent and straightforward in a homogeneous medium and refracted (reflected) on the boundaries of media with different optical properties. The apparatus of geometrical optics makes it possible to describe the process of the image formation, to estimate some parameters of the optical system (the angular field of view, linear magnification, etc.). However, the use of the geometrical optics does not make it possible to consider energy, statistical, spectral characteristics of optical signals.

The formation of signals and images in the systems can be viewed from the position of the wave theory, which is based on the system of differential Maxwell's equations describing electric and magnetic fields strength, electric displacement, magnetic induction and the electric charge density. The system of Maxwell's equations also includes constitutive relations that characterize the behavior of different media in the electromagnetic field. Taking into account the constitutive relations and boundary conditions, the system of Maxwell's equations is a complete one and it allows describing all the properties of the electromagnetic field and many of the processes of interaction of a field with a matter. From the standpoint of the wave theory it is convenient to describe diffraction phenomena, interference, spectral and polarization properties of optical signals. However, the wave theory application to description of the optical signals of low intensity leads to the results contradictory to the experimental data, when a small number of photons is recorded during the observation time.

Also the processes of signals and images formation in the systems can be considered from the position of the corpuscle (photon) optics, when light is represented as a stream of discrete particles (photons). Using the photon representation of the optical radiation, it is possible to describe phenomena that have a probabilistic nature (the generation of radiation, absorption in the neutral filters and in the substance of the photocathode). The corpuscular representation enables the analysis of the statistical properties of optical signals and applying methods of the theory of solutions when solving problems of detection, reception of signals and their parameters measurement.

The quality of a particular physical-mathematical model and its applicability limits depend on how fully other factors that characterize the observation conditions and their influence on the parameters of the described signals are taken into account. When analyzing the applicability limits of various models it is advisable to use the criteria for evaluating the efficiency of the optoelectronic systems. These criteria include: the value of the signal/noise ratio, the conditional probability of correct and false detection of objects, etc.

The choice of the basic physical principles and approaches, being the basis for creating mathematical models of signals under different conditions, should be based on the characteristics of the described signals (e.g., the range of intensities) and the background noise and interference condition. So, the concepts of the wave theory of light can be utilized when describing the high intensity signals. In the description of weak signals (for example, weak signals from space objects, signals under low light conditions) the corpuscular structure of the optical radiation should be considered.

Methods of primary and secondary data processing in the optoelectronic systems use the analog and digital methods, which include the spatial-coordinate and the spatial-spectral transformations. Each stage of transformations is aimed at improving the characteristics of signals and images in the optical-electronic systems, for example, spectral redistribution of energy, linear and nonlinear methods of filtering and differentiation. However, the stochastic signal being registered can have different statistical characteristics. This can lead to changes in the accuracy of signals' parameters determination and will not be sufficient to issue reliable results with the best probabilistic characteristics of the observed objects and their parameters. The inclusion of statistical properties of the optical radiation determines the need to develop more complete mathematical models and optimization of algorithms for signal receiving and image processing.

Having reviewed the main approaches to the description of the individual signals, it can be concluded that the mathematical description of signals in optoelectronic systems can be built using a variety of ideas about the structure of the optical radiation and the mechanisms of its interaction with elements of the optical path. So to estimate the spatial position of the signals in the plane of the photodetector the concepts of geometrical optics can be used. The spatial distribution of the signal components from the object is described by the wave representation of the structure and properties of the optical radiation and the theory of electromagnetic waves diffraction. To describe the statistical characteristics of optical signals it is necessary to use elements of the corpuscular (photon) theory of light and a notion of the optical radiation as a random flow of photons. To describe the interaction of light with an absorbent, the process of photocarriers formation, one should use the elements of quantum mechanical representations.

2. FORMATION OF IMAGES IN OPTOELECTRONIC SYSTEMS

The structure of the optoelectronic system is determined by the range of problems solved with its help. Depending on the purpose and objectives the system can incorporate various technical devices that perform the signals receiving, converting and processing for the purpose of the most effective, justified decision making about the presence of the observation object in the field of view or the value of the signal parameters.

The structural diagram of the optoelectronic system includes: an optical system, an optical receiver, post-detector signal processing shown in Figure 1.

The optical signal in the form of a light field or beam is propagating in the channel, gathered by the system optics and detected by the photodetector. In the gen-

Figure 1. Structural scheme of the generalized optoelectronic system

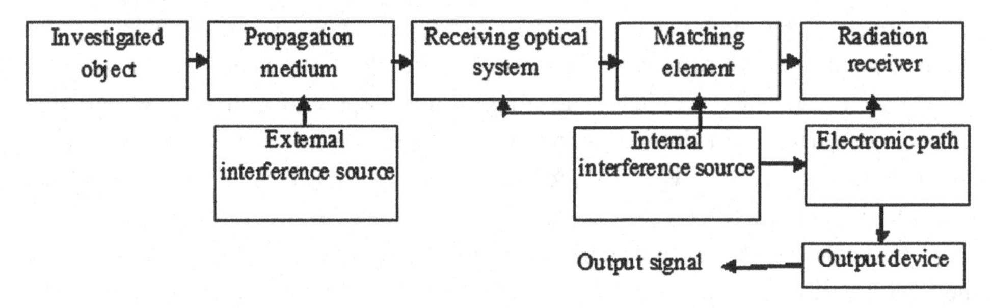

eral case, the optical field of the useful signal is distorted by the medium of propagation, diffraction and interference effects. The background or noise optical field caused by an external light arrive to the input of the optical receiver. The optical receiver noise is summed to the additive mixture of the useful optical fields and background radiation in the electronic path in the process of converting.

The process of signal formation on the photosensitive elements is not a stationary process. The histograms of the amplitudes of the signals generated by the pixels under the action of optical radiation are presented in Figure 2. Gradual changes of intensity are caused by the uneven lighting, and the abrupt changes in the intensity are stipulated by the quantum structure of optical radiation and dark noise.

The schematic representation of a fragment of one line in a frame is shown in Figure 3.

U_N is the noise component of background, which varies in space (CCD matrix);
U_S is the signal component (constant in time and varying in space);

Figure 2. Histogram of amplitudes of the light flux intensities

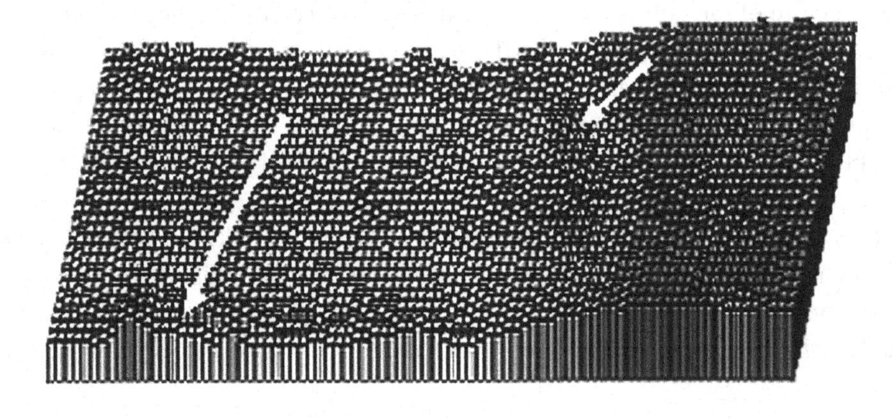

Figure 3. Fragment of a row in the frame

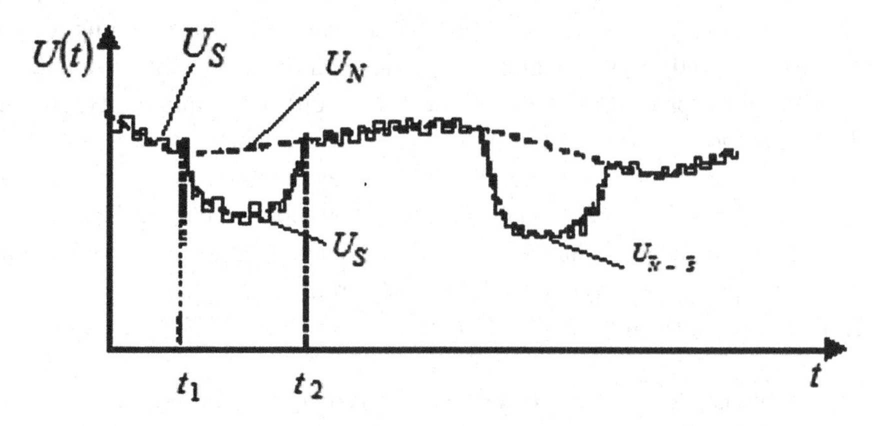

$U_{\tilde{N}}$ is the fluctuation component of noise of the background (varying in time and space);

$U_{\tilde{S}}$ is the fluctuation component of the signal (varying in time and space).

From the analysis of Figure 3 it follows that the quality of information in images processing depends both on a low-frequency (slowly varying) background noise component, and high-frequency fluctuation components of noise $U_{\tilde{N}}$ and signal $U_{\tilde{S}}$, caused by the corpuscular nature of the light flux and the internal noise of the receiver.

3. STATISTICAL PROPERTIES OF SIGNALS IN OPTOELECTRONIC SYSTEMS

Normal, log-normal, exponential distributions are used when making statistical models of the received optical radiation in the optoelectronic systems (Cunninghom & Shaw, 1999; Fedoseev, 2011; Lytyuga, 2009; Strelkov et al., 2007; Strelkova, 2014a; Cokes & Lewis, 1969; Yang et al., 2012; Nikitin et al., 2008; Strelkova, 2014b; Strelkov et al., 1998; Strelkova, 2014c; Johnson, 2013; Seitz, 2011). Statistical models of the received signals, based on Poisson statistics have gained a large widespread (Nowak & Kolaczyk, 2000; Timmermann & Nowak, 1999; Willett & Nowak, 2007; Zhang et al., 2002). These models take into account the corpuscular structure of the photon flow, the conditions of the optical radiation formation, the lack of interaction between photons in the stream, as well as the stochastic properties of optical signals with the quantum noise analysis. However, when compiling the statistical signal models based on the Poisson statistics, some assumptions are made, as a rule, and the optical signals properties limitations are accepted, reducing them to three basic properties of the simplest (Poisson) flow: stationarity, ordinary and the lack of the aftereffect. This allows using the invariance property of the Poisson flow when describing its interaction with the elements of the system. The use of such mathematical apparatus makes it possible to obtain analytical expressions for the main performance indicators of the system and to create high-performance algorithms for processing signals in the optoelectronic systems.

Let us introduce the following notation:

$\mu(t)$ is the intensity of the photoelectrons flow (the average number of photoelectrons escaping per a unit of time);

$\bar{n}$ is the average number of photoelectrons escaping in the interval τ.

These values are coupled by the relation

$$\bar{n} = \int_0^\tau \mu(t)dt$$

The intensity of the photoelectrons flow is proportional to the light intensity, i.e. $\mu(t) \sim I(t)$. The moments of photoelectrons escape and the number of photoelectrons knocked out during the time τ will be random due to the quantum nature of the signal and quantum nature of the interaction of light with a photodetector.

According to the quantum mechanics statements the possibility of the photoelectron escape in a small time interval is proportional to the duration of this interval and intensity of the signal $I(t)$. Therefore, the characteristics of the signal at the output of the photodetector is introduced taking into account the temporal dependence of the intensity of the light flux.

If the intensity of the light signal at the input of the photodetector is unchanged (does not fluctuate) in time $I(t) = I_0$, then the probability of the electrons escape in the interval τ depends only on its duration and does not depend on the location of the interval on the time axis. The flow of photoelectrons is stationary. The subsequent escapes of photoelectrons do not depend on the previous ones (absence of an after-effect). In this case, the distribution of the electrons number n in the interval of any duration is described by the Poisson law

$$P(n) = \frac{\overline{n}^{n}}{n!} e^{-\overline{n}} \tag{1}$$

where $P(n)$ is the probability of escape of n photoelectrons in the interval τ .

Mathematical expectation of the photoelectrons number in the interval

$$M(n) = \sum_{n=0}^{\infty} nP_{(n)} = \overline{n}$$

Dispersion in the number of photoelectrons escapes

$$\sigma^2 = <\left(n - \overline{n}\right)^2> = \sum_{n=0}^{\infty} \left(n - \overline{n}\right)^2 P(n) = \overline{n}$$

Thus, both dispersion and mathematical expectation are equal to the average number of photoelectrons in the observation interval. Along with stationarity and absence of an after-effect the Poisson flow is characterized by the ordinariness (the pursuit of photoelectrons singly and not in pairs or threes).

An important property of the Poisson law is that the sum N of the processes with the Poisson distributions is also the Poisson process with a mean value equal to the sum of average values of N processes.

When $n >> 10$ the Poisson distribution approaches the normal distribution:

$$P(n) = \frac{1}{\sqrt{2\pi \overline{n}}} e^{-\frac{(n-\overline{n})^2}{2\overline{n}}} \tag{2}$$

Consider the case where $\mu(t)$ is a random function that occurs in the process of the fluctuations of the light flux received at the input of the photodetector. Thus the probability of the photoelectron emergence in a small interval τ will depend on the value of $\mu(t)$, i.e. it will vary in time in a random way.

For easier description, let us introduce the normalized random function $\xi(t)$ such that $\mu(t) = \xi(t)\overline{\mu}$, where $\overline{\mu} = \lim_{T\to\infty} \frac{1}{T}\int_0^T \mu(t)dt$, $\overline{\xi} = \lim_{T\to\infty} \frac{1}{T}\int_0^T \xi(t)dt = 1$.

A random function $\xi(t)$ can be characterized by the correlation interval τ_C. The statistical characteristics of the signal at the output of the photodetector depend on the ratio of intervals τ and τ_C.

If the time of observation (duration of implementation) $\tau << \tau_C$, then the value $\xi(t)$ can be considered constant and not time-dependent in this interval. The average number of photoelectrons in the interval τ is equal to $n_1 = \xi \overline{n}$, where $\overline{n} = \mu\tau$.

The distribution of electrons in the interval at a fixed value of ξ corresponds to the Poisson distribution:

$$P\left(n / \xi\right) = \frac{\left(\xi \overline{n}\right)^n}{n!} e^{-\xi \overline{n}} \tag{3}$$

To find $P(n)$ distribution, the joint distribution of n and $\xi P(n,\xi)$ should be averaged over $\xi P(n,\xi)$. In this case $P(n,\xi) = P(n / \xi)P(\xi)$.

Usually the intensity of the signal reflected from the object and the background at the receiver input has normal fluctuations. This is because the signal from the object is created as a result of the interference of statistically independent fields reflected from a large number of "shiny spots" of the scattering surface. Background radiation is also created by a large number of randomly located scatterers (particles). Under normal fluctuations of tension the instantaneous intensity of the signal and, hence, the intensity of the photoelectrons flow, characterized by the value ξ fluctuates according to the exponential law

$$P(\xi) = e^{-\xi} \tag{4}$$

From (3) and (4) we get

$$P(n) = \int_0^\infty P(n / \xi)P(\xi)d\xi = \int_0^\infty \frac{(\xi \overline{n})^n}{n!} e^{-\xi \overline{n}} e^{-\xi} d\xi = \frac{1}{\overline{n}+1}\left(\frac{\overline{n}}{\overline{n}+1}\right)^n \tag{5}$$

This distribution is called geometric distribution, or distribution of Bose-Einstein. The mathematical expectation of this distribution is $M(n) = \overline{n}$, and dispersion

$$\sigma^2 = \overline{n}\left(1 + \overline{n}\right) \tag{6}$$

Another practically important case occurs when the observation time is $\tau >> \tau_C$. The values of the intensity of the flow of electrons $\mu(t)$ through the time intervals

$\Delta t >> \tau_C$ are statistically independent. If we split the observation time τ into $m = \tau / \tau_C$ intervals, then in each interval we will have a distribution $P(n)$ that obeys the law of the Bose-Einstein with a mean value of $\overline{n_m} = \mu\tau_C = \overline{n} / m$. Over the interval τ the addition m of statistically independent values occurs, each having the distribution of Bose-Einstein. Thus we turn to the negative binomial distribution

$$P(n) = \frac{(m+n-1)!}{(m-1)!n!}\left(\frac{\overline{n}_m}{\overline{n}_m + 1}\right)^n \cdot \left(\frac{1}{\overline{n}_m + 1}\right)^m \qquad (7)$$

The mathematical expectation of this distribution is equal to the average number of photoelectrons escaping per the interval τ:

$$M(n) = \overline{n} = \overline{n}_m m,$$

and dispersion

$$\sigma^2 = \overline{n}\left(1 + \frac{\overline{n}}{m}\right) \qquad (8)$$

The number of photons in the light flux is proportional to the modulus square of the electric component of the electromagnetic wave. The photocathode converts the signal photons flow in the flow of signal carriers of N_S charges, the number of which is proportional to the number of photons in the light flux.

In addition to the signal flow of N_S charges the interfering flow of the N_N carriers are formed caused by the action of the background optical radiation quanta and internal noise of the photodetector. Resolution elements of the photodetector produce an accumulation of flows N_S and N_N for the accumulation time T_a.

Let's break the plane of the photodetector into the elementary areas of the size $\Delta S_{ij} = \left(u_i - u_{i-1}\right)\left(v_j - v_{j-1}\right)$, where $i = 1,\ldots,k$; $j = 1,\ldots,m$ (Figure 4).

Each ΔS_{ij} elementary area has n_{ij} charges for the accumulation time T_a. Let's write the average number of charges in the cases, when only the interference or only the signal components are present in the implementation, as

Figure 4. Plane of photodetector

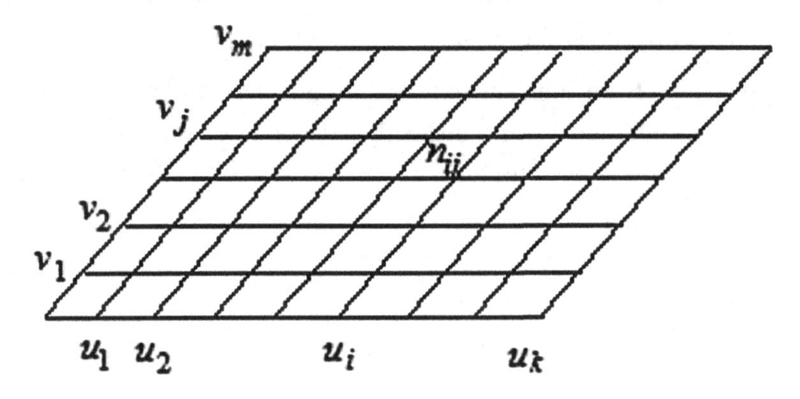

$$\bar{n}_{ij} = N_N(u,v) = N_N T_a \Delta S_{ij},$$
$$\bar{n}_{ij} = N_S(u,v) = N_S T_a \Delta S_{ij}, \tag{9}$$

where N_S and N_N are the average count rate, respectively, of the signal and interference charges per unit time on a unit square.

The spatial distribution of the number of charges N_S over the photocathode:

$$N_S(u,v) \approx N_{S_0} \left[\frac{\sin\left(c(u-u_0)\frac{a}{2}\right)}{c(u-u_0)\frac{a}{2}} \right]^2 \left[\frac{\sin\left(d(v-v_0)\frac{b}{2}\right)}{d(v-v_0)\frac{b}{2}} \right]^2 \tag{10}$$

where N_{S_0} is the number of charges, formed under the action of light radiation of the intensity $|E_0|^2$.

As the flows of charge carriers N_S and N_N have the Poisson statistics, the probability of the n_{ij} charges appearance on the ΔS_{ij} site for the accumulation time T_a

$$P(n_{ij}) = \frac{(\bar{n}_{ij})^{n_{ij}}}{n_{ij}!} \exp[-\bar{n}_{ij}] \tag{11}$$

In the case when the number of charges n_{ij} is conditioned only by the noise component, the expression (11) with regard to (9) takes the form:

$$P_N\left(n_{ij}\right)=\frac{\left(N_N T_a \Delta S_{ij}\right)^{n_{ij}}}{n_{ij}!}\exp\left[-N_N T_a \Delta S_{ij}\right] \tag{12}$$

Let's write the multivariate probability density of n_{ij} values for the noise component:

$$P_N\left(\widehat{n}\right)=\prod_{i=1}^{k}\prod_{j=1}^{m}\frac{\left(N_N T_a \Delta S_{ij}\right)^{n_{ij}}}{n_{ij}!}\exp\left[-N_N T_a \Delta S_{ij}\right] \tag{13}$$

Multidimensional probability density of n_{ij} values for an additive mixture of noise and signal components is determined by the expression:

$$P_{S+N}\left(\widehat{n}\right)=\prod_{i=1}^{k}\prod_{j=1}^{m}\frac{\left(T_a \Delta S_{ij}\left(N_S+N_N\right)\right)^{n_{ij}}}{n_{ij}!}\exp\left[-N_N T_a \Delta S_{ij}\right] \tag{14}$$

Using expressions (13) and (14) we write the logarithm of the likelihood ratio for the case when

$$\frac{N_S}{N_N}<<1 \tag{15}$$

Under the condition (15) it can be assumed that $\ln\left(1+\frac{N_S}{N_N}\right)\sim\frac{N_S}{N_N}$. Then

$$\ln L=\sum_{i=1}^{k}\sum_{j=1}^{m}\left[n_{ij}\frac{N_S}{N_N}-N_S T_a \Delta S_{ij}\right] \tag{16}$$

Substituting in (16) N_S value we will receive:

$$\ln L = \frac{N_{C_0}}{N_{\Pi}} Y\left(\hat{n}\right) - W_C \tag{17}$$

For brevity sake, the following expressions are used in the expression (17):

$$Y\left(\hat{n}\right) = \sum_{i=1}^{k} \sum_{j=1}^{m} n_{ij} \left[\frac{\sin\left[c\left(u_i - u_0\right)\frac{a}{2}\right]}{c\left(u_i - u_0\right)\frac{a}{2}} \right]^2 \left[\frac{\sin\left[d\left(v_j - v_0\right)\frac{b}{2}\right]}{d\left(v_j - v_0\right)\frac{b}{2}} \right]^2 \tag{18}$$

where $\hat{n} = n_{1j}, n_{2j}, \ldots, n_{kj}, n_{i1}, n_{i2}, \ldots, n_{km}$ is the assumed implementation.

It is obvious that the value W_S is proportional to the energy of the received signal.

$$W_S = \sum_{i=1}^{k} \sum_{j=1}^{m} N_{C_0} \left[\frac{\sin\left[c\left(u_i - u_0\right)\frac{a}{2}\right]}{c\left(u_i - u_0\right)\frac{a}{2}} \right]^2 \left[\frac{\sin\left[d\left(v_j - v_0\right)\frac{b}{2}\right]}{d\left(v_j - v_0\right)\frac{b}{2}} \right]^2 T_a \Delta S_{ij} \tag{19}$$

From the expression (16) it follows that the logarithm of the likelihood ratio with a precision of constants is determined by the value of the function $Y\left(\hat{n}\right)$. Therefore, the algorithm of detection of a signal of low intensity using an optoelectronic system can be represented in the form shown in Figure 5.

The $Y\left(\hat{n}\right)$ function values are a random variable and depend on the adopted implementation of $\hat{n}$. The distribution law, which $Y\left(\hat{n}\right)$ value from the expression (17) is subject to, is not obvious. Although n_{ij} values involved in (17) have the

Figure 5. The algorithm of detection of a signal

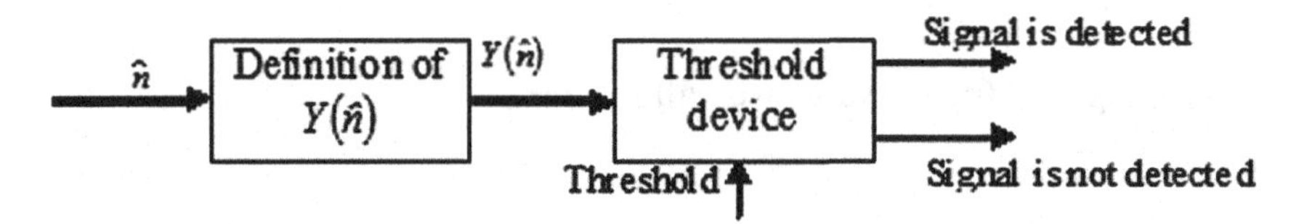

Poisson statistics, the sum of $k \cdot m$ terms, each of which is the product of independent Poisson variables by different factors, in a certain way dependent on the spatial coordinates, obviously, will not obey the Poisson law.

However, if k and m are large enough to satisfy the condition $\Delta S_{ij} \to 0$, for large n_{ij} values, due to the law of large numbers, the law of $Y(\hat{n})$ value distribution will tend to the normal law.

Therefore, in the first approximation $Y(\hat{n})$ value will be assumed to be distributed by the normal (Gaussian) law.

To estimate the effectiveness of the proposed algorithm it is possible to use the criterion of the value of the signal/noise ratio. As it is known from (Левин, 1968), this criterion is the universal one in evaluating the effectiveness of the known signal detection on the background of Gaussian noise. Let us write down the expression for the signal-to-noise ratio ϕ in the form similar to that given in (Strelkova, 1999; Gal'jardi & Karp, 1978):

$$\phi = \frac{\bar{\alpha}}{\sqrt{D(\alpha)}} \tag{20}$$

where $\bar{\alpha}$ is the average value of the measured value α; $D(\alpha)$ is the dispersion α.

In our case the measured value is the value of the function $Y(\hat{n})$ defined by expression (18). To calculate the value of the signal/noise ratio (20) it is necessary to determine the mean value $\overline{Y(\hat{n})}$ and the dispersion $D(Y(\hat{n}))$. Since the information about the signal component is contained in its additive mixture with noise, it is necessary to center the noise component. Then the expression (20) can be written in the form:

$$\phi = \frac{\bar{Y}_{S+N}(\hat{n}) - \bar{Y}_N(\hat{n})}{\sqrt{D(Y_{S+N}(\hat{n})) + D(Y_N(\hat{n}))}} \tag{21}$$

Let us calculate the values appearing in (21). Let us write the expression for $\overline{Y_{\Pi}(\hat{n})}$ average value:

$$\overline{Y_{\Pi}\left(\widehat{n}\right)} = \sum_{i=1}^{k}\sum_{j=1}^{m}\overline{n_{ij}}\left[\frac{\sin\left[c\left(u_i - u_0\right)\frac{a}{2}\right]}{c\left(u_i - u_0\right)\frac{a}{2}}\right]^2\left[\frac{\sin\left[d\left(v_j - v_0\right)\frac{b}{2}\right]}{d\left(v_j - v_0\right)\frac{b}{2}}\right]^2 \tag{22}$$

Given that the linear size of the main lobe of the intensity distribution is considerably less than the linear size of the photodetector, and replacing the sum by the integral in infinite limits, we get an expression for the average value of noise:

$$\overline{Y_N\left(\widehat{n}\right)} = N_N T_a \int_{-\infty}^{\infty}\left[\frac{\sin\left[c\left(u_i - u_0\right)\frac{a}{2}\right]}{c\left(u_i - u_0\right)\frac{a}{2}}\right]^2 du \int_{-\infty}^{\infty}\left[\frac{\sin\left[d\left(v_j - v_0\right)\frac{b}{2}\right]}{d\left(v_j - v_0\right)\frac{b}{2}}\right]^2 dv \tag{23}$$

or after integration:

$$\overline{Y_N\left(\widehat{n}\right)} = \frac{4\pi^2}{cadb} N_N T_a \tag{24}$$

The expression for the noise component dispersion:

$$D\left(Y_N\left(\widehat{n}\right)\right) = N_N T_a \int_{-\infty}^{\infty}\left[\frac{\sin\left[c\left(u_i - u_0\right)\frac{a}{2}\right]}{c\left(u_i - u_0\right)\frac{a}{2}}\right]^2 du \int_{-\infty}^{\infty}\left[\frac{\sin\left[d\left(v_j - v_0\right)\frac{b}{2}\right]}{d\left(v_j - v_0\right)\frac{b}{2}}\right]^2 dv \tag{25}$$

or after integration:

$$D\left(Y_{\Pi}\left(\widehat{n}\right)\right) = \frac{16\pi^2}{9cadb} N_{\Pi} T_a \tag{26}$$

Similar arguments will be held for the determination of the mean value and dispersion of the additive mixture of signal and noise:

$$\overline{Y_{S+N}\left(\widehat{n}\right)} = N_N T_a \int_{-\infty}^{\infty} \left[\frac{\sin\left[c\left(u_i - u_0\right)\frac{a}{2}\right]}{c\left(u_i - u_0\right)\frac{a}{2}} \right]^2 du \int_{-\infty}^{\infty} \left[\frac{\sin\left[d\left(v_j - v_0\right)\frac{b}{2}\right]}{d\left(v_j - v_0\right)\frac{b}{2}} \right]^2 dv +$$

$$+N_{S_0} T_a \int_{-\infty}^{\infty} \left[\frac{\sin\left[c\left(u_i - u_0\right)\frac{a}{2}\right]}{c\left(u_i - u_0\right)\frac{a}{2}} \right]^4 du \int_{-\infty}^{\infty} \left[\frac{\sin\left[d\left(v_j - v_0\right)\frac{b}{2}\right]}{d\left(v_j - v_0\right)\frac{b}{2}} \right]^4 dv = \tag{27}$$

$$\frac{4\pi^2}{cdab} N_N T_a + \frac{16\pi^2 T_a}{9cdab} N_{S_0} T_a,$$

$$D\left(Y_{S+N}\left(\widehat{n}\right)\right) = N_N T_a \int_{-\infty}^{\infty} \left[\frac{\sin\left[c\left(u_i - u_0\right)\frac{a}{2}\right]}{c\left(u_i - u_0\right)\frac{a}{2}} \right]^2 du \int_{-\infty}^{\infty} \left[\frac{\sin\left[d\left(v_j - v_0\right)\frac{b}{2}\right]}{d\left(v_j - v_0\right)\frac{b}{2}} \right]^2 dv +$$

$$+N_{S_0} T_{=} \int_{-\infty}^{\infty} \left[\frac{\sin\left[c\left(u_i - u_0\right)\frac{a}{2}\right]}{c\left(u_i - u_0\right)\frac{a}{2}} \right]^6 du \int_{-\infty}^{\infty} \left[\frac{\sin\left[d\left(v_j - v_0\right)\frac{b}{2}\right]}{d\left(v_j - v_0\right)\frac{b}{2}} \right]^6 dv = \tag{28}$$

$$\frac{16\pi^2}{9cdab} N_N T_a + \frac{121\pi^2 T_{=}}{100cdab} N_{S_0} T_a.$$

The expression for the signal-to-noise relation value will be:

$$\phi = \frac{16\pi^2 T_a N_{S_0}}{9cdab\sqrt{\frac{121\pi^2}{100\,cdab} T_a N_{S_0} + 2\frac{16\pi^2}{9\,cdab} T_a N_N}} \tag{29}$$

4. IMAGE PROCESSING IN OPTOELECTRONIC SYSTEMS

The problem of signal detection on the background of additive noise in the image field formed by the optical system is the primary objective for most optoelectronic systems. The results of digital processing of images depend strongly on the quality of information contained in the resulting image, which is determined by the magnitude of the contrast ratio κ and the value of the signal/noise ratio ϕ. In practice, the post- detector accumulation method is the most widely used when recording low-contrast objects characterized by small values of the signal/noise ratio. The registration time has a significant impact on the quality of the signals detection in the optoelectronic systems, and it is one of the parameters determining the energy ratio between the signal and noise components of the response of the system's photosensitive element to the optical exposure. With the increase in the accumulation time T_a the radiation energy of the object and the background detected by the photodetector increases, average values and dispersion of the signal and noise components are increasing in proportion to T_a. According to expression (29) the process of energy accumulation use will result in the increase in the magnitude of the signal/noise ratio at the output of the detector in proportion to $\sqrt{T_a}$, and, therefore, in the increase in the penetrating ability of the optoelectronic system.

Traditional methods of detecting signals from the objects in optoelectronic systems are based on the threshold signal processing, that is, on comparing the magnitude of the photosensitive elements response to the impact of the additive mixture of signal and background components with a fixed threshold, the value of which is conditioned by the selected criterion of the decision making quality. The decision about the signal detection is making, when the amplitude of the electrical signal generated by one resolution element of the matrix of photodetectors, exceeds the threshold value (single count method).

In (Lytyuga, 2009; Левин, 1968; Gal'jardi & Karp, 1978) it is shown that the quasi-optimal solution for the problem of detecting optical signals on the background of additive noise can be implemented by comparing the received implementation of the "signal + noise" additive mixture with the threshold. The decision about the presence of the signal component in the received implementation (TV frame) is made when exceeding a certain threshold valuably the response amplitude of the element of the photodetector matrix resolution. The so-called energy detection of a signal is realized.

Experimental studies have been devoted to the research on the effect of registration time on the quality of the signals detection in optoelectronic systems. A video clip was used that contained an image of the Beehive (the Manger) constellation recorded under conditions of compensation of the Earth daily rotation. Video material was provided by the staff of the Nikolaev Astronomical Observatory (A.V. Shulga, etc.). The method of successive accumulation, based on the logical summation of a sequence of images, was implemented in the course of the experiment.

The original image and the result of processing of the video are shown in Figure 6.

1) the image formed for the accumulation time of 40 MS; 4) the resulting image; 2), (5) the parts of the image containing the detected point objects; 3), 6) histograms of the intensity of the photodetector output signal. Arrows show a point object for which the value of the signal/noise ratio was calculated.

The gloss value of the investigated object in the catalog Redshift – $9^{m}_{,}6$

The dependence of the signal-to-noise ratio value in the resulting image on the number of original frames is shown in Figure 7. The calculations were performed for the object indicated by arrows in the figure. The penetrating power of optical-electronic systems is increased by $3^{m}_{,}0$.

With the accumulation of the sequential images in the general case it is necessary to carry out element-by-element summation of the digital images. For the convenience of the digital processing and reducing the information processing time, it

Figure 6. Result of experimental verification of signal processing

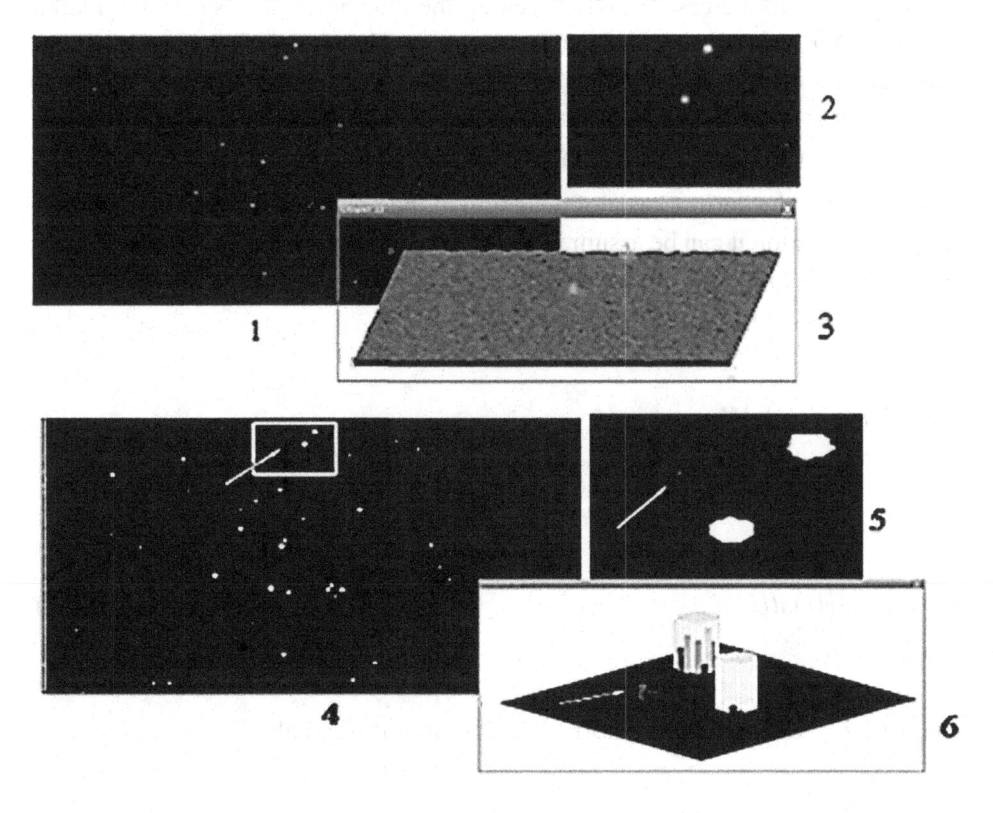

Figure 7. Dependence of signal/noise ratio on the number of realizations chosen for processing

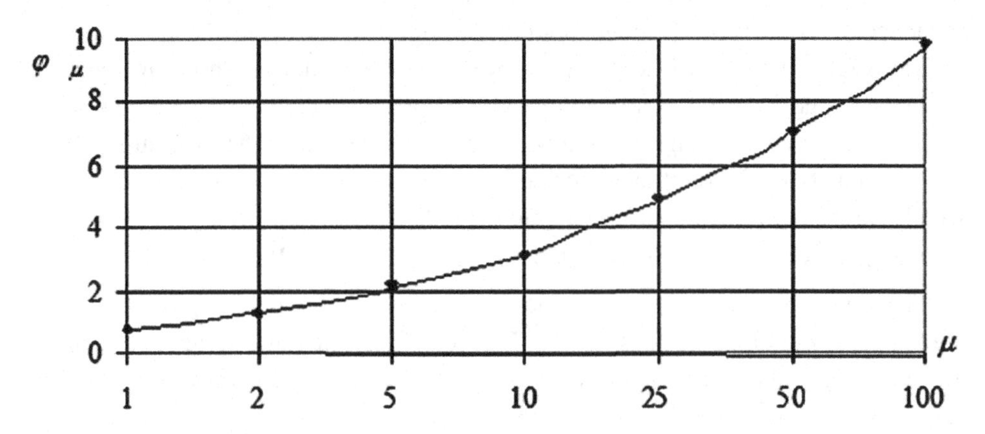

may be best to perform a binary quantization of the video signal. The result of this processing is a sequence of numeric binary arrays corresponding in dimension to the size of a television frame in which the marks "1" correspond to the case of exceeding permission of a certain threshold by the element response, and the marks "0" correspond to the case of not exceeding the threshold. In this case the marks "1" can be formed under the influence of both the signal and background components. Therefore, at a co-processing of the sequence of binary quantized frames it is necessary to assess and take into account the change of the conditional probabilities of the correct detection (C) and the emergence of false marks (F).

To evaluate the performance of detection in the binary quantized frames in the first approximation it can be assumed that the centered noise component is distributed according to the normal law. Then detection performance in a single binary quantized frame will be equal to:

$$C_1 = \int_{U_0}^{\infty} w_{N+S}(U)\, dU \tag{30}$$

$$F_1 = \int_{U_0}^{\infty} w_N(U)\, dU \tag{31}$$

where C_1 is the probability of correct detection of the signal;

F_1 is the probability of the false marks emergence;

$w_{N+C}\left(U\right)$ is the probability density of the total fluctuation component of the signal and background;

$w_N\left(U\right)$ is the probability density of the fluctuation component of the background.

To obtain the resulting image the logical summation of the binary quantized frames will be held, for example, for two frames, according to the logical rule "And":

$$\begin{cases} 1+1=1 \\ 1+0=0 \\ 0+1=0 \\ 0+0=0 \end{cases} \tag{32}$$

To summarize let's select a finite number n of binary frames. Let's calculate the detection characteristics for the resulting image (C_R and F_R). The probability of occurrence of the marker in the resulting frame can be defined as the probability k of favorable outcomes in a series of n tests with the likelihood of success in a single trial p. The probability of a correct detection and the probability of a false marks emergence can be described by the binomial distribution:

$$C_R = C_n^k p^k q^{n-k} \tag{33}$$

$$F_R = C_n^k p^k q^{n-k} \tag{34}$$

where k is the number of frames to be marked;

n is the number of frames in the series;

$p = C_1$ and is determined using (30) to calculate the probability of correct detection;

$p = F_1$ and is determined by the expression (31) for the probability of occurrence of false marks;

$q = 1 - p$; $0 \le k \le n$.

Figure 8 shows the dependence C_R and F_R calculated in accordance with (33) and (34), from k for $n = 15$ for the value of the signal-to-noise ratio $\phi = 1$; $C_1 = 0,5$; $F_1 = 0,16$.

From the analysis of the curves shown in Figure 8, it follows that in the summation of n frames the detection characteristics in the total image greatly depend on k. It is advisable to choose k values within certain limits to ensure the optimum ratio of probabilities of the correct detection of the signal and an emergence of the false marks.

The possibility emerges to stabilize one of the detection characteristics when optimizing the other at the co-processing of a sequence of binary quantized images. For example, the stabilization of the probability of a correct detection and reduction of the probability of false marks in the resulting frame. This requires the quantization of the original video frames with an adaptive threshold.

The adaptive choice of the threshold should be conducted in such a way as to provide the desired stabilized value of the probability in the resulting image.

Taking into account the above, it is proposed a method of joint processing of the frames to ensure a given probability of correct detection and minimization of a probability of false marks occurrence in the resulting image.

The method consists in the sequential execution of the following operations. Upon reception of the first frame, the noise component should be centered and dispersion should be measured. The probability of correct detection and the prob-

Figure 8. Probability of correct detection

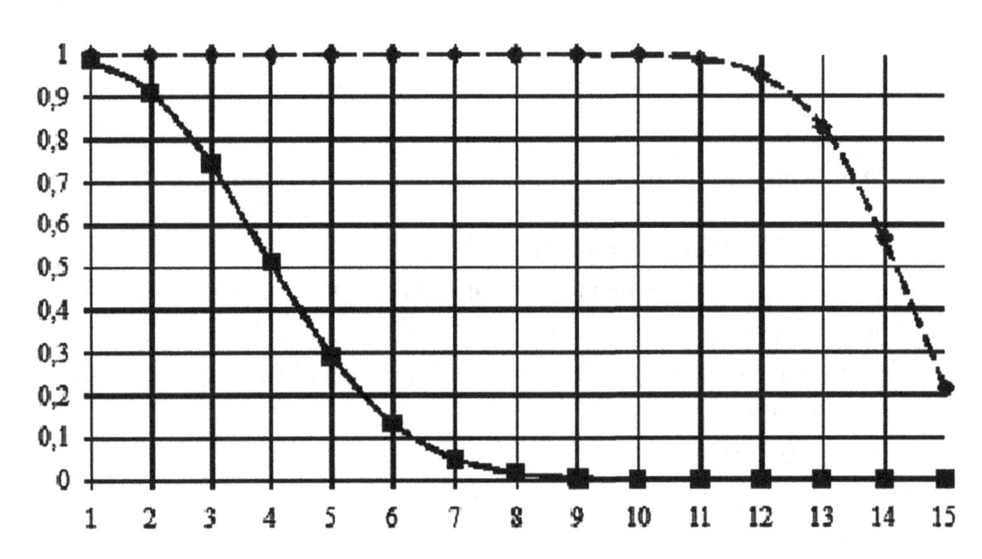

ability of false marks should be calculated for a specific value of the signal component (signal/noise ratio). Based on the values of the detection characteristics, the number of frames for summing n should be selected. It is possible to use the values of the integral distribution function for the normalized random variable, distributed according to the normal law for computing. The value of an adaptive threshold, providing a specified value of C_R and F_R, should be calculated according to the expressions (33) and (34).

A binary quantization of the summed frames should be performed taking into account the selected threshold. The binary quantized frames should be summed up according to the logical rule (32). The dependence of the detection characteristics on the number of the summed frames for the case of stabilization of the correct detection probability at the level $C_R = 0,9$ and the value of the signal/noise relation in a single frame $\phi_1 = 2$ are shown in Figure 9.

From the analysis of the curves in the figure it is seen that the probability of the false alarm in the resulting image will be of a minimal value when $k = 9$ for the given initial conditions. This procedure makes it possible to detect the objects with conditional probabilities $C_R = 0,9$ and $F_R = 3{,}23 \cdot 10^{-7}$ in the resulting image.

The method for selecting an adaptive threshold is implemented in the optical-electronic system, which is used for morphological studies in microscopy. Experimental investigations were devoted to the research on the action exerting by the process of optimization of time accumulation on the quality of signals detection in optoelectronic systems. A video clip, containing an image of erythrocytes in the

Figure 9. Value of signal/noise relation in a single frame

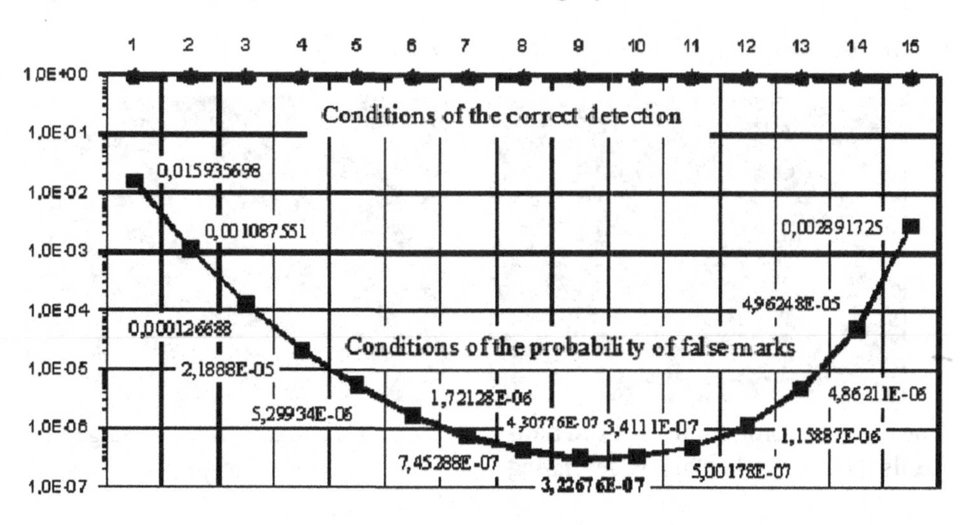

process of hemolysis, was used. In the process of lysis, hemolytic destroys erythrocytes and the hemoglobin releases into the blood plasma, the cells become low-contrast, complicating greatly the process of determining the shape of erythrocytes in microscopy in Figure 10.

The results of processing are presented in Figure 11. The application of this method has made it possible to estimate accurately the time intervals for which the morphological changes of cells were taking place, to recognize automatically the shape of erythrocytes.

5. CONCLUSION

The development of theoretical methods of image processing in optical-electronic systems includes, both deepening of knowledge about the physical processes of propagation and reception of optical radiation and the development of methods, algorithms for image processing based on the obtained knowledge, and the creation of software to implement them.

Processing of the source (input) signals recorded on the noise background takes place at the first stage of image processing in optoelectronic systems. The main purpose of the primary processing involves the increase in the probability of a correct signal detection in the additive mixture of the useful signal and noise against a background of the noise received at the input of the system. Known methods of filtering a useful signal, such as the accumulation method, frequency filtering method, correlation method, nonlinear and linear filtering method, are based on the properties of the useful signal and noise. The peculiarities of spatial-temporal pat-

Figure 10. The image of erythrocytes before and after lysis

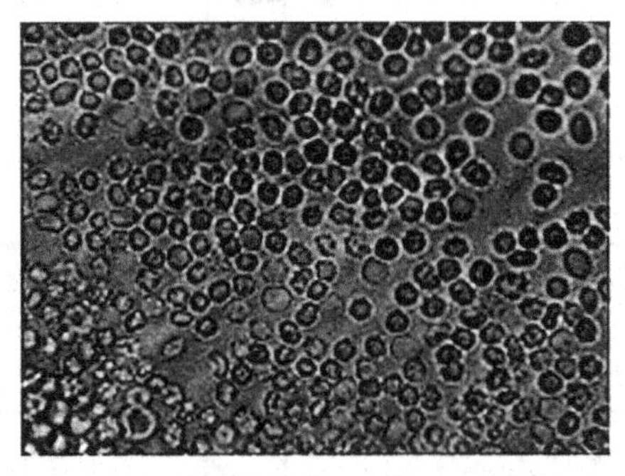

The first television frame, the red blood cells prior to the hemolysis beginning

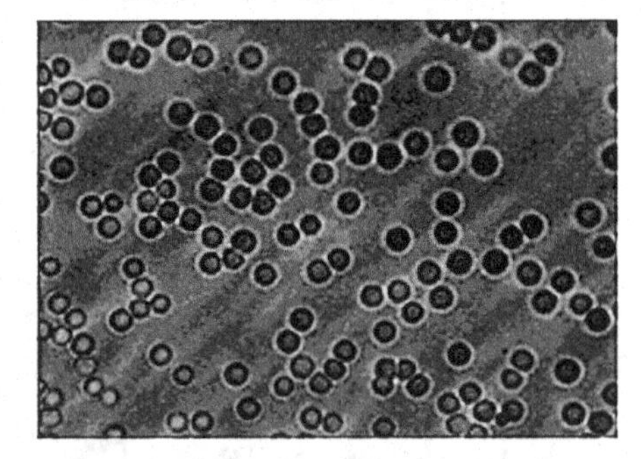

Last TV frame, the end-stage of hemolysis

Figure 11. Results of experimental verification of image enhancement

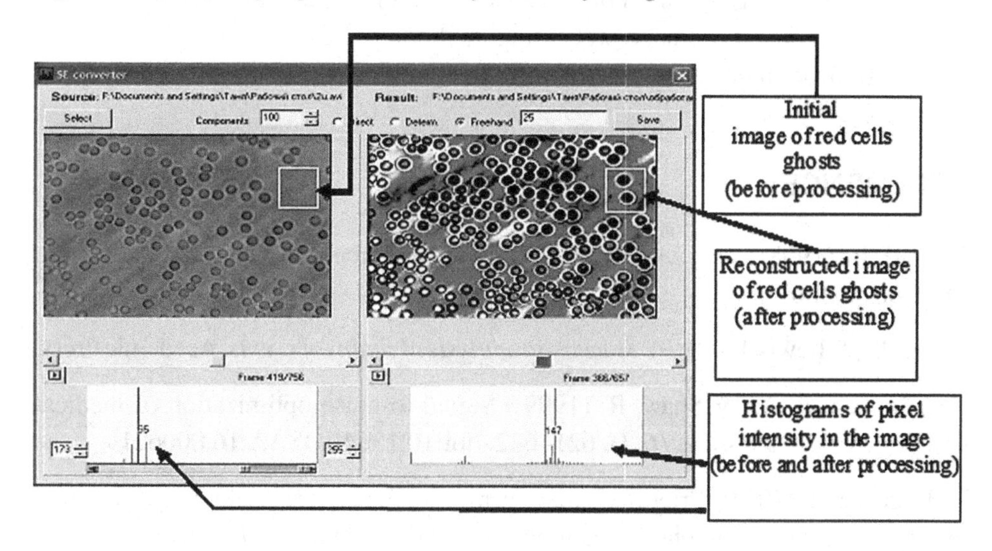

terns, corpuscular and wave properties of optical signals underlie the development of more detailed mathematical and statistical models of output signals.

The principle stage in the statistical description of signals involves determining of their statistical characteristics. The use of the proposed theoretical method of estimating the mean value and dispersion of the received useful signal and noise, based on the joint use of the corpuscular and wave descriptions of signals and noise, makes it possible to describe the energy parameters of the received signals more fully, the expression Equations 23-28.

The aim of the second stage of image processing in optoelectronic systems is the provision of increasing the magnitude of the signal/noise ratio Equation 29. The basis for these methods is the statistical analysis of the system's output signals. Taking into account Poisson and sub-Poisson statistics of the output signals the image processing algorithms are developed. Criteria of efficiency of the proposed algorithms and methods for image processing are probabilistic characteristics of signals and noise calculated according to the expressions Equations 30-31. Optimization of the described characteristics allows using methods of frame-by-frame and inter-frame processing with the parameters that are best for different conditions of the experiments.

Thus, theoretical methods of processing can be determined by:

1. Definition of statistical characteristics both of the useful signal and noise, taking into account physical processes of their formation.

2. Selection and calculation of efficiency criteria of the optoelectronic systems in development of image processing methods.
3. Optimization of image processing algorithms based on statistical analysis.

REFERENCES

ЛевинБ.Р. (1968). *Теоретические основы статистической радиотехники*. М.: Сов. радио.

Cokes, D., & Lewis, P. (1969). *Stochastic analysis of chains of events*. Academic Press.

Cunningham, I. A., & Shaw, R. (1999). Signal-to-noise optimization of medical imaging sytems. *JOSA A*, *16*(3), 621–632. doi:10.1364/JOSAA.16.000621

Fedoseev, V. I. (2011). Priem prostranstvenno-vremennyh signalov v optiko-jelektronnyh sistemah (puassonovskaja model'). *Universitetskaja kniga, 232*.

Gal'jardi, R., & Karp, S. (1978). Opticheskaja svjaz. Moscow: Svjaz.

Johnson, D. H. (2013). *Statistical signal processing*. Retrieved from http://www.ece.rice.edu/~dhj/courses/elec531/notes. pdf

Lytyuga, A. (2009). Mathematical model of signals in television systems with low-orbit space objects observation in daytime. *Collected Works of Kharkiv University of Air Force*, *4*(22), 41–46.

Nikitin, V. M., Fomin, V. N., Borisenkov, A. I., Nikolaev, A. I., & Borisenkov, I. L. (2008). *Adaptive noise protection of optical-electronic information systems*. Belgorod.

Nowak, R. D., & Kolaczyk, E. D. (2000). A statistical multiscale framework for Poisson inverse problems. *Information Theory. IEEE Transactions on*, *46*(5), 1811–1825.

Seitz, P. (2011). Fundamentals of noise in optoelectronics. In *Single-photon imaging* (pp. 1–25). Springer Berlin Heidelberg. doi:10.1007/978-3-642-18443-7_1

Strelkov, A. I., Stadnik, A. M., Lytyuga, A. P., & Strelkova, T. A. (1998). Comparative Analysis of Probabilistic and Determinate Methods for Attenuating Light Flux. *Telecommunications and Radio Engineering*, *52*(8), 54–57. doi:10.1615/TelecomRadEng.v52.i8.110

Strelkov, A. I., Zhilin, Y. I., Lytyuga, A. P., Kalmykov, A. S., & Lisovenko, S. A. (2007). Signal Detection in Technical Vision Systems. *Telecommunications and Radio Engineering, 66*(4).

Strelkova T. A. (1999). The Potentialities of Optical and Electronic Devices for Biological Objects Investigations. *and Radio Engineering, 53*, 190-194.Telecommunications

Strelkova, T. A. (2014). Influence of video stream compression on image microstructure in medical systems. *Biomedical Engineering*, *47*(6), 307–311. doi:10.1007/s10527-014-9398-1

Strelkova, T. A. (2014). Statistical properties of output signals in optical-television systems with limited dynamic range. *Eastern-European Journal of Enterprise Technologies*, *2*(9 (68)), 38–44. doi:10.15587/1729-4061.2014.23361

Strelkova, T. A. (2014). Studies on the optical fluxes attenuation process in optoelectronic systems. *Semiconductor Physics Quantum Electronics & Optoelectronics*, *17*(4), 421-424.

Timmermann, K. E., & Nowak, R. D. (1999). Multiscale modeling and estimation of Poisson processes with application to photon-limited imaging. *Information Theory. IEEE Transactions on*, *45*(3), 846–862.

Willett, R. M., & Nowak, R. D. (2007). Multiscale Poisson intensity and density estimation. *Information Theory. IEEE Transactions on*, *53*(9), 3171–3187.

Yang, F., Lu, Y. M., Sbaiz, L., & Vetterli, M. (2012). Bits from photons: Oversampled image acquisition using binary poisson statistics. *Image Processing. IEEE Transactions on*, *21*(4), 1421–1436.

Zhang, X. J., Fan, X. Z., Liao, J., Wang, H. T., Ming, N. B., Qiu, L., & Shen, Y. Q. (2002). Propagation properties of a light wave in a film quasiwaveguide structure. *Journal of Applied Physics*, *92*(10), 5647–5657. doi:10.1063/1.1517731

Chapter 7
Machine Vision Optical Scanners for Landslide Monitoring

Javier Rivera Castillo
Autonomous University of Baja California, Mexico

Oleg Sergiyenko
Autonomous University of Baja California, Mexico

Moises Rivas-Lopez
Autonomous University of Baja California, Mexico

Julio Cesar Rodríguez-Quiñonez
Autonomous University of Baja California, Mexico

Wendy Flores-Fuentes
Autonomous University of Baja California, Mexico

Daniel Hernandez-Balbuena
Autonomous University of Baja California, Mexico

ABSTRACT

An application of landslide monitoring using optical scanner as vision system is presented. The method involves finding the position of non-coherent light sources located at strategic points susceptible to landslides. The position of the light source is monitored by measuring its coordinates using a scanner based on a 45° sloping surface cylindrical mirror. This chapter also provides a background about the concept of landslides and technologies for monitoring. finally, the results of experiments of position light source monitoring in laboratory environment using the proposed method are presented.

DOI: 10.4018/978-1-5225-0632-4.ch007

INTRODUCTION

Landslide is a term associated to downslope movements of rock, debris or earth under the influence of gravity, which may cover a wide range of spatial and temporal scales. Physical factors, human activities and natural environmental changes like earthquakes, volcanic activity, heavy rainfalls and changes in ground water are typical natural triggering mechanisms for landslides, which amplifies the inherent weakness in rock or soil (Savvaidis, 2003).

Landslides are a major hazard worldwide causing fatal disasters, property losses and degradation of the environment, so it is necessary to understand the mechanisms that triggers them. To understand the dynamics of the landslides it is necessary to monitor de surface displacements including accurate inventories of past slides, their location, extent, type and triggering mechanism (Kerle, Stump, & Malet, 2010). The magnitude, velocity and acceleration of displacements can provide an indication of the stability of the slope. These movements, if detected early enough, can indicate impeding catastrophic failure of a slope mass. (Savvaidis, 2003)

The long term monitoring of instability phenomena has become relatively inexpensive and affordable due to the recent technological growth in the field of data storage and transmitting. Measurements of landslide movements that integrate hydrological and geological data have greatly improved the knowledge of landslide mechanics and the integration of different techniques allows for a better understanding of this kind of phenomena and thus better protecting human settlements and infrastructures. (Anderson & Tallapally, 1996; Rentschler & Moser, 1996)

There are many landslide types depending on the slope movement and the type of terrain. Slope movement may be fall, topple, lateral spreading, slide and flow and the earth material could be rock or soil. In past decades, photograph sensors, including extensometers, in-place inclinometers, tiltmeters, pressure transducers and rain gauges, were used to landslides monitoring. However, researchers had several difficulties for accurate assessments and evaluations, monitoring times were long and the measurement process very expensive.

In this chapter a machine vision application to landslide monitoring tasks is presented. Machine vision is a new alternative that uses different optoelectronics devices to as vision systems; nowadays the most used devices include photograph and video cameras. However, Camera images are detailed and therefore contain a large amount of information. Our application uses an optical scanner with incoherent light emitter as vision system (Figure 14).

The scanner used in this application was also proposed for structural health Monitoring tasks (Rentschler & Moser, 1996), (Rivas López, Flores Fuentes, Rivera Castillo, Sergiyenko, & Hernández Balbuena, 2013), but in this chapter we will show the advantages in landslide monitoring using incoherent light. Several position measurements were carried out with a laser emitter and incoherence light emitter. Position measurements were carried out during the scanning process moving the position of light emitter source from the nearest distance to the optical aperture sensor to the farthest position from the optical aperture sensor along the same angle line. Experiments show that with laser emitter, alignment difficulties were presented in all positions. When the laser emitter was moved to 10mm up, down, right or left, the light beam was not detected in the scanner aperture.

On the other hand, an incoherence light emitter was positioned to distances of 2m, 4m, 6m, and 8m. In each position the emitter was moved 10cm, 20cm, 50cm, to up, down, right and left and the light was detected in the scanner aperture without problem in all positions. In this way we opted to use optical scanning with incoherent light for measurements of coordinates and calculate displacements.

The Optical Scanner System (OSS) of this study consists of the following elements:

1. An incoherent light source emitter (non-rotating) mounted on the structure under monitoring.
2. A passive rotating optical aperture sensor designed with a 45°-sloping mirror, and embedded into a cylindrical micro rod. The beam of light is deviated by the 45°-sloping mirror to a double convex lens and filtered, in order to remove any interference and enhance the focus.
3. A photodiode to capture the beam of light while the cylindrical micro rod mounted on a dc electrical motor shaft is rotating.

This last element generates the targeted signals to be analyzed by the proposed method.

The light emitter source is set at a distance from the receiver; the receiver includes a mirror, which spins with an angular velocity ω. The beam emitted arrives with an incident angle β with respect to the perpendicular mirror, and is reflected with the same angle β, according to the reflecting principle to pass through the lens that concentrates the beam to be captured by the photodiode, which generates a signal with a shape similar to the Gaussian function. When the mirror starts to spin, a sensor is synchronized with the origin generating a pulse that indicates the starting of measurement that finishes when the sensor releases the next start signal. This signal is captured by a data acquisition module to start the signal processing method. (Rivas López, Flores Fuentes, Rivera Castillo, Sergiyenko, & Hernández Balbuena, 2013)

A complete description of the application will be explained in the section 5 and experimental results will also be presented.

1. BACKGROUND

Sohn and Farrar established that Structural Health Monitoring problem is fundamentally one of a statistical pattern recognition paradigm consisting of a four-part process: (1) operational evaluation, (2) data acquisition, fusion and cleansing, (3) feature extraction and information condensation, and (4) statistical model development for feature discrimination.

During the operational evaluation four questions must be answered: (1) how damage is defined for the system to be monitored? (2) What are de operational end environmental conditions under which functions? (3) What are the limitations on acquiring data in the operational environment? And (4) What are the economic and/or life safety motives for performing the monitoring?

In the data acquisition, fusion end cleansing step it is necessary to establish the type, number and placement of sensors, the data acquisition system, the storage and transmission type and the data cleansing methods.

In the feature extraction and information condensation process it is possible to distinguish between undamaged and damaged structure by means of the analysis of the measured data and to do some kind of compression data.

Statistical model development involves the selection of the algorithms to be used. Usually they fall into two categories: (1) supervised learning when there are available data form undamaged and damaged structure. (2) Unsupervised learning refers to algorithms applied to data that do not contain data from the damage structure. (Hoon Sohn, 2002)

According to USGS in 2001 the economic cost of landslides in United States was $ 3.5 billion dollars and 20 to 50 people per year are killed during these catastrophes. Rock falls, rock slides and debris flows are the most important events involving people deaths.

On December 29th, 2013, a collapse of 30 centimeters in the Tijuana-Ensenada Scenic Road became the collapse of up to 40 meters along a stretch of 300 meters at kilometer 93. This damage is attributable to the San Andreas fault, and although there were no deaths it will take at least one year to be repaired at a high economic and social cost (see Figures 1 and 2).

There are various types of landslides depending on the earth material and the mode of movement where the Rotational and Translational are the two major types. The next list is a brief description of such a landslides taken from the USGS paper "Landslide Types and processes".

1. **Rotational Slide:** This is a slide in which the surface of rupture is curved concavely upward and the slide movement is roughly rotational about an axis that is parallel to the ground surface and transverse across the slide.

Figure 1. Collapse in the Scenic Road, Ensenada, Mexico

Figure 2. Another view of the collapse in the Scenic Road, Ensenada, Mexico

2. **Translational Slide:** In this type of slide, the landslide mass moves along a roughly planar surface with little rotation or backward tilting. A block slide is a translational slide in which the moving mass consists of a single unit or a few closely related units that move downslope as a relatively coherent mass.
3. **Falls:** Falls are abrupt movements of masses of geologic materials, such as rocks and boulders that become detached from steep slopes or cliffs. Separation occurs along discontinuities such as fractures, joints, and bedding planes, and movement occurs by free-fall, bouncing, and rolling. Falls are strongly influenced by gravity, mechanical weathering, and the presence of interstitial water.
4. **Topples:** Toppling failures are distinguished by the forward rotation of a unit or units about some pivotal point, below or low in the unit, under the actions of gravity and forces exerted by adjacent units or by fluids in cracks.
5. **Flows:** There are five basic categories of flows that differ from one another in fundamental ways.
    a. **Debris Flow:** A debris flow is a form of rapid mass movement in which a combination of loose soil, rock, organic matter, air, and water mobilize as a slurry that flows downslope. Debris flows include <50% fines. Debris flows are commonly caused by intense surface-water flow, due to heavy precipitation or rapid snowmelt, that erodes and mobilizes loose soil or rock on steep slopes. Debris flows also commonly mobilize from other types of landslides that occur on steep slopes, are nearly saturated, and consist of a large proportion of silt- and sand-sized material. Debris-flow source areas are often associated with steep gullies, and debris-flow deposits are usually indicated by the presence of debris fans at the mouths of gullies. Fires that denude slopes of vegetation intensify the susceptibility of slopes to debris flows.
    b. **Debris Avalanche:** This is a variety of very rapid to extremely rapid debris flow.
    c. **Earthflow:** Earthflows have a characteristic "hourglass" shape. The slope material liquefies and runs out, forming a bowl or depression at the head. The flow itself is elongate and usually occurs in fine-grained materials or clay-bearing rocks on moderate slopes and under saturated conditions. However, dry flows of granular material are also possible.
    d. **Mudflow:** A mudflow is an earthflow consisting of material that is wet enough to flow rapidly and that contains at least 50 percent sand-, silt-, and clay-sized particles. In some instances, for example in many newspaper reports, mudflows and debris flows are commonly referred to as "mudslides."

e. **Creep:** Creep is the imperceptibly slow, steady, downward movement of slope-forming soil or rock. Movement is caused by shear stress sufficient to produce permanent deformation, but too small to produce shear failure. There are generally three types of creep: (1) seasonal, where movement is within the depth of soil affected by seasonal changes in soil moisture and soil temperature; (2) continuous, where shear stress continuously exceeds the strength of the material; and (3) progressive, where slopes are reaching the point of failure as other types of mass movements. Creep is indicated by curved tree trunks, bent fences or retaining walls, tilted poles or fences, and small soil ripples or ridges.

6. **Lateral Spreads:** Lateral spreads are distinctive because they usually occur on very gentle slopes or flat terrain. The dominant mode of movement is lateral extension accompanied by shear or tensile fractures. The failure is caused by liquefaction, the process whereby saturated, loose, cohesionless sediments (usually sands and silts) are transformed from a solid into a liquefied state. Failure is usually triggered by rapid ground motion, such as that experienced during an earthquake, but can also be artificially induced. When coherent material, either bedrock or soil, rests on materials that liquefy, the upper units may undergo fracturing and extension and may then subside, translate, rotate, disintegrate, or liquefy and flow. Lateral spreading in fine-grained materials on shallow slopes is usually progressive. The failure starts suddenly in a small area and spreads rapidly. Often the initial failure is a slump, but in some materials movement occurs for no apparent reason. Combination of two or more of the above types is known as a complex landslide.

There are a lot of Landslide causes. It can be Geological causes like Weak or sensitive materials, Weathered materials, Sheared, jointed, or fissured materials, Adversely oriented discontinuity (bedding, schistosity, fault, unconformity, contact, and so forth), Contrast in permeability and/or

stiffness of materials. Also it can be caused by Morphological causes like Tectonic or volcanic uplift, Glacial rebound, Fluvial, wave, or glacial erosion of slope toe or lateral margins, Subterranean erosion (solution, piping), Deposition loading slope or its crest, Vegetation removal (by fire, drought), Thawing, Freeze-and-thaw weathering, Shrink-and-swell weathering.

Human activities can be a cause for landslides, for example: Excavation of slope or its toe, Loading of slope or its crest, Drawdown (of reservoirs), Deforestation, Irrigation, Mining, Artificial vibration, Water leakage from utilities.

2. OPTOELECTRONIC DEVICES FOR LANDSLIDE MONITORING

P. D. Savvaidis made a summary of the systems and techniques for landslide monitoring including remote sensing or satellite techniques, photogrammetric techniques, geodetic techniques or observational techniques, and geotechnical or instrumentation or physical techniques (Savvaidis, 2003).

I.Baron and R. Supper applied a Questionnaire on National State on Landslide Site Investigation and monitoring which was disseminated among European institutes and representatives within the frame of the Safeland project with the following objectives: (1) Assessing general state of the slope instability investigation and monitoring in different European countries; (2) Assessing effectiveness/reliability of each method for slope instability investigation and monitoring; and (3) Applicability of the monitoring techniques for early warning.

They found that active rotational and translational slides with recent moving rates less than 10 mm/month are the most observed and the most frequently applied investigation methods are geological, geomorphic and engineering-geological and core drilling, testing of strength properties / deformability and clay mineralogy, studying of aerial photographs, LiDAR airborne Laser scans (ALS), radar interferometry, resistivity measurements and refraction seismic.

Aerial photographs, satellite optical very high resolution (VHR) imagery, LiDAR ALS, radar interferometry and measurement of resistivity, reflection and refraction seismic, time domain electromagnetic, passive acoustic emission, geophysical logging were the most reliable investigation methods.

They also found that the monitoring of movement and deformation was most frequently done by repeated orthophotos, radar interferometry, differential LiDAR ALS, webcam, dGPS, total station, inclinometer and wire extensometers. Most frequently monitored hydro-meteorological factors were precipitation amount, pore water pressure and air temperature; the most frequently monitored geophysical parameters were passive seismic/acoustic emissions, electromagnetic emissions and direct current resistivity (I. Baron, 2010)

In a more recent study, Michoud in "New classification of landslide-inducing anthropogenic activities" introduces a table with the techniques available for landslide monitoring. Such a table includes passive optical sensors (ground based imaging, aerial imagery and satellite imaging, active optical sensors (distance meters, terrestrial Lidar (TLS), and airborne Lidar (ALS)), active microwave sensors (interferometric radar distance meter, differential InSAR, Advanced InSAR, ground based InSAR and Polsar and Polinsar, ground-based geophysic sensor (seismic, electricity, electromagnetic (low frequency), ground penetrating radar, gravimetry and borehole geophysics), offshore sensors (2D and 3D seismic, sonar and multifaisseau), geotechnical sensors (extensometers, inclinometers, piezometers, contact earth pressure cells and multiparametric in place systems), and other sensors like global navigation satellite systems (GNSS) and core logging (Michoud, 2012).

As can be seen in the above section, there are many techniques to monitor areas with probability of a catastrophic landslide. In terms of optoelectronic devices for landslide monitoring are the next ones: Passive and active optical sensors. As an example of the first one is ground based and aerial imaging working on the basis of photogrammetric techniques where the combination of aerial photography and infrared imagery results in a more accurate and complete portrayal of terrain conditions. Infrared imagery provides information that can lead to landslide susceptible terrain, like conditions in drainage, moisture in surface and near surface, the presence of bedrocks at shallow deeps, and distinction between loose materials. At this time, aerial photographs can be georeferenced and taking photos at different times and comparing them with the use of a computer can show changes in the terrain. One problem with this technique occurs when there are heavy changes in vegetation between epochs.

As an example of active optical sensors is the Electronic distance meter instruments widely used in landslide monitoring due to the ability to measure distances with high accuracy. They are electro-optical instruments with visible or near infrared continuous radiation and the accuracy are in the range of mm. Electronic theodolites and electronic distance meters are widely used in some countries for angle and distance measurements in high accuracy deformation surveying so triangulation and trilateration horizontal networks are employed. For the height determination, differential levelling is the traditional technique. Vertical position can be determined with these instruments with very high accuracy ranging in the $\pm$ 0.1 – 1 mm over distances of 10 – 100 m. High accuracy electronic theodolites and electronic distance meters replaces geodetic levelling instruments in a more economical trigonometric height measurements obtaining accuracy better than 1 mm in the determination of height between two targets 200 m apart. This combination equipment is now called "total surveying station and as with all the optical instruments working over the earth surface, the refraction error is still the major problem. CCDs are now installed in accuracy theodolites so the centroid can be located in relation to the cross-hairs as well as robotic systems to automate the process. The main advantage of using total station instruments is that they provide 3D coordinate information of the points measured and some disadvantages are that it is required an unobstructed line of sight between the instrument and the targeting prisms and the price.

LIDAR technologies are used to create a point cloud with latitude, longitude and height information by means of a laser beam, a scanner and a GPS. Two types of LIDAR are topographic and bathymetric. Topographic LIDAR typically uses a near-infrared laser to map the land, while bathymetric lidar uses water-penetrating green light to also measure seafloor and riverbed elevations (NOAA, 2015). The operating principle of the LIDAR is by means of emitting a laser beam and measuring the reflection from the earth surface to get the distance measurement and combining that information with position and orientation data generated by GPS and Inertial measurement units integrated. Basically are two types of LIDAR: airborne and terrestrial. Airborne has the advantage of covering a large area but it needs a plane or satellite to correct operation, so the time between scanning is large. On the other hand, Terrestrial LIDAR has the advantage of continuous monitoring versus small area coverage.

Digital cameras and video cameras are now used to measure slow motion landslides creating a lot of data to process by way of pattern recognition software to recognize topography and landform changing.

Ming-Chih Lu proposed an image based landslide monitoring system using a laser projector placed in a non-landslide area, a grid plate and a wireless camera to measure land displacements (Ming-Chih, Su-Chin, Yu-Shen, & Cheng-Pei, 2015).

3. TYPICAL OPTOELECTRONIC SCANNERS

Nowadays optoelectronic scanners are widely used for multiple applications; most of the position or geometry measuring scanners use the triangulation principle or a variant of this measurement method. There are two kinds of scanners for position measuring tasks: scanners with static sensors and scanners with rotating mirrors. Optical triangulation sensors with CCD or PSD are typically used to measure manufactured goods, such as tire treads, coins, printed circuit boards and ships, principally for monitoring the target distance of small, fragile parts or soft surfaces likely to be deformed if touched by a contact probe.

3.1. Scanners with Position Triangulation Sensors Using CCD or PSD

A triangulation scanner sensor can be formed by three subsystems: emitter, receiver, and electronic processor as shown in Figure 3. A spot light is projected onto the work target; a portion of the light reflected by the target is collected through the lens by the detector which can be a CCD, CMOS or PSD array. The angle (α) is calculated, depending on the position of the beam on the detectors CCD or PSD array, hence the distance from the sensor to the target is computed by the electronic processor. As stated by Kennedy William P. in (Kennedy, 2012), the size of the spot

Figure 3. Principle of triangulation

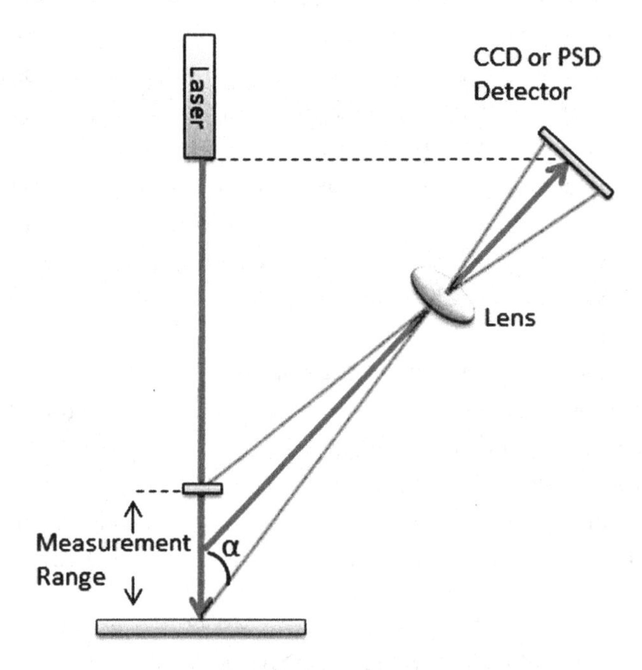

is determined by the optical design, and influences the overall system design by setting a target feature size detection limit. For instance, if the spot diameter is 30 μm, it will be difficult to resolve a lateral feature <30 μm.

Techniques are then used to determine the location of the object. For situations requiring the location of a light source on a plane, a position sensitive detector (PSD) offers the potential for better resolution at a Many devices are commonly utilized in different types of optical triangulation position scanners and have been built or considered in the past for measuring the position of light spot more efficiently. One method of position detection uses a video camera to electronically capture an image of an object. Image processing lower system cost (Vahelal, 2010). However, there are other kinds of scanners used commonly in large distances measurement or in structural health monitoring tasks, these scanners will be explained in the next section.

3.2. Scanners with Rotating Mirrors and Remote Sensing

In the previous section, we described the operational principle of scanners for monitoring the distance of small objects, now we will describe the operational principle of scanners with rotating mirrors for large distances measurement or in structural health monitoring tasks.

There are two main classification of optical scanning: remote sensing and input/output scanning. Remote sensing detects objects from a distance, as by a space-borne observation platform. For example, an infrared imaging of terrain. Sensing is usually passive and the radiation incoherent and often multispectral. Input / output scanning, on the other hand, is local. A familiar example is the document reading (input) or writing (output). The intensive use of the laser makes the scanning active and the radiation coherent. The scanned point is focused via finite-conjugate optics from a local fixed source; see (Bass, 1995).

In remote sensing there are a variety of scanning methods for capturing the data needed for image formation. These methods may be classified into framing, push broom, and mechanical. In the first one, there is no need for physical scan motion since it uses electronic scanning and implies that the sensor has a two-dimensional array of detectors. At present the most used sensor is the CCD and such array requires an optical system with 2-D wide-angle capability. In push broom methods a linear array of detectors are moved along the area to be imaged, e. g. airborne and satellite scanners. A mechanical method includes one and two dimensional scanning techniques incorporating one or multiple detectors and the image formation by one dimensional mechanical scanning requires the platform with the sensor or the object to be moved in order to create the second dimension of the image.

In these days there is a technique that is being used in many research fields named Hyperspectral imaging (also known as imaging spectroscopy). It is used in remotely sensed satellite imaging and aerial reconnaissance like the NASA's premier instruments for Earth exploration, the Jet Propulsion Laboratory's Airborne Visible-Infrared Imaging Spectrometer (AVIRIS) system. With this technique the instruments are capable of collecting high-dimensional image data, using hundreds of contiguous spectral channels, over the same area on the surface of the Earth, as shown in Figure 4 where the image measures the reflected radiation in the wavelength region from 0.4 to 2.5 μm using 224 spectral channels, at nominal spectral resolution of 10 nm. The wealth of spectral information provided by the latest generation hyperspectral sensors has opened ground breaking perspectives in many applications, including environmental modelling and assessment; target detection for military and defence/security deployment; urban planning and management studies, risk/hazard prevention and response including wild-land fire tracking; biological threat detection, monitoring of oil spills and other types of chemical contamination (Plaza, 2009).

While remote sensing requires capturing passive radiation for image formation, active input/output scanning needs to illuminate an object or medium with a ''flying spot,'' derived typically from a laser source. some examples divided into two principal functions are shown below: input (when the scattered radiation from the scanning spot is detected) and output (when the radiation is used for recording or

Figure 4. The concept of hyperspectral imaging illustrated using NASA's AVIRIS (Plaza, 2009)

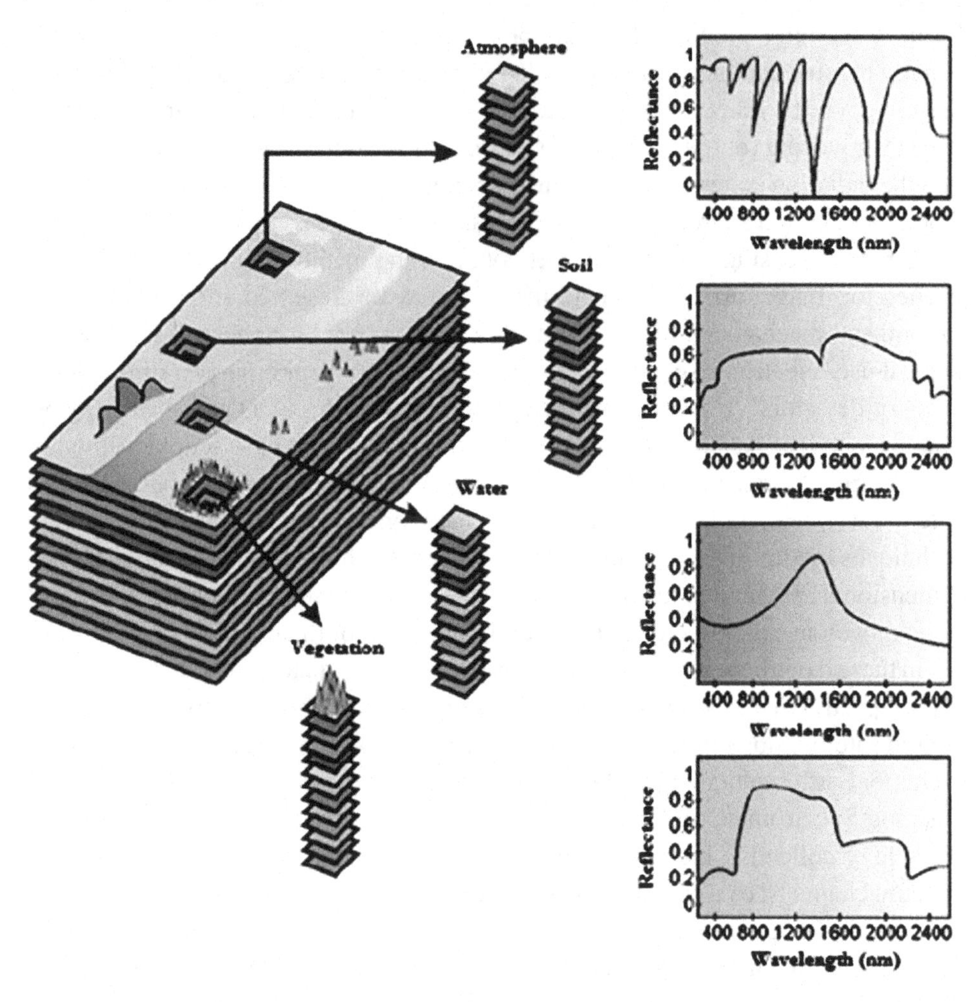

displaying). Therefore, it can be said that in input scanning the radiation is modulated by the target to form a signal and in the output scanning it is modulated by a signal.

3.3. Examples of Input / Output Scanning

3.3.1. Polygonal Scanners

These scanners have a polygonal mirror rotating at constant speed by way of an electric motor and the radiation received by the lens is reflected on a detector. The primary advantages of polygonal scanners are speed, the availability of wide scan angles, and velocity stability. They are usually rotated continuously in one direction at a fixed speed to provide repetitive unidirectional scans which are superimposed in the scan field, or plane, as the case may be. When the number of facets reduces to one, it is identified as a monogon scanner, Figure 5 illustrates a hexagonal rotating mirror scanner.

3.3.2. Pyramidal and Prismatic Facets

In these types of scanners, the incoming radiation is focused on a regular pyramidal polygon with a number of plane mirrors facets at an angle, rather than parallel, to the rotational axis. This configuration permits smaller scan angles with fewer facets than those with polygonal mirrors. Principal arrangements of facets are termed prismatic or pyramidal. The pyramidal arrangement allows the lens to be oriented close to the polygon, while the prismatic configuration requires space for a clear passage of the input beam.

Figure 5. Polygon scanner

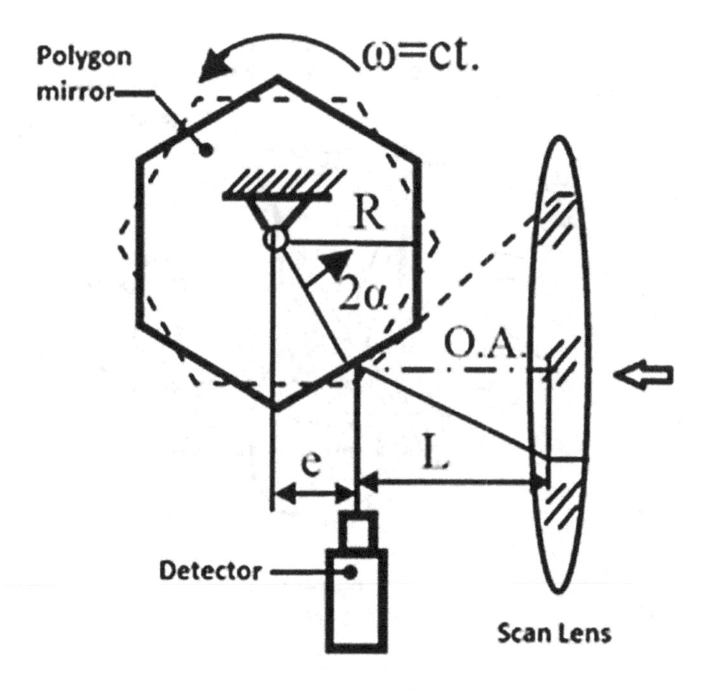

3.3.3. Holographic Scanners

Almost all holographic scanners comprise a substrate which is rotated about an axis, and utilize many of the characterising concepts of polygons. An array of holographic elements disposed about the substrate serves as facets, to transfer a fixed incident beam to one which scans. As with polygons, the number of facets is determined by the optical scan angle and duty cycle, and the elemental resolution is determined by the incident beam width and the scan angle. In radially symmetric systems, scan functions can be identical to those of the pyramidal polygon. Meanwhile there are many similarities to polygons, there are significant advantages and limitations (Figure 6).

3.3.4. Galvanometer and Resonant Scanners

To avoid the scan nonuniformities which can arise from facet variations of polygons or holographic deflectors, one might avoid multifacets. Reducing the number to one,

Figure 6. Polygonal scanner
(From http://beta.globalspec.com/reference/34369/160210/chapter-4-3-5-4-scanner-devices-and-techniques-postobjective-configurations)

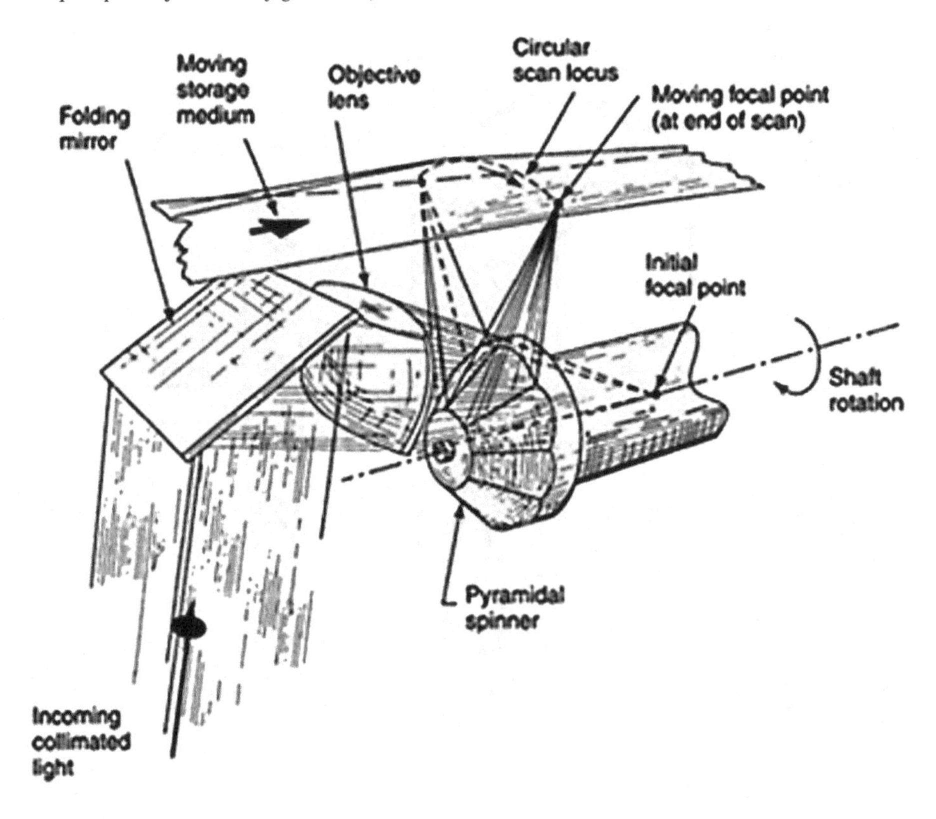

the polygon becomes a monogon. This adapts well to the internal drum scanner, which achieves a high duty cycle, executing a very large angular scan within a cylindrical image surface. Flat-field scanning, however, as projected through a flat-field lens, allows limited optical scan angle, resulting in a limited duty cycle from a rotating monogon. If the mirror is vibrated rather than rotated completely, the wasted scan interval may be reduced. Such components must, however, satisfy system speed, resolution, and linearity. Vibrational scanners include the familiar galvanometer and resonant devices and the least commonly encountered piezoelectrically driven mirror transducer as shown in Figure 7.

3.3.5. Scanner with 45° Cylindrical Mirror

Optical scanning systems can use coherent light emitting sources, such as laser or incoherent light sources like the lights of a vehicle. In the use of laser as light emitting source, the measurements are independent of environment lighting, so it is possible to explore during day and night, however, there are some disadvantages such as the initial cost, the hazard due to its high energy output, and that they cannot penetrate dense fog, rain, and warm air currents that rise to the structures, interfering the laser beam, besides, it is difficult to properly align the emitter and receiver.

Figure 7. Galvanometer scanner
(From http://www.yedata.com)

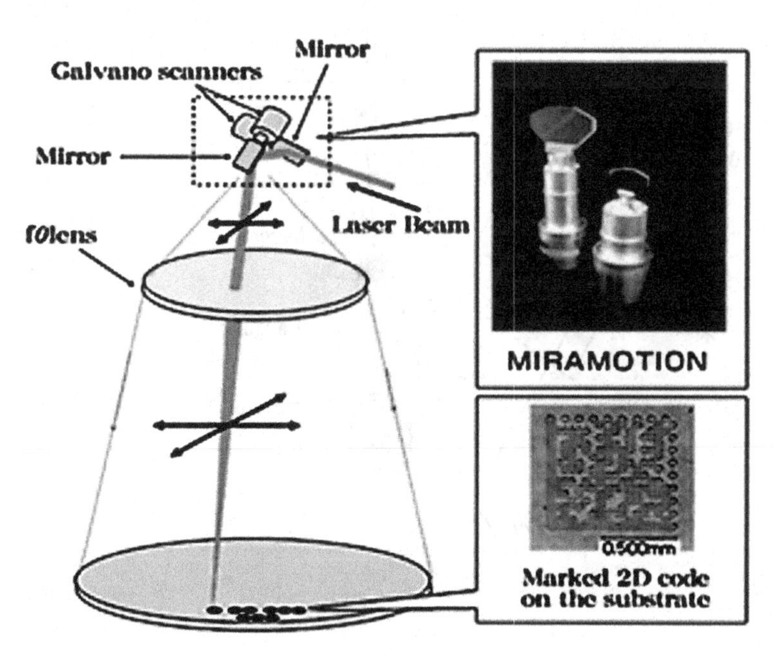

A passive optical scanning system for SHM can use conventional light emitting sources placed in a structure to determine if its position changes due to deteriorating. Figure 6 illustrates a general schematic diagram with the main elements of the optical scanning aperture used to generate the signals to test the proposed method.

As can be seen in Figure 8, the optical system is integrated by the light emitter source set at a distance from the receiver; the receiver is compound by the mirror E, which spins with an angular velocity ω. The beam emitted arrives with an incident angle β with respect to the perpendicular mirror, and is reflected with the same angle β, according to the reflecting principle to pass through a lens that concentrates the beam to be captured by the photodiode, which generates a signal "f" with a shape similar to the Gaussian function. When the mirror starts to spin, the sensor "s" is synchronized with the origin generating a pulse that indicates the starting of measurement that finishes when the photodiode releases the stop signal. This signal is released when the Gaussian signal energetic centre has been detected.

Figure 8. Scanner with 45° cylindrical mirror

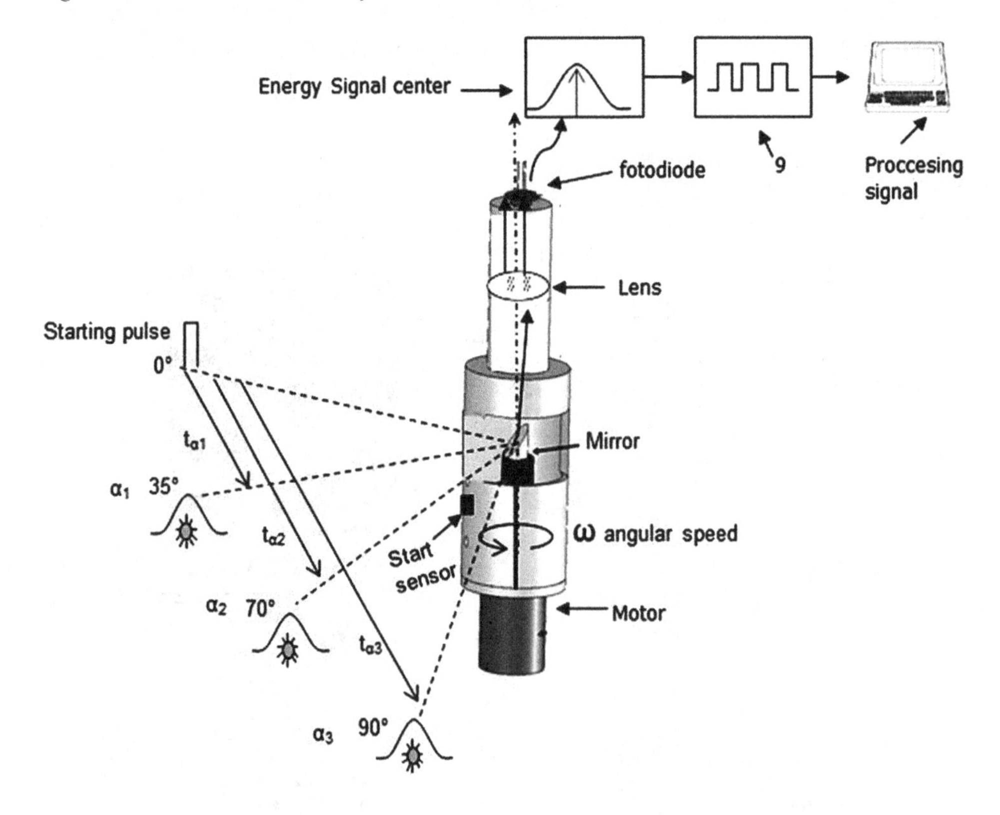

Figure 9 shows that light intensity increments in the centre of the signal generated by the scanner. The sensor "s" generates a starting signal when $t\alpha = 0$, then the stop signal is activated when the Gaussian function geometric centre has been detected.

The distance $T2\pi$ is equal to the time between m1 and m1, that are expressed by the code $N2\pi$ as defined in equation 1.

$$N_{2\pi} = T2\pi \cdot f_0 \tag{1}$$

On the other hand, the time $t\alpha$ is equal to the distance between m1 and m2, could be expressed by the code defined in equation 2.

$$N\alpha = t_{\alpha} \cdot f_0 \tag{2}$$

where f0 is a standard frequency reference. With this consideration the time variable could be eliminated from equation 2 obtaining equation 3.

Figure 9. Signal generated by a 45° cylindrical mirror scanner

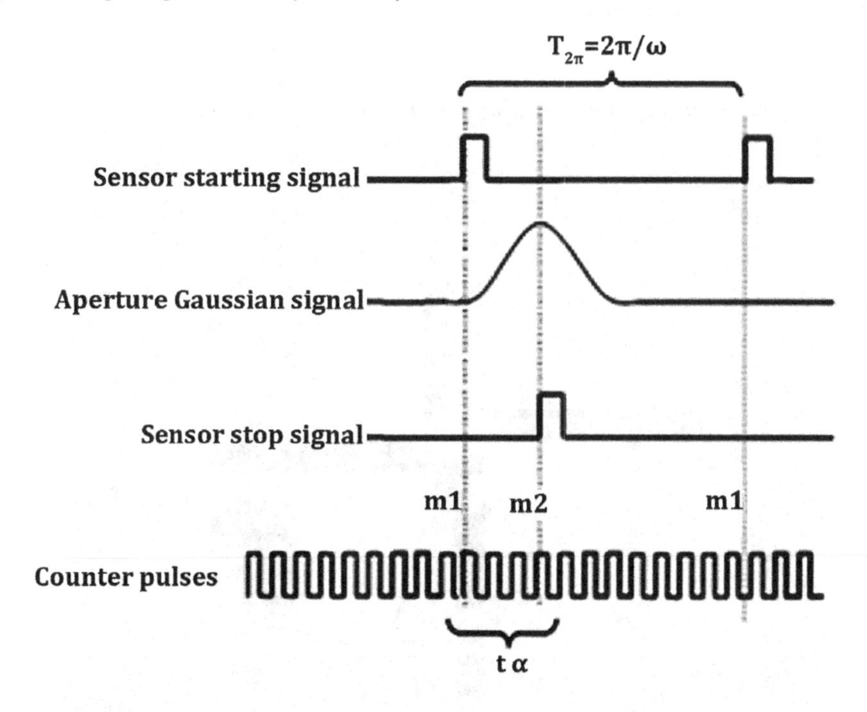

$$\alpha = 2\pi \cdot N_{\alpha} / N_{2\pi} \tag{3}$$

4. INCOHERENT VS. COHERENT LIGHT IN SCANNERS FOR LANDSLIDE MONITORING

The objective of this study was to compare the effectiveness of laser and incoherent light used in a scanning system for landslide monitoring. A total of 63 measurements were taken, with the purpose of characterize the scanner development. Position measurements were carried out during the scanning process from 1 to 6 m, moving the position of light emitter source from the nearest distance to the optical aperture sensor to the farthest position from the optical aperture sensor, varying the lateral distance and the height, 5 measurements were taken at each 2m.

The response of system with a laser was better than incoherent light only when the laser was aligned with the scanning aperture. In the other hand, when incoherent light was used, the scanner has satisfactory development in all check points.

For this experimentation, the set up consisted of placing a source of incoherent light onto a surface marked with a scale in four directions as illustrated in Figure 10. Figure 11 shows the system response in four different check points located at 6m of the scanning aperture.

Figure 10. Set up for experimentation with incoherent light source

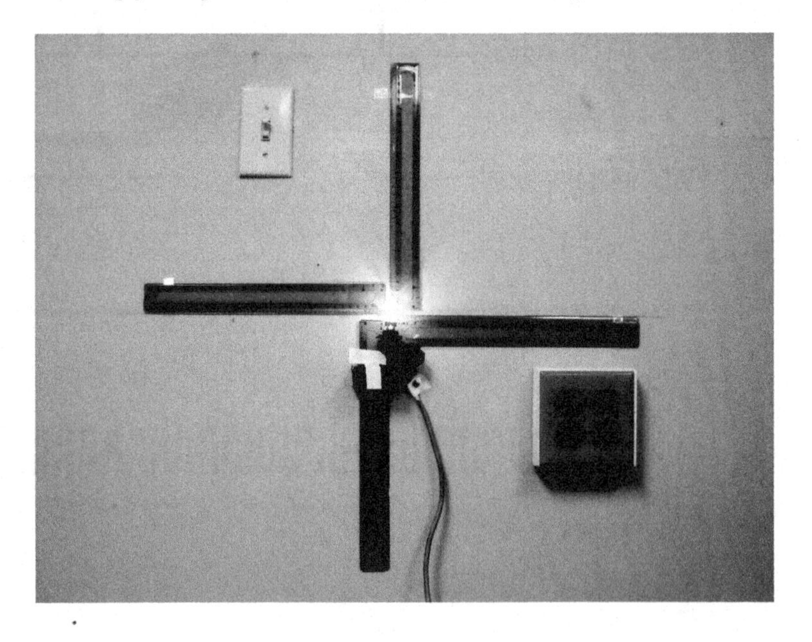

Figure 11. System response in four different check points located at 6m of the scanning aperture

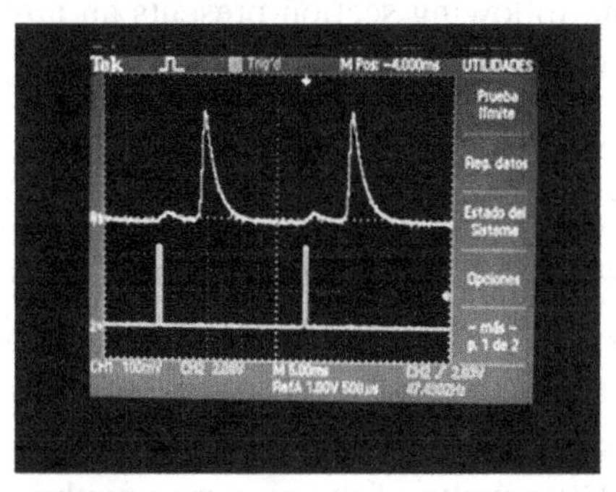

a) Response of scanner with incoherent source light located 30cm above of center.

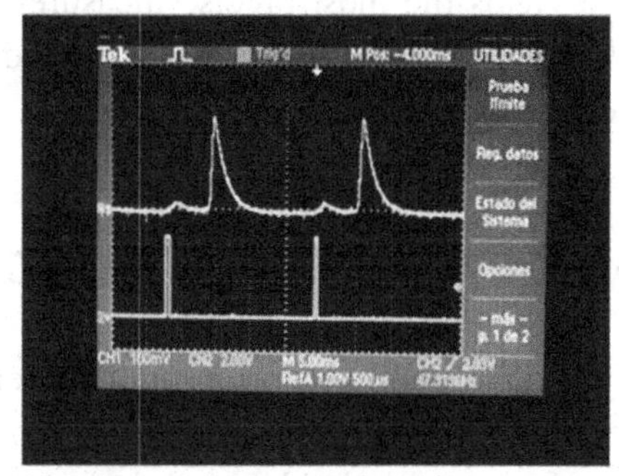

b) Response of scanner with incoherent source light located 30cm below of center.

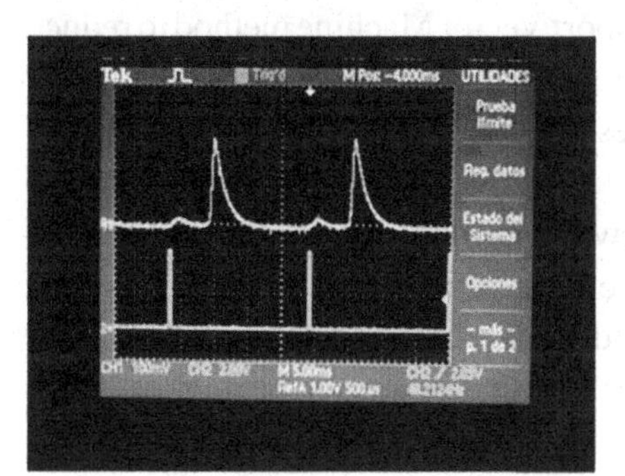

c) Response of scanner with incoherent source light located 30cm to rigth of center.

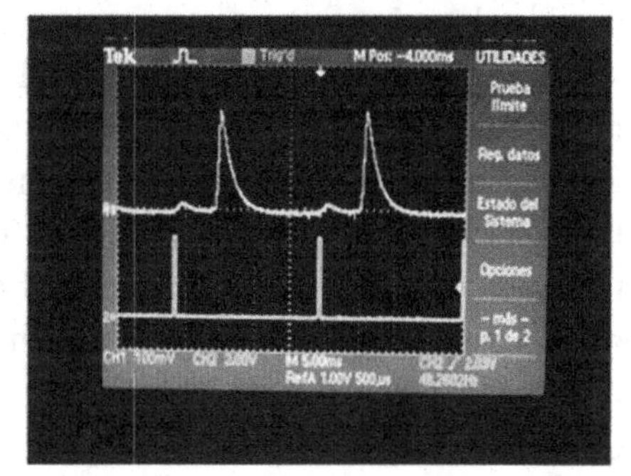

a) Response of scanner with incoherent source light located 30cm to left of center

As can be seen in Figure 11, the response to each check point did not differ significantly between them. In all cases the maximum value of measurement was about 300mV.

These results show the advantage of using incoherence light with optical scanners. However, the application of such scanners in landslides monitoring involves measurements containing the coordinates of the light-emitting source to determine the amount of displacement.

Experimental readings showed that, in order to find the position of a light source, the targeted signal resemblances a Gaussian shaped signal. This is mainly observed when the light source searched by the optoelectronic scanning is a punctual light source. This last assertion corresponds to the fact that when the punctual light source expands its radius, a cone-like or an even more complex shape is formed depending on the properties of the medium through which the light is travelling. To reduce errors in position measurements, the best solution is taking the measurement in the energy center of the signal generated by the scanner. The Energy Centre of the signal concept employs different mathematical methods as a way to assess which one yields the most precise measurement. The following section presents an application of an optical scanner in landslides monitoring, with experimentation in laboratory environment.

5. A NEW APPROACH BASED IN OPTICAL SCANNERS

Typically, a landslide prone area is a remote one and the sensors need to be rugged, weather resistant, portable, with low power consumption, with a low cost and easy interfacing with the data acquisition system. The novelty of the proposed method is the combination of optical scanner system with incoherent light detection and machine learning techniques, specifically Support Vector Machine method to reduce error in measurements. This method was previously validated with successfully results, taking measurements at distances less than 1.5m (Flores Fuentes, et al., 2014). The prototype used to test the method was significantly improved to increase the distance range of measurements. The new experiment (figure 10) and results, taking measurements at a distance of 6m, are showed in this section. The experiment was developed with low cost sensors and incoherent light emitter. Additional advantage of using inexpensive sensors is that allow more units to be deployed. For example, Tungsten Quartz Halogen Lamps may be used as a beacon for the 45° mirror scanner explained above.

Consider the emitted power of a light bulb filament as a function of wavelength λ and given by Planck's radiation formula: (Corke, 2013)

$$E\left(\lambda\right)=\frac{2hc^{2}}{\lambda^{5}\left(e^{\frac{hc}{k\lambda T}}-1\right)}\ \mathrm{W\ m^{-2}\ m^{-1}} \tag{4}$$

where T is the absolute temperature (K) of the source, h is the Planck's constant, k is the Boltzmann's constant end c is the speed of light. This is the power emitted per unit area per unit wavelength.

The total amount of power radiated (per unit area) is the area under the blackbody curve and is given by the Stefan-Boltzmann law:

$$E=\frac{2\pi^{5}k^{4}}{15c^{2}h^{3}}T^{4}\ \mathrm{W\ m^{-2}} \tag{5}$$

if

$$\sigma = \frac{2\pi^5 k^4}{15c^2 h^3} \tag{6}$$

Then

$$E = \sigma T^4 \text{ W m}^{-2} \tag{7}$$

For lighting applications, typical filament temperature is between 2800°K to 3400°K. On the other hand, for infrared heating applications, tungsten temperature could be anywhere from 2000°K to 3200°K, most commonly used temperature being 2500°K. As the filament temperature is raised, the radiation spectrum changes. For a given filament temperature T(°K), the wavelength at which the radiant energy is maximum can be estimated by the Wien's Law:

$$\lambda_{max} = \frac{2.8978\times10^{-3}}{T} \tag{8}$$

The filament temperature can be estimated from the ratio of the hot resistance (RH) and cold resistance (RC @ 25°C) of the filament. Since the data of resistance of tungsten as a function of the temperature is accurately known, the ratio RH / RC gives a fairly accurate estimate of filament temperature. Emissivity of tungsten at blue end is higher than at red end and therefore, the apparent colour temperature of a tungsten filament is higher than the actual filament temperature by 50 to 100°K around 3000°K (Litex Electricals Pvt. Ltd., 2015).

As can be seen in Figure 12, Halogen_Spectrum, the greatest Energy of a Tungsten Quartz Halogen Lamp is in the infrared section with a peak between 700-800 nm, leaving a small part in the visible spectrum. If this type of lamp is coupled with an optical sensor like the OPT 301 with peak sensitivity between 700-800 nm it is possible to get a very good system response (see Figure 13). With this in mind, the system shown in Figure 8 was constructed and tested.

Measurements were performed scanning from 45° to 135°, positioning the light emitter source at near distance to the optical aperture sensor and displacing it away from the optical aperture sensor through the same angle. Ten measurements were taken at each point –i.e. angle, distance, repeating this measurement process every 5°. Figure 14 depicts this process. A total of 6020 measurements built the dataset that was used to train and to test the SVM algorithm in the error prediction.

Figure 12. Tungsten quartz halogen lamp spectrum

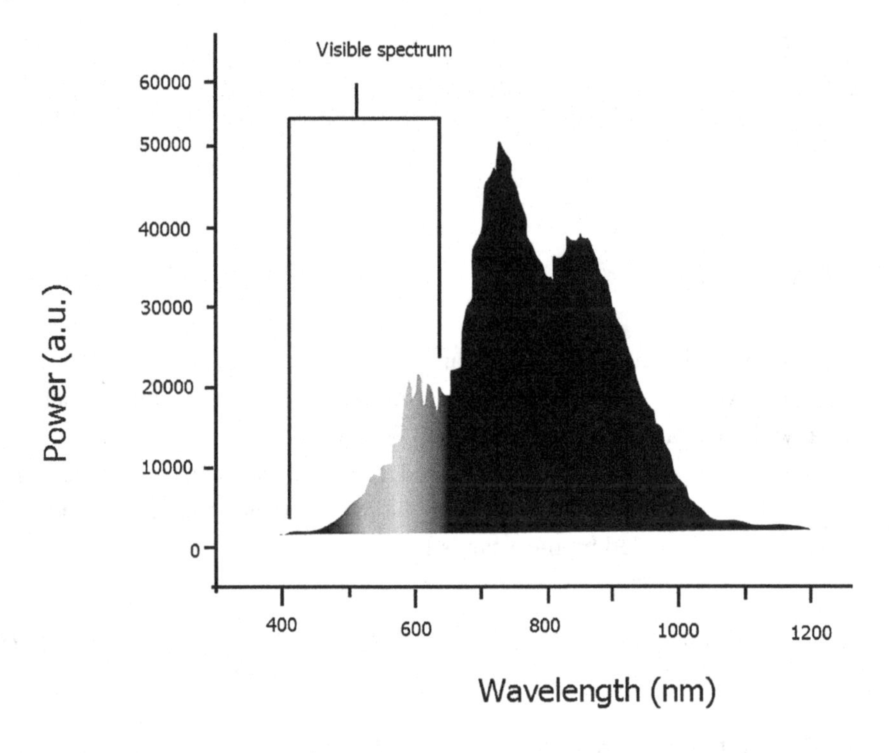

Figure 13. Spectral responsivity of OPT 301

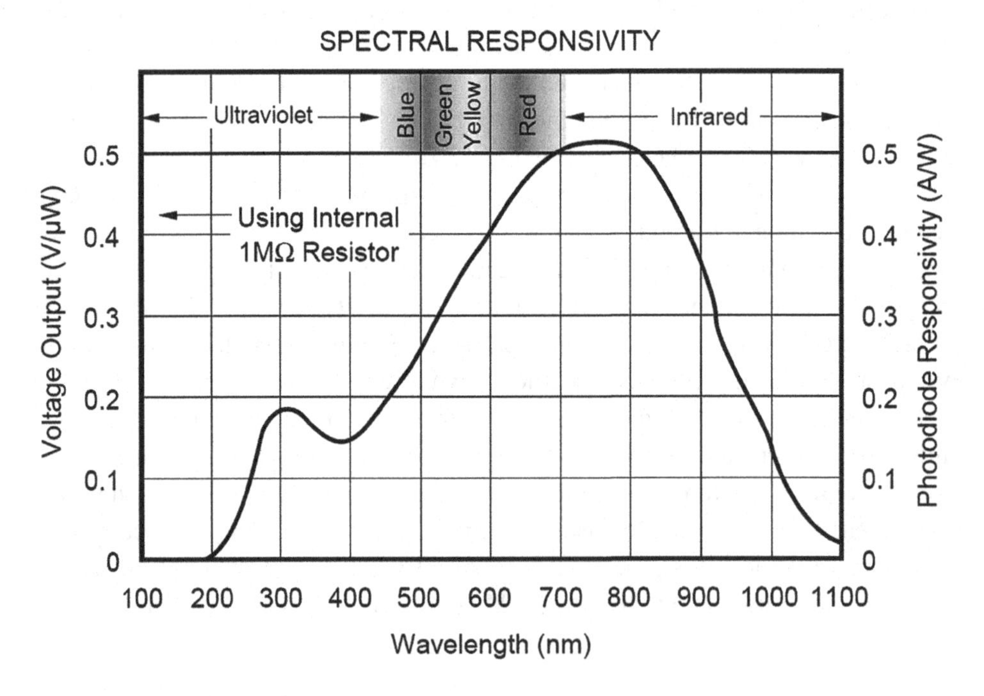

Figure 14. Measurements angle and distance representation

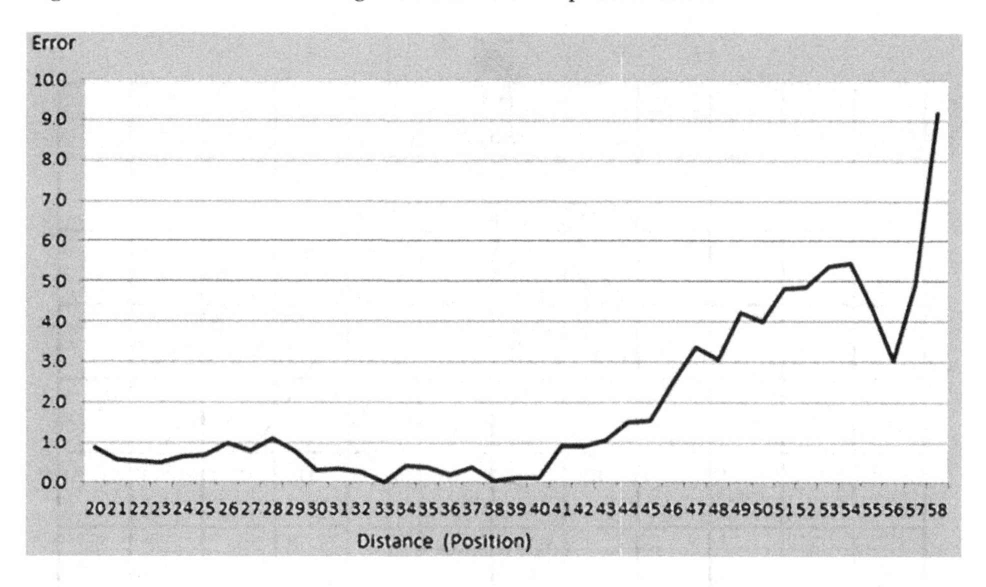

Figures 15 and 16, gives an error profile by position and angle. It is seen that the error does not follows a well-defined behavior or any known function.

As far as the light emitter source is moved away from the optical aperture sensor, the error increments. Also, it was noticed that the best scanning window angle was close the 90°.

Measurement errors have been calculated by (9) without error adjustment and by (10) with error adjustment using the most appropriate method to be implemented on optical scanning system for better measurement accuracy.

$$E = \alpha_m - \alpha_t \tag{9}$$

$$E_P = \alpha_{mc} - \alpha_t \tag{10}$$

where:

E is the real Error,
Ep is the predicted Error,
αm is the angle measured by the system,
αt is the target angle measured,
αmc is the angle measured by the system corrected (digital rectified).

Figure 15. Measurement error average by position

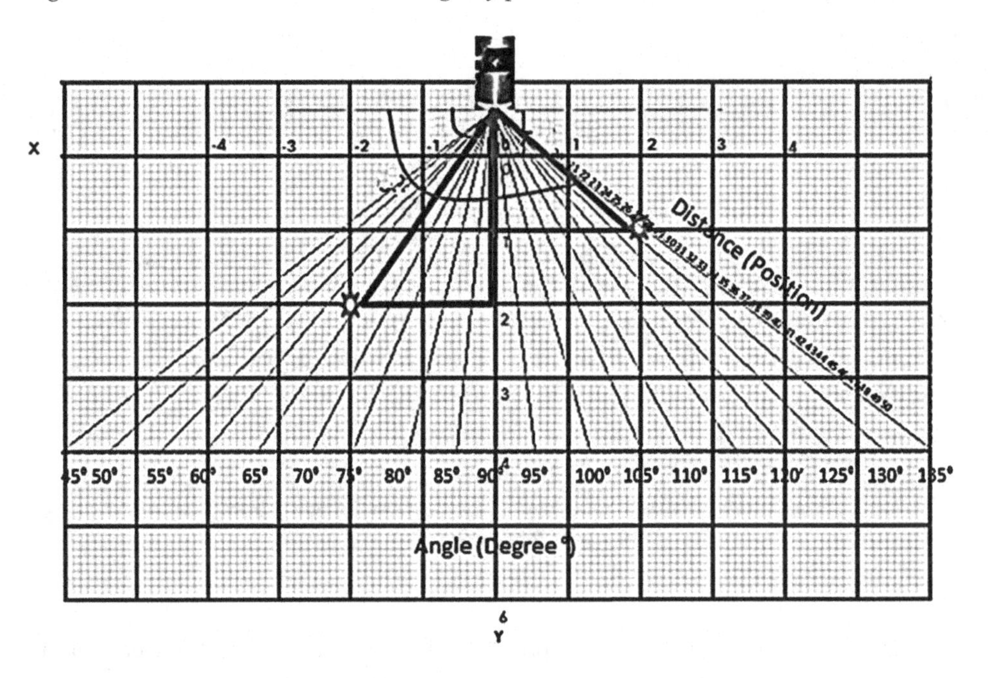

Figure 16. Measurement error average by angle°

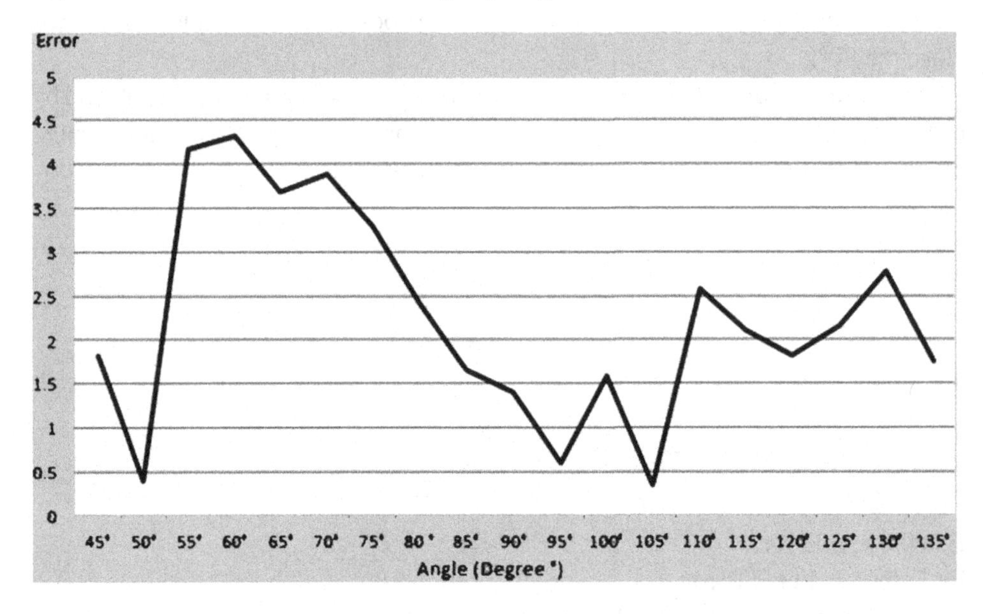

Before correcting the measurements, 53.39% of measurement had an error less than 2.99°.

Each set of measurements could be affected by different error sources generated due to environmental conditions or even errors due to the mechanism by itself. Hence, systematic and random errors do not follow a linear function, since their behavior is by the position -i.e. angle and distance, scanning frequency. For this reason, error compensation by a linear function is not suitable to the task at hand. Therefore, in this work it is proposed to use error approximation functions (artificial intelligence algorithms under assessment) toper form the error compensation.

To predicting and compensating error seven artificial intelligence algorithms were evaluated taking the measurements in the energy center of the signal generated by the scanner. The seven algorithms are: Principal Components Regression (PCR); Partial Leas Square Regression (PLSR); Ridge Regression (RR); Least Absolutely Shrinkage and Selection Operator (LASSO); and Generalized Linear Models Fitting (GLMFIT). To select the one that best suit the data model with the goal to predict the value of the outcome error measure based on a number of input measures to compensate the error measurement, caused by the few disadvantage on rotatory mirror scanners and explore the big benefits of advantages as simplicity, low cost, suitable to any kind of coherent or incoherent light detection and focus in the search of only the features of interest K-Fold validation was made obtaining the best results with support vector machine regression. (Flores Fuentes, et al., 2014; Gascon-Moreno, et al., 2011).

Support Vector Machines (SVM) was developed to solve classification and regression problems in machine learning or pattern recognition area. The statistical learning theory goal in modelling is to choose a model from the hypothesis space, which is closest (with respect to some error measure) to the underlying function in the target space as described on (Gunn, 1998) (Gascon-Moreno, et al., 2011). According to Qingsong Xu, SVM is capable of modelling nonlinear systems by transforming the regression problem into a convex quadratic programming (QP) problem and solving it with a QP solver (Xu, 2013), basically, support vector machines regression can be applied to: a) Linear regression by the introduction of an alternative loss function (as Quadratic, Laplace, Huber and ε-insensitive); and b) Non-linear regression by a non-linear mapping of the data into a high dimensional feature space where linear regression is performed.

In this chapter we show an application of prediction measurement error by means of a SVM regression to perform the digital rectification by measurement correction, adding the predicted error to each measurement.

Figure 17 shows measurement improvement through a rotatory mirror scanner applying Support Vector Machine as algorithm for error compensation. Darker points indicate error bigger than 6.00°. White points indicate error bigger than 3.00°

Figure 17. Results after the compensating error

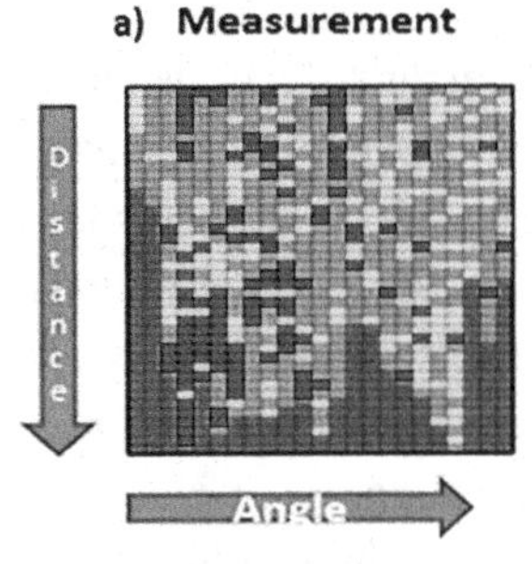

53.39%

Error Range Quantitative Analysis(Without Error Adjustment)

Error Percentage Classification	*Percentage*
Error > 6.00°	17.80%
3.00° < *Error* < 5.99°	28.81%
Error < 2.99°	53.39%

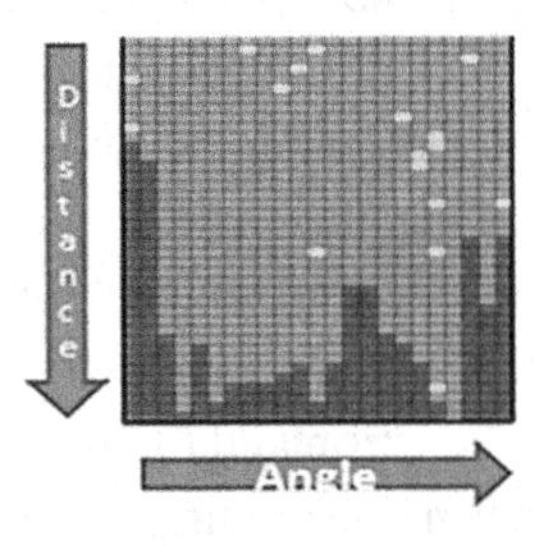

98.3%

Error Range Quantitative Analysis(With Error Adjustment)

Error Percentage Classification	*Percentage*
Error > 6.00°	0.00%
3.00° < *Error* < 5.99°	1.70%
Error < 2.99°	98.3%

and less than 5.99°. Gray points indicate error less than 2.99°. So, error bigger than 6.00° are reduced to 0, error bigger than 3.00° and less than 5.99° are reduced from 28.81% to 1.70%, and error less than 2.99° are increased from 53.39% to 98.30%.

The selection of compensation method was published by the authors in (Flores Fuentes, et al., 2014), (Flores Fuentes, et al., 2014).

As shown in Figure 17, the error is reduced to less than 3° in 98% of measurements. However, it is necessary to reduce this error even more. The authors have been visualized other improvements for further research, on the mechanical-electrical optical system, as in the energy signal center, digital processing and its rectification by the analysis of the motor rotation frequency effects.

6. FUTURE RESEARCH DIRECTIONS

Improvements to the proposed system include the development of a dual aperture system to implement the triangulation method to distance and angle measurements at the same time as well as an increment in the range. With the use of two aperture scanners separated by a known distance it is possible to measure angle and distance at the same time by means of dynamic triangulation calculations. This coordinate calculation method is called dynamic due to the rotation ability of the scanning apertures allowing to have angles that can form triangles with different shapes for a very short period of time and when a triangle is formed thanks to the light emitter detected by two apertures, it is obtained all the necessary information to calculate the coordinate of the light emitter. Also, other improvements have been visualized for further research to reduce optical noise using optical and digital filters to enhance accurate of measurement in operational real environment. Maybe a system with height measurement capabilities could be possible so it can measure x and y axis angles.

7. CONCLUSION

In this chapter was presented an alternative method to landslide monitoring. The focus of this application involves finding the position of non-coherent light sources located at strategic points susceptible to landslides. The results of experiments of position light source monitoring in laboratory environment showed that the incoherent light is better than laser scanners for such applications because alignment problems are not critical.

The experiments also showed that the error is significantly reduced when using artificial intelligence techniques like support vector machine to compensate it. Thus the proposed method has been proved successfully and was shown to be an inexpensive way for detecting non-coherent light sources. However, the scope of the features explored in this research is limited, therefore the authors have been visualized other improvements for further research, on the mechanical-electrical optical system, as in the energy signal center, digital processing and its rectification by the analysis of the motor rotation frequency effects.

REFERENCES

Anderson, S. A., & Tallapally, L. K. (1996). Hydrologic response of a steep tropical slope to heavy rainfall. 7th International Syposium on Landslides (pp. 1489-1498). Trondheim, Norway: Brookfield, Rotterdam.

Baron, I. R. S. (2010). State of the Art of Landslide Monitoring in Europe: Preliminary results of the Safeland Questionnaire. In R. S. Baron (Ed.), Landslide Monitoring Technologies & Early Warning Systems (pp. 15-21). Vienna: Geological Survey of Austria.

Bass, M. (1995). Handbook of Optics, Vol. II- Devices, Measurements and Properties. New York: McGraw-Hill, Inc.

Corke, P. (2013). *Robotics, Vision and Control*. Berlin: Springer.

Flores Fuentes, W., Rivas López, M., Sergiyenko, O., González Navarro, F. F., Rivera Castillo, J., Hernández Balbuena, D., & Rodríguez Quiñonez, J. (2014). Combined Application of Power Spectrum Centroid and Support Vector Machines for Measurement improvement in Optical Scanning Systems. *Signal Processing, 98*, 37–51. doi:10.1016/j.sigpro.2013.11.008

Flores Fuentes, W., Rivas López, M., Srgiyenko, O., González Navarro, F. F., Rivera Castillo, J., & Hernández Balbuena, D. (2014). Machine Vision Supported by Artificial Intelligence Applied to Rotary Mirror Scanners. *Proceedings of the 2014 IEEE 23th International Symposium on Industrial Electronics* (pp. 1949-1954). Istambul, Turkey: ISIE. doi:10.1109/ISIE.2014.6864914

Gascon-Moreno, J., Ortiz-Garcia, E., Salcedo-Sanz, S., Paniagua-Tineo, A., Saavedra-Moreno, B., & Portilla-Figueras, J. (2011). Multi-parametric Gaussian Kernel Function Optimization for Ɛ-SVMr Using a Genetic Algorithm. In J. Cabestany, I. Rojas, & G. Joya (Eds.), *Advamces in Computational Intelligence* (pp. 113–120). Berlin: Springer. doi:10.1007/978-3-642-21498-1_15

Gunn, S. R. (1998). *Support vector machines for classification and regression*. University of Southhampton.

Hoon Sohn, C. F. (2002). *A Review of Structural Health Monitoring Literature 1996-2001*. Retrieved from www.lanl.gov.projects/damage

Kennedy, W. P. (2012). *The Basics of triangulation sensors*. Retrieved from http://archives.sensorsmag.com/articles/0598/tri0598/main.shtml

Kerle, N., Stump, A., & Malet, J.-P. (2010). Object oriented and cognitive methods for multidata event-based landslide detection and monitoring. *Landslide Monitoring Technologies & Early Warning Systems*.

Litex Electricals Pvt. Ltd. (2015). Retrieved from http://www.litexelectricals.com/halogen_lamp_characteristics.asp

Michoud, , M. J.-H. (2012). New classification of landslide-inducing anthropogenic activities. *Geophysical Research Abstracts*.

Ming-Chih, L., Su-Chin, C., Yu-Shen, L., & Cheng-Pei, T. (2015, March). *Machine Vision in the realization of Landslide monitoring applications*. Retrieved from INTERPRAEVENT: http://www.interpraevent.at/?tpl=startseite.php&menu=41

NOAA. (2015). Retrieved from http://oceanservice.noaa.gov/facts/lidar.html

Plaza, J. P. (2009). Multi-Channel Morphological Profiles for Classification of Hyperspectral Images Using Support Vector Machines. *Sensors (Basel, Switzerland)*, 197. PMID:22389595

Rentschler, K., & Moser, M. (1996). Geotechniques of claystone hillslopes-properties of weathered claystone and formation of sliding surfaces. *7th International Symposium on Landslides* (pp. 1571-1578). Tronheim, Norway: Brookfield, Rotterdam.

Rivas López, M., Flores Fuentes, W., Rivera Castillo, J., Sergiyenko, O., & Hernández Balbuena, D. (2013). A Method and Electronic Device to Detect the Optoelectronic Scanning Signal Energy Centre. In S. L. Pyshkin, & J. M. Ballato (Eds.), Optoelectronics (pp. 389-417). Crathia: Intech.

Savvaidis, P. D. (2003). Existing Landslide Monitoring Systems aand Techniques. School of Rural and Surveying Engineering, The Aristotle University of Thessaloniki.

Vahelal, A. (2010). *Digitization.* Gujarat, India: HasmukhGoswami College of Engineering.

Xu, Q. (2013). Impact Detection and Location for a Plate Structure using Least Squares Support Vector Machines. *Structural Health Monitoring*.

Chapter 8
3D Imaging Systems for Agricultural Applications:
Characterization of Crop and Root Phenotyping

Frédéric Cointault
University of Burgundy, France

Simeng Han
University of Burgundy, France

Gilles Rabatel
IRSTEA, France

Sylvain Jay
IRSTEA, France

David Rousseau
CREATIS, France

Bastien Billiot
Roullier Company, France

Jean-Claude Simon
University of Burgundy, France

Christophe Salon
University of Burgundy, France

ABSTRACT

The development of the concepts of precision agriculture and viticulture since the last three decades has shown the need to use first 2D image acquisition techniques and dedicated image processing. More and more needs concern now 3D images and information. The main ideas of this chapter is thus to present some innovations of the 3D tools and methods in the agronomic domain. This chapter will particularly focus on two main subjects such as the 3D characterization of crop using Shape from Focus or Structure from Motion techniques and the 3D use for root phenotyping using rhizotron system. Results presented show that 3D information allows to better characterize crucial crop morphometric parameters using proxy-detection or phenotyping methods.

DOI: 10.4018/978-1-5225-0632-4.ch008

INTRODUCTION

In order to optimize crop management and take into account the intra-parcellar variability, precision agriculture has been developed over the past thirty years. It consists in a localized crop management using new technologies such as computing, electronics and imaging. Two types of imagery can thus be used: proxy-detection and remote sensing. The conception of a proxy-detection system is motivated by the need of better resolution, accuracy, temporality and lower cost, compared to remote sensing. The use of computer vision techniques allows obtaining this information automatically with objective measurement in contrast with the difficulty and subjectivity of visual or manual acquisition. However, information is often obtained in two dimensions which does not allow to provide some particular characteristics of plants such as growth or early yield estimation, determination of foliar volume … Thus, the use of 3D information appears as an essential point.

Two main 3D representation families can be distinguished: the surface and the volume representations.

The first one is composed with 1) the depth maps, in which the value of each pixel corresponds to a distance between the corresponding point in the scene and the acquisition system; 2) the surfels (for surface element), composed of attributes describing a local sample of the surface like its coordinates, texture …; and 3) the meshing, largely used for 3D representation, group of points defined by their 3D coordinates. The volume representation contains 1) the voxels (for volumetric elements), appellation of 3D pixel constituted with colorimetric information associated with spatial coordinates; and 2) the spherical harmonics allowing a frequency-domain representation of the 3D model in spherical coordinates.

Unlike the active vision, no adding illumination is needed for the "passive" methods to work. In this category of 3D reconstruction methods, we find triangulation methods using several cameras, methods based on image analysis and methods needing particular optics.

The objective of this chapter is not to detail all the existing 3D techniques but rather to explore the use of some of these methods in precision agriculture or viticulture. Thus we propose to present innovations on 3D tools and methods used in agriculture for different applications such as 3D characterization of crop using Depth imaging systems (Billiot et al., 2013; Chéné et al., 2012) or Structure from Motion techniques (Jay et al., 2015; Santos & de Oliveira, 2012) and 3D use for root phenotyping using rhizotron systems (Clark et al., 2011). Results for the different applications, in terms of 3D tools and image processing are presented, discussed, and improvement proposals are given before to detail future research directions.

Background

In agriculture, viticulture or orchard, a lot of researches have been done on the use of 3D information and images (Figure 1).

The design of a 3D acquisition system is required in order to obtain new parameters related to crops. Numerous 3D acquisition techniques exist and they have been the subject of active research for several decades. The overall principle is to determine the shape and structure of a scene from the analysis of the acquired images. The representation of depth information will depend on the type of acquisition technique used, e.g., 3D mesh representation or depth map. In all cases, the imaging methods aim at retrieving numerous parameters that characterize the crop row structure. They must be non-destructive since these measurements have to be carried out all along the plant life. They must also be fast and automatic so that a maximum of data may be processed with a minimum of user interactions. Plant structural information can be accurately retrieved from three-dimensional images that can be computed in several ways.

The most common 3D approach in computer vision is the stereovision principle, also called Shape from Stereo. This 3D imaging technique consists in the acquisition of a pair of images of the same scene by two cameras from different angles. Then, based on the pinhole camera model and epipolar geometry (Hartley, 2013), the

Figure 1. Left: Kinect image; right: use of motion information
Sources: Chéné, 2012; SenseFly.

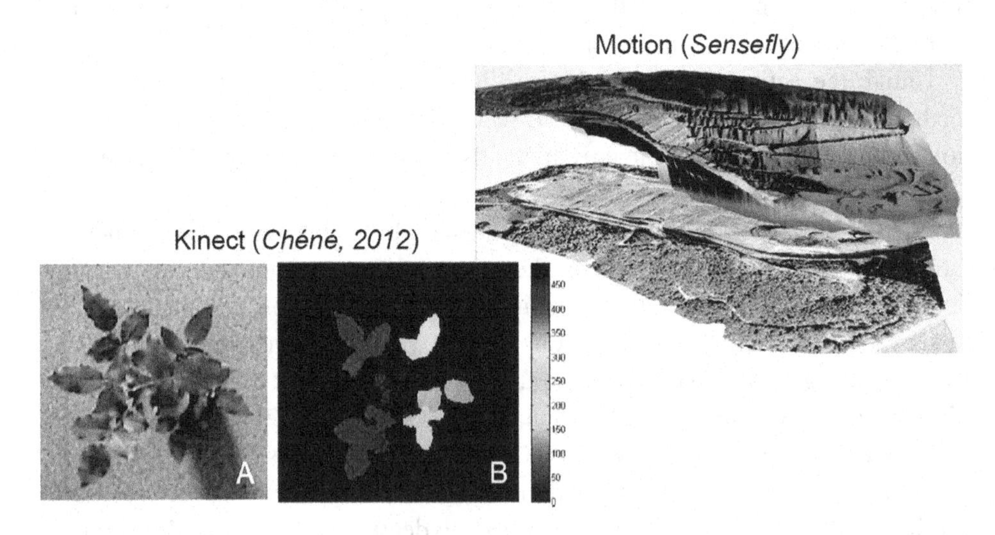

depth is determined from the disparity (difference between the position of an object viewed from multiple angles). This measure of disparity is the main difficulty for smooth functioning of this technique and depends on the choice of the base between cameras and their tilt angles. Indeed, the larger the base is, the more accurate the measure will be, but there will be more occlusions (a point on the scene viewed by a camera is not necessarily viewed by the other).

These occlusion problems do not allow to obtain good results due to the kind of scene where this phenomenon often happens (crops). A 3D reconstruction technique that frees itself from occlusion problems is necessary. We can group 3D reconstruction techniques into three large families: geometric approaches, photometric approaches and those based on the physical properties of the acquisition system. Geometrical approaches are based on the knowledge of the scene structure and the internal and external parameters of the cameras used. Stereovision technique is part of this approach. In the case of photometric approaches, the principle is the evaluation of a pixel's intensity to obtain 3D information as in the case of the method known as Shape from Shading. Finally, many techniques of the previous techniques are based on the pinhole model; the third approach uses a real optical system. The main difference is that instead of considering a perfect projection of all points of the scene onto the image plane, only some of these points are projected correctly. This phenomenon comes from a limited depth of field that will be explained later.

Two main activities in precision agriculture are of great interest for farmers and researchers: the early estimation of yield (for better yield knowledge and reduction of manual work) and the plant phenotyping (for comparison of varieties and relation establishment between genotypes and phenotypes). Thus in this chapter, we detailed first Structure from Motion and Shape from Focus methods used for crop characterization, which constitutes new techniques generally used in industrial or microscopic domains. The previous methods avoid the drawbacks due to stereovision technique allowing better accuracy. In a second section, we propose the use of laser-scanner for plant root image acquisition in order to characterize RSA (Root System Architecture). This is based on the use of rhizotrons, which are containers devoted to grow plants and equipped with outer transparent sides so that roots can be visualized in situ, non-invasively and dynamically.

Each section is constituted similarly: first a presentation of the method used, then details on the experiments done and the results obtained, and finally some discussions on the techniques proposed. The next sections will deal with future research directions before to conclude on this chapter.

3D USE FOR CROP CHARACTERIZATION

3D Plant Modelling by Structure from Motion

Method

Structure from Motion (SfM) is a computer vision technique allowing recovering the 3D structure of a scene by combining the pictures acquired from different points of view with a moving camera. In the domain of remote sensing and UAV (Unmanned Aerial Vehicle) imagery, this technique is also referred to as photogrammetry. In the present study, SfM was used for recovering the 3D structure of a crop scene from a camera positioned above at a short distance (typically one meter). In a second phase, the crop was separated from the ground using a segmentation algorithm based on both height and color information, in order to compute structural characteristics such as plant height and leaf area.

3D Model Extraction

SfM is based on stereo reconstruction: as soon as a scene point is seen from two different camera positions, and provided these positions and camera intrinsic parameters are known, the 3D position of this scene point can be recovered. The intrinsic parameters of a camera are its internal parameters such as the focal length, the image enlargement factors, the coordinates of the projection of the optical center of the camera on the image plane. In practice, it means that for SfM 3D modelling, every point in the scene must be seen at least twice, thus requiring a large image overlapping (more than 50%).

Basically, the SfM algorithm, which consists in analyzing these overlapping images, is made of three steps:

1. For every couple of overlapping images, a set of matching points is searched. Feature points are detected in each image using SIFT algorithm (Lowe, 1999) or equivalent, and then matched based on their descriptor similarity and RANSAC selection.
2. Using a global optimization procedure (bundle adjustment), these sets of matching points allow determining not only the corresponding point 3D positions, but also the camera 3D position at each image acquisition and the camera distortion parameters. It results in a sparse 3D point cloud (Figure 2).
3. Finally, based on camera positions, a dense matching is made using cross-correlation (the knowledge of camera positions allows the restriction of cross-correlation searching to epipolar lines). It results in a dense 3D point cloud, defined in arbitrary coordinates. A reference object placed in the scene

Figure 2. Example of 3D sparse point cloud, where tetrahedrons on the top represent the successive camera positions and orientations
Jay et al., 2015.

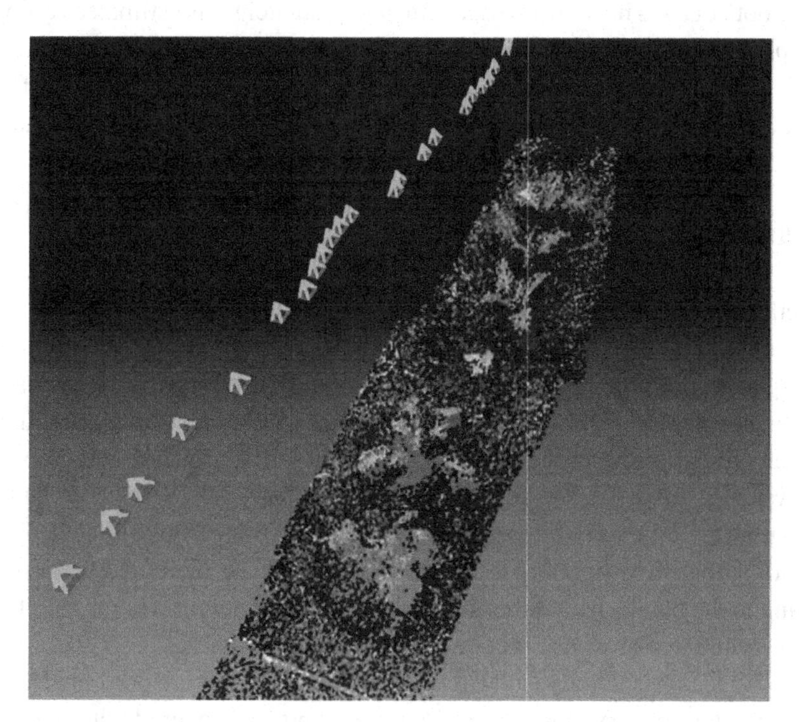

(typically a tape) then allows rescaling this dense point cloud in real metric coordinates.

The whole SfM process was implemented using MicMac, an open source photogrammetric software suite (MicMac, IGN).

3D Model Segmentation and Leaf Area Computation

In order to further analyze the 3D structure of the crop, it is first required to separate it from the background. A usual approach for vegetation segmentation consists in using the Excess Green (ExG) index, defined as:

$$ExG = (2G-R-B)/(R+G+B) \text{ (normalized form)} \quad (1)$$

where G, R and B are respectively green, red and blue colors. However, in presence of various possible backgrounds (soil, sand, grass, etc.), the color information alone is not robust enough. Therefore, a 2D clustering strategy based on ExG and height and using k-means algorithm has been implemented. The label "crop" is then assigned to the class containing the greenest and highest points. Figure 3 shows an extreme example with green background where classification is nearly based on height only.

Once both classes have been discriminated, plant height is estimated by computing the plant highest height minus the mean ground height in a local neighborhood around the plant. Finally, a triangular meshing of the plant point cloud is performed using ball-pivoting algorithm (Bernardini et al., 1999) and leaf area is estimated by summing every triangle area.

Experiments and Results

The method was applied on a data set collected under outdoor conditions using a digital single-lens reflex camera. The latter was orientated towards nadir and images were acquired approximately every 10 cm by manually moving the camera at about 1 m above the ground level. Various crop structures were considered, ranging from sunflowers to Brussels sprouts (23 plants in total). After image acquisition, plant height was measured. Leaf blades were then harvested and placed on a blank sheet of known size. They were scanned, and after extraction of leaf-related pixels by thresholding the corresponding Excess Green index map, their area was estimated according to the blank sheet dimensions. Overall, plant height was ranging from 10 to 65 cm, while leaf area was ranging from 27 to 1201 cm^2.

Some 3D reconstruction results of three sunflower plants are presented in Figure 4. In this case, as observed in Figure 4a, the ground was covered with weeds, thus

Figure 3. Crop classification: example of a green background with grass and weeds (a) and the corresponding representation using the Excess Green index

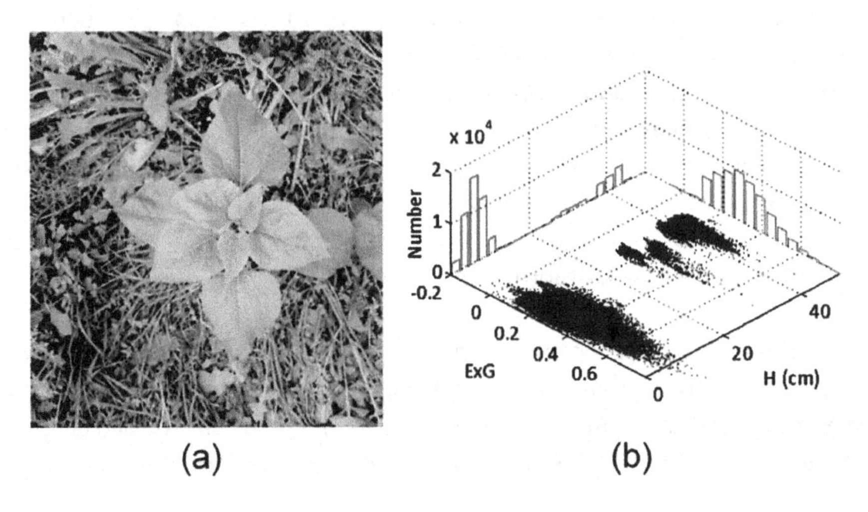

Figure 4. (a) Some of the 16 RGB images acquired above the sunflower row, and (b) associated 3D model

making discrimination between crop and ground more difficult. These RGB images confirm that each targeted plant was seen from various view angles, which is an essential prerequisite when implementing SfM. The more view angles we have, the more accurate the 3D reconstruction. However, using lots of images can dramatically increase the processing time. Therefore, a compromise has to be reached between the 3D model accuracy and processing time.

The obtained 3D model presented in figure 4b was accurate, both in terms of estimated leaf orientation and ground shape. The specific structure of sunflower, i.e., flat leaves at different heights, was indeed well retrieved. Stems were not reconstructed because, as observed in figure 4a, they were not seen by the sensor. This clearly shows that SfM is an efficient and convenient tool for 3D modelling because its implementation does not require any specific imaging system (e.g., unlike a calibrated stereocamera).

After discrimination of plant- and ground-related pixels, such a 3D model allows the characterization of crop row structure through several parameters, ranging from plant height to leaf area or mean leaf inclination, the latter being a critical parameter for vegetation remote sensing (Jacquemoud et al., 2009). In this chapter, we focused on the retrieval of plant height and leaf area. Overall, height estimation was extremely accurate for every considered species since the RMSE and measurement error were similar and close to 1 cm ($R^2 = 0.99$; RMSE = 1.1 cm). As a result, realistic height maps could be obtained from the 3D model as shown in Figure 5 for the Savoy cabbage row.

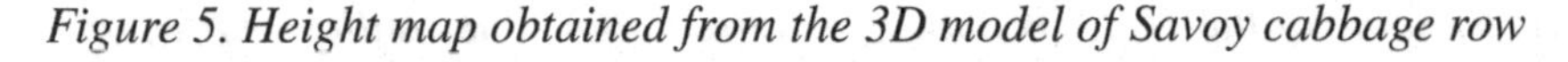

Figure 5. Height map obtained from the 3D model of Savoy cabbage row

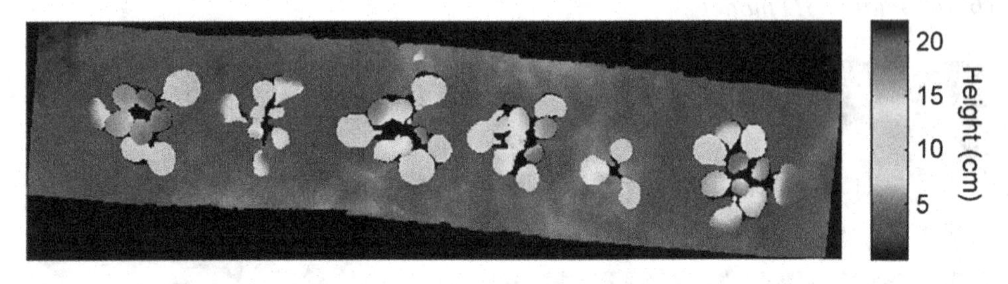

Similarly, we show in Figure 6 that a strong correlation was obtained between estimated and actual leaf areas ($R^2 = 0.94$; RMSE = 85 cm^2). The estimation results were good for every species, therefore proving that the proposed method can handle various plant structures. Interestingly, sunflower leaf area was well estimated even if both the ground and sunflower plants were green (Figure 4). This shows that combining color and height is an effective and robust way to discriminate the plants of interest from the ground.

As a summary, these results indicate that SfM is an interesting tool for crop 3D characterization. Combining it with simple image processing techniques allows the retrieval of various plant structural parameters such as height and leaf area. Both the accuracy and ease of use of this method makes it potentially suitable for phenotyping applications.

Figure 6. Leaf area estimation results

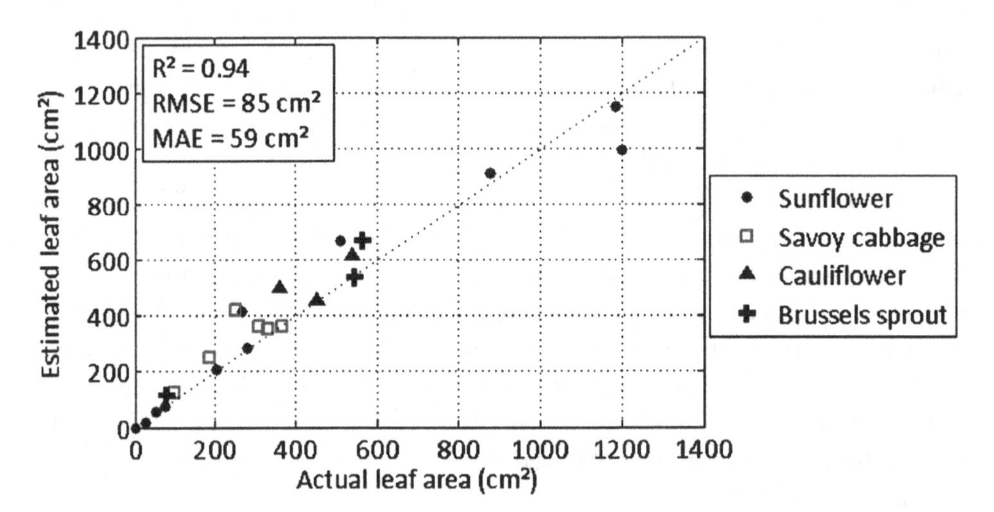

3D Plant Modelling by Shape from Focus

Method

The second 3D image acquisition system presented is based on Shape from Focus (SFF) technique, for the design of a 3D image acquisition system dedicated to natural complex scenes composed of randomly distributed objects with spatial discontinuities. In agronomic sciences, the 3D acquisition of natural scene is difficult due to the complex nature of the scenes. The whole system proposed is based on the monocular and passive 3D Shape from Focus technique initially used in the microscopic domain for confocal images (Figure 7).

The camera use is a CCD color camera with Bayer filter (IDS UI-2280SE), associated with a lens of 50 mm of focal. In order to minimize the effect of illumination conditions on the image acquisition process, two 3W-power LEDs are used and controlled. To avoid the problem of displacement of the optical part, the scene is scanned using the variation of the distance setting and a motorized lens (Qioptiq MeVis-Cm).

Adaptation of this technique to the macroscopic domain has been done combined with specific and novel image processing used to perform such technique. Two new focus measures are applied on a 2D image stack previously acquired by the system. When these focus measures are performed, the depth map of the scene can be created with high accuracy.

Figure 7. Acquisition process (left) and apparatus conceived with camera and power LEDs for the illumination (middle and right)

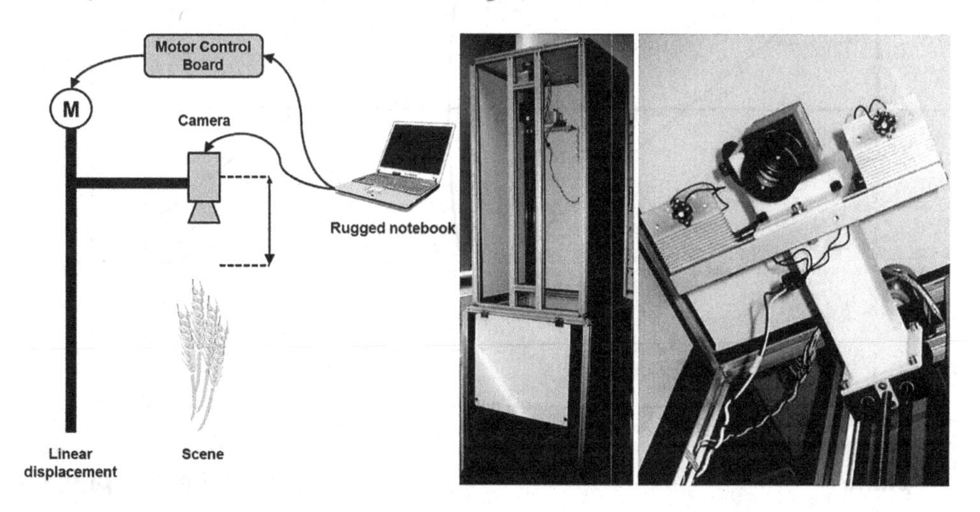

To better understand the physical principles governing the creation of sharp or blurred image and the acquisition process of image stack, a brief reminder of the optical properties is proposed. In Figure 8, all the rays emitted by the point P of an object and intercepted by the lens are refracted by this one and converge at point Q in the image plane.

The equation for the focal lens depending on the camera/object distance and lens/image plane is:

$$\frac{1}{f} = \frac{1}{o} + \frac{1}{s} \tag{2}$$

Each point of the object is projected onto the image plane at a single point and leads to the formation of the image Is(x,y). If the image plane does not merge with the sensor plane, where the distance between them is δ, the energy received from the object by the lens is distributed on the sensor plane in a circular shape. However, the shape of this energy distribution depends on the shape of the diaphragm aperture, considered circular. The radius of this shape can be calculated by:

Figure 8. Sharp and unsharp image formation

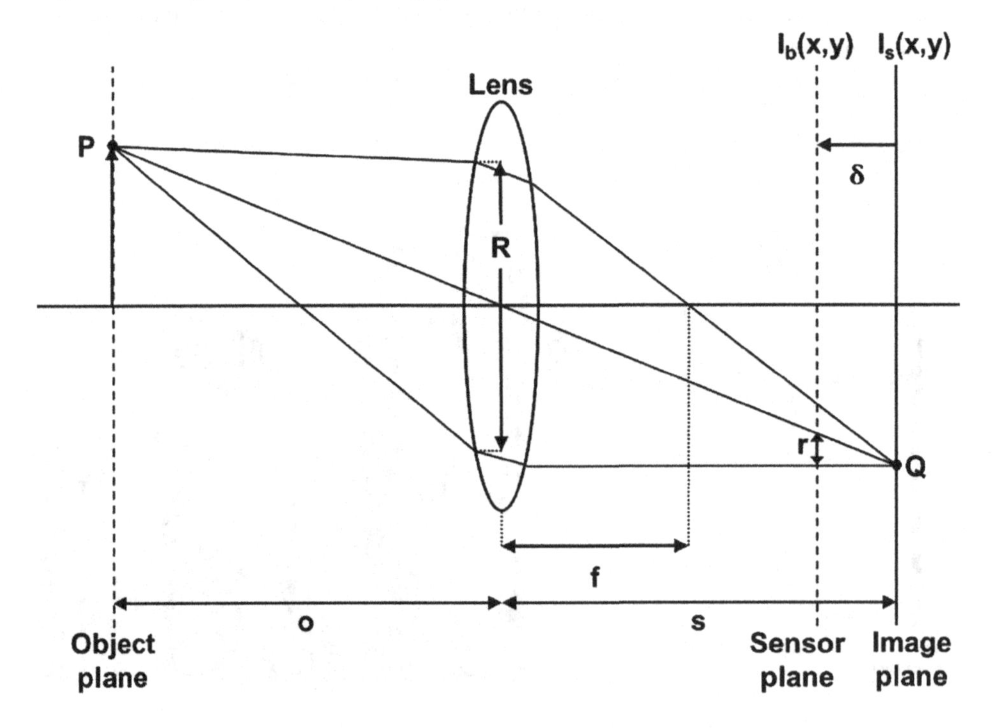

$$r = \frac{\delta.R}{s} \tag{3}$$

where R is the aperture of the lens.

The blurred image Ib(x,y) formed on the sensor plane can be considered as the result of a convolution between a sharp image Is(x,y) and a blur function h(x,y):

$$I_b(x,y) = I_s(x,y) * h(x,y) \tag{4}$$

This blur function can be approximated by the low pass filter:

$$h(x,y) = \frac{1}{2\pi\sigma_h^2} \exp^{-\frac{x^2+y^2}{2\sigma_h^2}} \tag{5}$$

The spread parameter σ_h is proportional to the radius r, thus the larger the distance δ between the image plane and the sensor plane is, and the higher frequencies are cut. In consequence, we obtain a blurred image. The depth of field depends on four parameters: camera/object distance, aperture, focal length and radius of the circle of confusion. The choice of all these parameters will affect not only the depth of field (DoF) but also the field of view (FoV) available. In conclusion, the depth of field decreases when the focal length or the aperture value increases. In the same way, it increases when the diameter of the circle of confusion or the camera/object distance increases. This depth of field is directly correlated with the depth resolution of the 3D reconstruction.

The first step of SFF technique concerns the image sequence acquisition: at each pixel corresponds an image in which this pixel is sharp. A measure of the local sharpness is applied at each pixel and its neighborhood allowing determining the image in which each point of the scene is the sharpest, as the autofocus principle. In the literature, several operators allow evaluating the sharpness and are also used in the 3D reconstruction algorithm SFF used locally at the pixel neighborhood. Generally, the sharpness measurements are based on the maximization of the high frequency information since a sharp image contains more intensity changes and primitives with high gradient (corners, edges, etc.) compared to a blur image. Obviously, the main drawback of this 3D reconstruction method is its application only to scenes with high frequency components that is with textured areas.

The more efficient and common operators are the following (Billiot et al., 2013):

- **FMtenvar:** Variance of the gradient range.
- **FMnrjgrad:** Energy of the gradient.
- **FMsml:** Sum of the modified laplacians.
- **FMvar:** Contrast based-variance of the grey levels.
- **FMentropy:** Measure of the local entropy.
- **FMcorr:** Measure of correlation.
- **FMorient:** Measure by adjustable filters.
- **FMwavratio:** Measure by wavelet ratio.

Two new powerful operators, based on Generalized Fourier Descriptors (GFD) (Gauthier et al., 1991), and described below, have also been proposed by Billiot (Billiot et al., 2013):

- **FMGFDpca:** Measure by linear regression of the GFD.
- **FMGFDsum:** Measure by the sum of the GFD components.

Since the SFF technique is commonly used for textured areas, the use of the frequential measures has great potential. We can notice that we found several focus measures based on textured analysis operators as Haralick operators (entropy, contrast, correlation ...). We used the Generalized Fourier Descriptors (GFD), which are defined as follows. Let f be a square summable function on the plane. The Fourier transform is:

$$\hat{f} = \int_{R^2} f(x)\exp(-j\xi x)dx \tag{6}$$

If (λ,θ) are polar coordinates of point ξ, we shall denote $\hat{f}(\lambda,\theta)$ the Fourier transform of *f* at point (λ,θ):

$$D_f(\lambda) = \int_0^{2\pi} \left|\hat{f}(\lambda,\theta)\right|^2 d\theta \tag{7}$$

Df is the GFD feature vector which describes each local neighborhood as in Figure 9.

When the vector of each pixel is computed, we propose two manners to retrieve the sharpest position of the pixel.

The first one is to sum all the values of each vector. Indeed, the biggest the sum of the energy of each frequency is, the sharpest the pixel is because a sharp area contains more high frequencies than a blurred area. The focus measure is defined by:

Figure 9. Principle of GFD feature vector creation

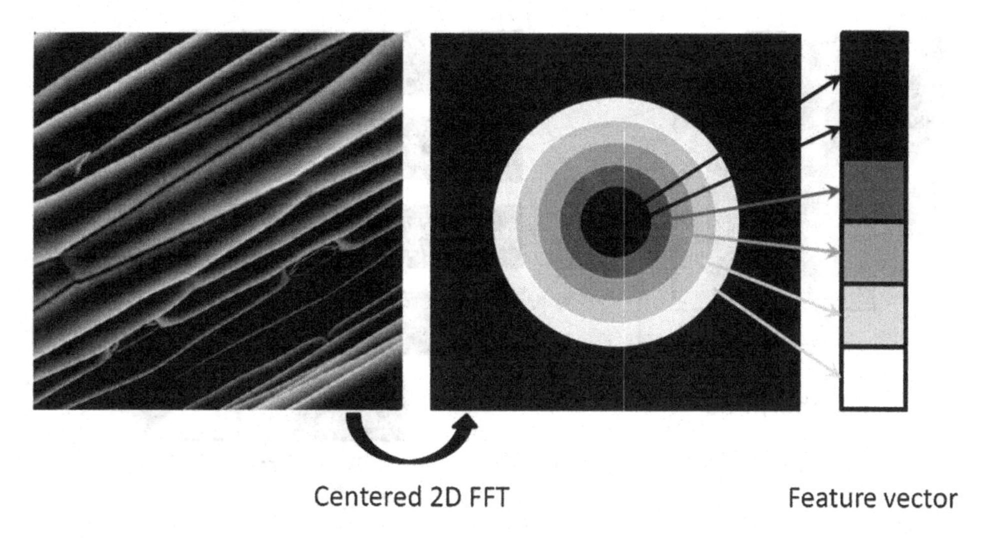

$$FM_{gfdsum} = \sum_{\lambda=1}^{N} D_f(\lambda) \tag{8}$$

with N is the size of the vector.

The second GFD based focus measure (FMgfdpca) uses a sequence of neighborhood of one pixel. We compute each vector of the same pixel in each image, we concatenate these vectors in a matrix. The matrix is then processed by PCA (Principal Component Analysis). Then, the sharpest pixel position is retrieved from the maximum factorial scores of the PCA first axis (Figure 10).

$$FM_{gfdpca} = position(\max(PCAvector)) \tag{9}$$

Finally, the 3D reconstruction of the scene contains several steps according to the situations. The common step concerns the creation of the depth map, constituted with different spatial locations of the points of the scene. If the system has been calibrated, these grey levels are significant of a metric location of each point. If a 3D representation of the scene is needed, the depth map is projected on three axes and the real texture of the scene is flattened on the projected surface. The main advantage of this method is tied on its functioning for scenes with several obstructions between objects. Moreover, depending on the optical properties of the imaging system, it allows accurate and dense reconstruction of the scene.

Figure 10. Scheme of FMgfdpca

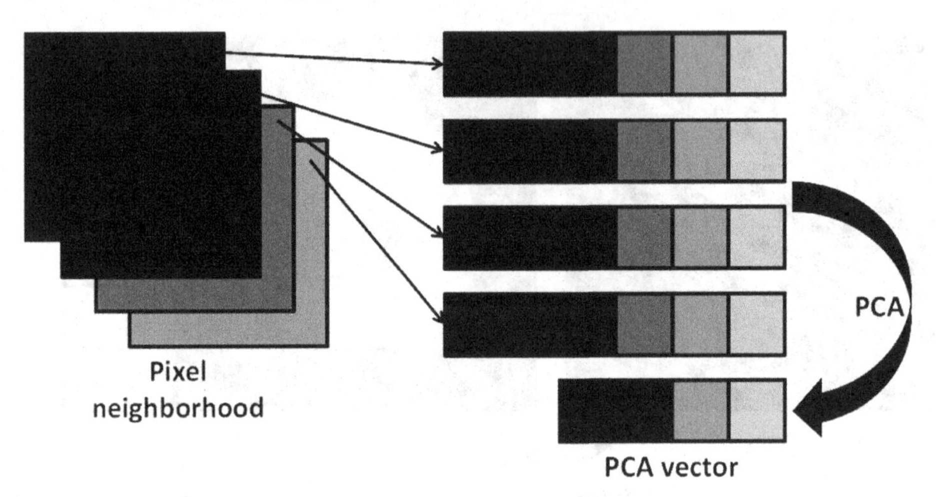

Experiments and Results

Since the method of 3D reconstruction will be applied on textural natural scenes, first validations on VisTex (VisTex, 2002) database is more pertinent than on other synthetic databases. Indeed, this database is constituted of natural scene images took under uncontrolled conditions and merged on different classes: barks of trees, walls, textiles, foods, metals, soils, plants, water, woods (Figure 11).

72 images of 512x512 pixels have been used. In order to evaluate the noise robustness of the different focus measure operators, the original images are noised according to 5 noisy levels (0, 5, 10, 15 and 20%). After, two successive blurs are applied on each image to simulate a depth image. The sizes of the neighborhood used for the focus measures of each pixel are squares with widths of 64, 32, 16, 8, and 4 pixels. Thus, for a couple (operator / image of the database), 25 depth maps are obtained for each combination of parameters (additive noise and neighborhood size which will be called "couple of parameters"). Finally, 19.800 depth maps are proposed. These maps are constituted of 3 grey levels for each focus band (band 1: original image; band 2: image with small blur; band 3: image with high blur). An ideal depth map is entirely constituted of grey levels of value 85 whereas an error induces a grey level of 170 or 255. These maps are created according to an evaluation of the focus block by block with a neighborhood of chosen size, and pixel by pixel.

Recurrent problematics in plant study are the detection of diseases, the follow-up of growth, the estimation of a foliar volume or the evaluation of a yield. Thus, the whole method, including 3D image acquisition, has been tested on several agro-

Figure 11. VisTex texture database
VisTex, 2002.

nomic scenes especially for wheat ear detection. This application is tied to a more global objective based on early yield determination. Such estimation could help farmers to improve their yield, to better manage their fields and to reduce their use of pesticides. The main component of the yield evaluation concerns the determination of the number of ears per square meter, which is currently done manually. The relevance on the use of image acquisition and processing for this specific task has been previously proved (Cointault et al., 2008). Even if the use of algorithms, such as skeletonization, allows discriminating the ears, the help of 3D depth information is more adapted. Indeed, ear covering implies that the ears are not in the same spatial or depth plan. Figure 12 presents an example of 3D reconstruction result obtained by SFF technique.

The Figure 13 presents an example of depth maps and 3D texture projection for three different sharpness operators.

Finally, the Figure 14 presents a result of superimposition between a sharpness image and its depth map for the same three sharpness operators previously mentioned.

Figure 12. 3D reconstruction by SFF apparatus: a. depth map. b. sharp image. c. false color projection. d. texture plating projection
Billiot, 2013.

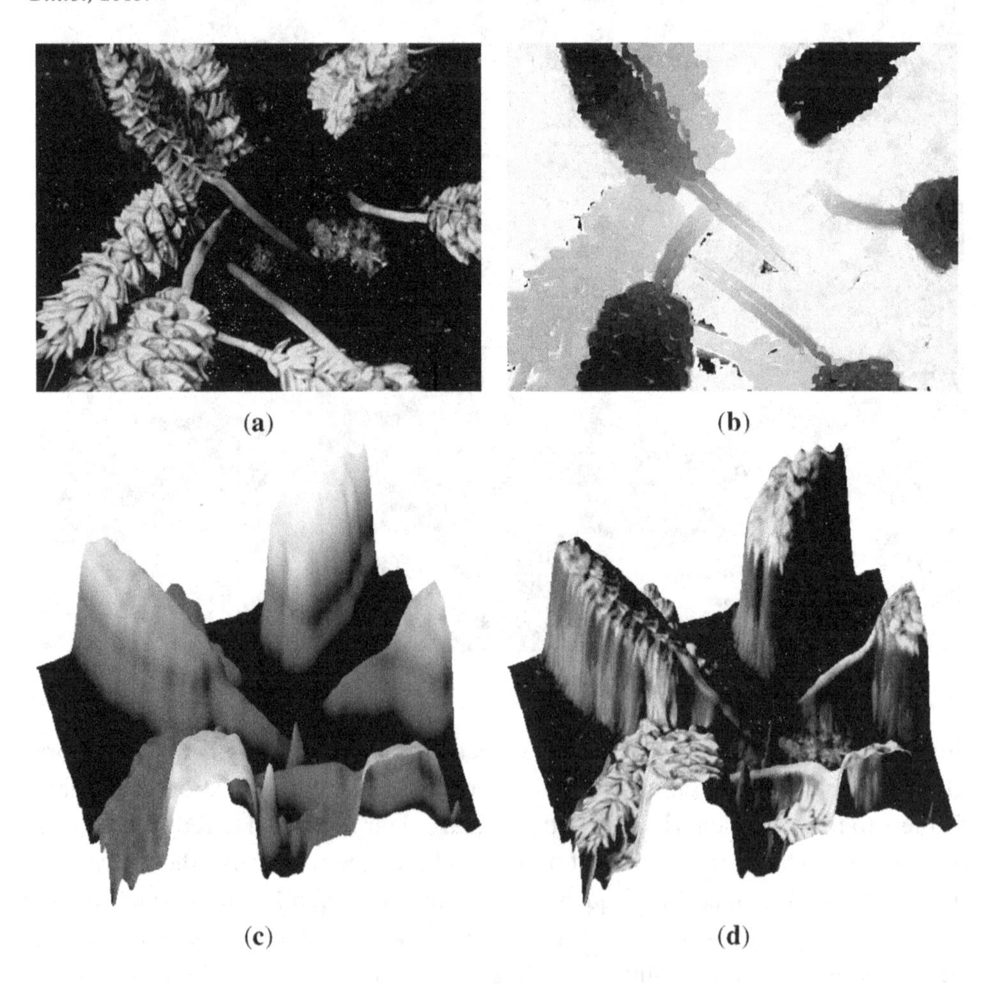

Figure 13. Depth maps and 3D projection of wheat ear scene for a 9x9 neighborhood: a. FMtenvar; b. FMGFDsum; c. FMentropy

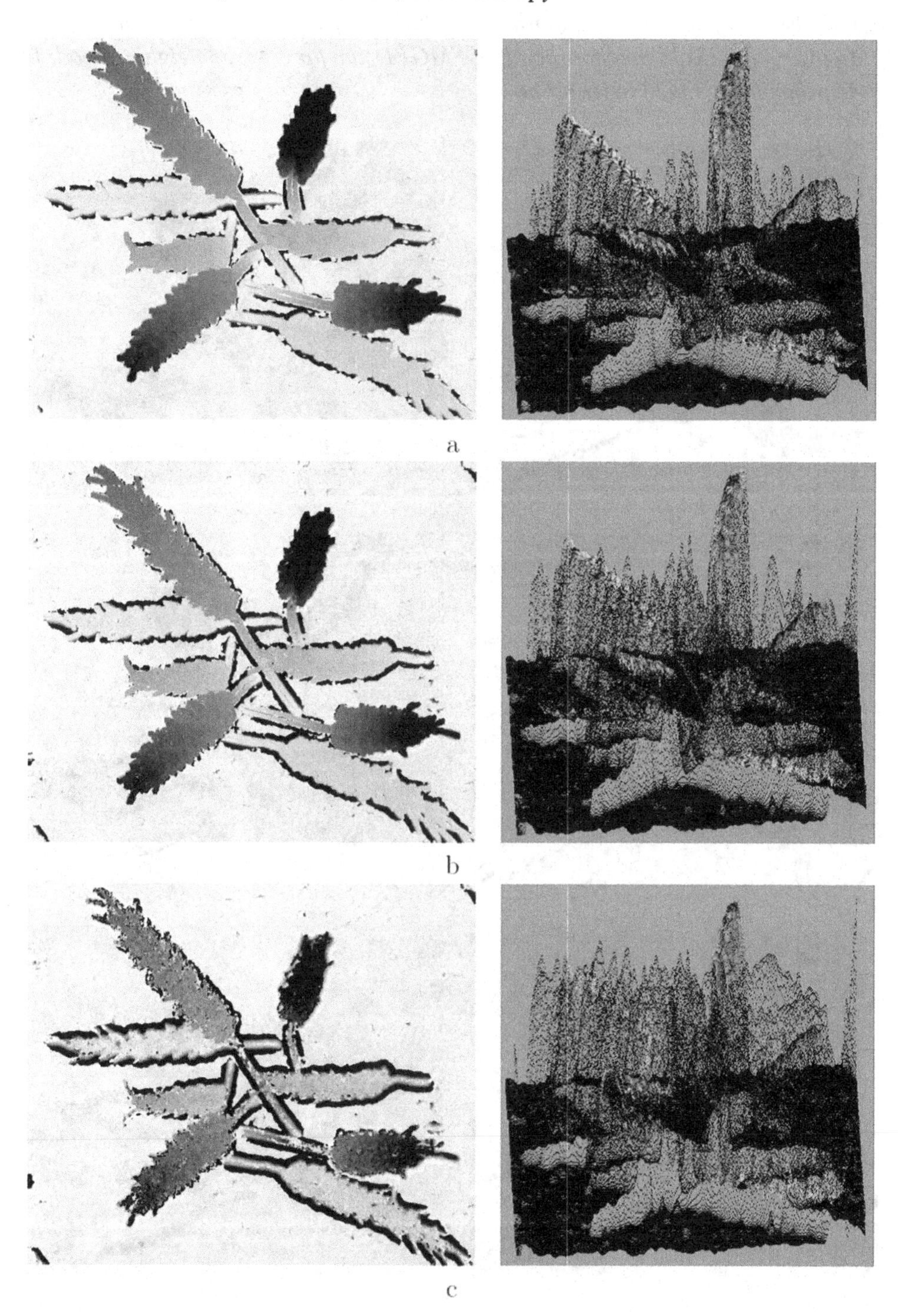

Figure 14. Superimposition between the sharp image and the depth map for 3 sharpness operators under different conditions: a. FMtenvar for a 9x9 neighborhood; b. FMGFDsum for a 9x9 neighborhood; c. FMentropy for a 9x9 neighborhood; d. FMtenvar for a 33x33 neighborhood; e. FMGFDsum for a 33x33 neighborhood; f. FMentropy for a 33x33 neighborhood.

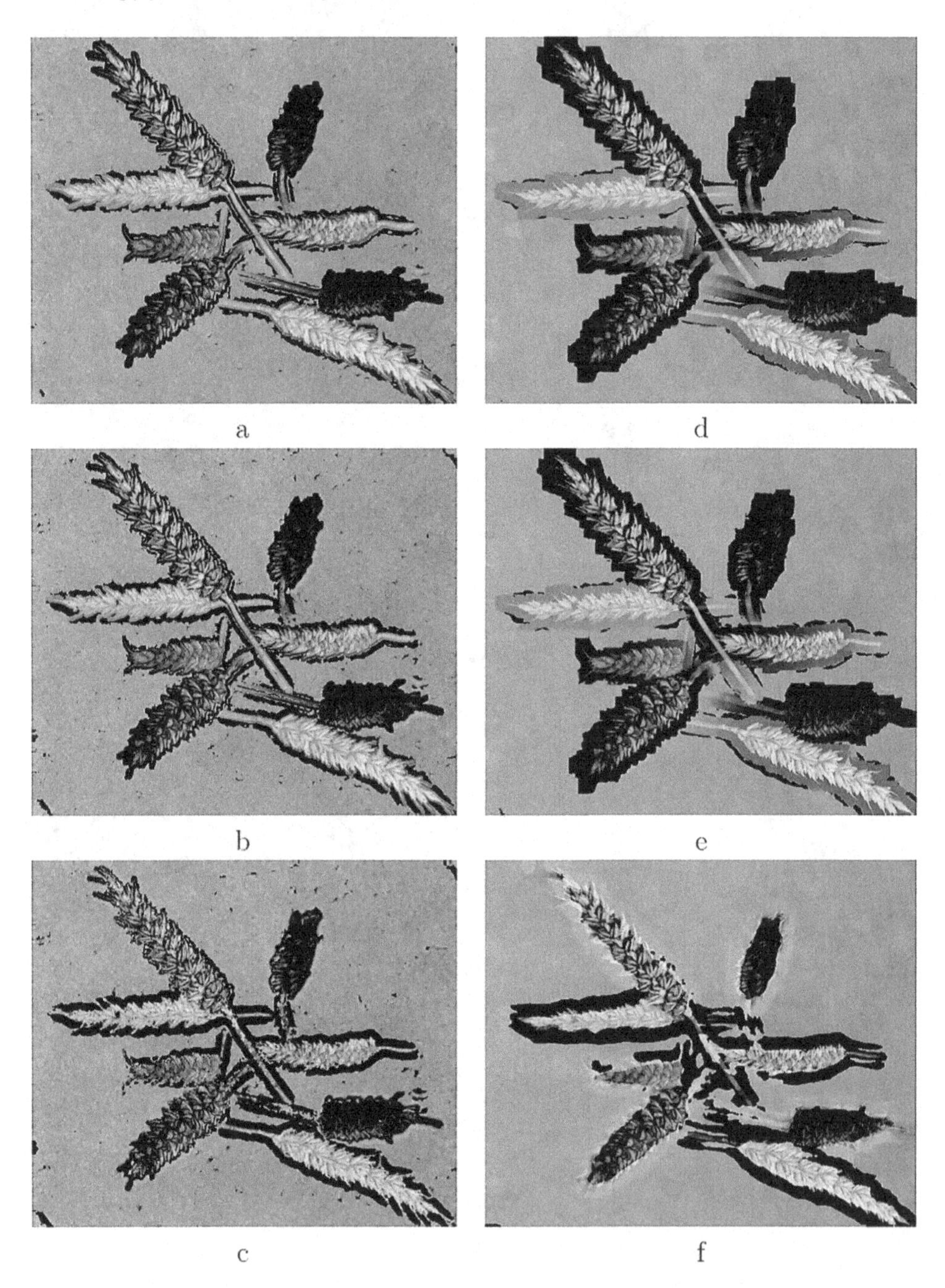

Whatever the results are according to the applications, the classification of the three previous operator studied is the same. Thus the operator based on the variance measurement of the gradient amplitude FMtenvar remains the operator with the better results, even if our operator based on GFD (FMGFDsum) is of great relevance for agronomical scenes. The results obtained for different scenes prove that the Shape from Focus technique is relevant for textured scenes. However, on no-textured images, the results are coherent with the scene analysis.

To resume, we recommend the different operators according to the 4 following cases:

- Large size of the neighborhood (large window analysis) and no additional noise: excepted the operators FMentropy and FMnrjgrad, all the other operators provide right results.
- Large size of the neighborhood and additional noise: the operators FMtenvar, FMvar, FMGFDpca and FMGFDsum are well indicated with similar results due to their noise robustness.
- Small size of the neighborhood and no additional noise: excepted FMcorr, FMentropy, FMnrjgrad and FMorient all the other operators can be used, especially FMsml and FMwavratio.
- Small size of the neighborhood and additional noise: this configuration is the most constraining for the sharpness measurement. The determination of the most appropriate operator implies the search of those not increasing the error for a simultaneous variation of the two parameters, such as FMtenvar, FMvar, FMGFDpca and FMGFDsum.

3D USE FOR ROOT PHENOTYPING

Agriculture faces the challenge to nourish an ever-growing population (Tilman et al., 2011) when at the meantime there is both an increasing pressure on arable land and more constrained resources for plant growth. This happens in the context of climate change, agriculture being both causal and exposed to detrimental environmental impacts. Various tools are being developed to increase breeding processes, select, create new varieties more adapted to the nowadays environmental constrained cropping systems. Among these, the exponential increase in high-throughput sequencing technologies provoked the dilemma to characterize the expression of the huge variety of underlying plant genes. The objective is to provide quantitative analyses of plant structure and function modifications in response to environmental factors modulations such as temperature, hydric stresses, low nutrients (nitrogen, sulphur and phosphorus), and soil availability, associated to both low input agriculture and resource-limited environments. As such, phenotyping becomes a major bottleneck in plant breeding.

Context and Methods

The majority of phenotyping efforts to the determination of plant phenotypic traits was up to now devoted to shoots (yield, leaf area), quality of the harvested products or pathogens resistance. This relied on the synergistic development of new sensor technology and automated management of phenotypic processes (for growing plants, manipulating them and acquiring the phenotypic traits using a variety of innovative sensors). Plant phenotyping can be done in the field or more generally on dedicated platforms. As such numerous applications were developed for monitoring within-field spatial and temporal variability of crop, soil water and nutrient distribution, disease epidemiology, and weed infestation (embedded on new agricultural machinery or airborne platforms equipped with onboard sensors). In controlled environment platform (i.e. climatic chambers or greenhouses), phenotyping systems are being developed, classified as sensor-to-plant or plant-to-sensor (Fiorani & Schurr, 2013).

Besides the numerous efforts towards plant shoot phenotyping, maintaining or improving crop yields while saving fossil energy inputs also implies to i) select plants with more efficient root system for acquisition of telluric resources (water, phosphorus, nitrogen, sulphur) and ii) improve our understanding of plant – soil rhizospheric microorganisms interactions. As such, root system architecture (RSA), which comprises a broad range of root morphological parameters (root length and density, root branching and total root surface), is a key component to optimally explore the soil for efficient nutrient uptake and it thus plays a major role in the adaptation of the plant to heterogeneous and fluctuating environmental conditions and gain higher crop yields in constrained agriculture regions (Wasson et al., 2012).

Despite this, and much lagging behind shoot, roots phenotyping received much less attention and success. This arises first from the technical difficulties to access roots directly in the field, which requires an important investment in terms of time and means and which consists on contact with the plants or destruction of the plants to observe the roots, as briefly described in Figure 15.

Moreover, RSA is highly plastic in response to variations of their environment (Malamy, 2005), fluctuations of temperature (Nagel et al., 2009), water availability (Bengough et al., 2011), soil structure (Cairns et al., 2011) and nutrients availability of such as nitrogen, phosphorus, iron and sulphate (Voisin et al., 2003; Manschadi et al., 2014). A major bottleneck for root phenotyping research is the development of devices suitable for high-throughput phenotyping of roots. A list of pre requisite for developing such resources is to visualize if possible the whole root system in its

Figure 15. (a) Manual method (Hoover, B., Cal Poly, 2010-2011) using a knife or a gouge, shovel the root, clean soil and measure by hand. (b) CI-600 In-Situ Root Imager (CID, Bio-Science, 2005): install acrylic tubes within the study area prior to the growing season, when the plant begins to build a network of roots, slide the scanner head within the tube at the desired depth and download images. (c) Minirhizotron Observation Tubes (Wilkinson, S., University of Lancaster): plants are installed in minirhizotron observation tubes which can be opened later.

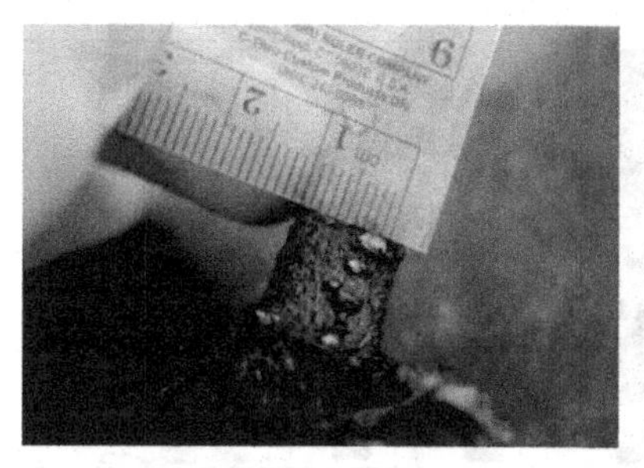

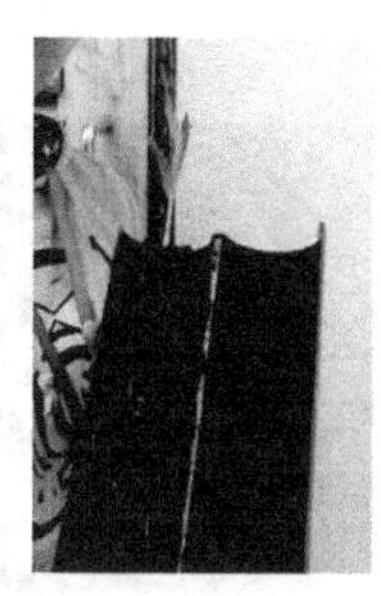

integrity. The less the amount of roots being identified, the less the tools will be able to differentiate among genotypes tiny or subtle variations in root phenotypic traits. Roots have also to be visualized possibly at high resolution, to identify changes in root system events. While it is obviously easier to characterize seedlings root architecture (Benoit et al., 2015, Benoit et al., 2013), phenotyping roots and acquiring traits at later developmental stages or highly branched root systems becomes tedious because the overlap or roots highly complicates root architecture disentangling. Lastly, it is highly desirable to perform plant root phenotyping dynamically and non-invasively, for a large number of plants, a variety of species and their genotypes, subjected to different environmental treatments.

Some root phenotyping systems already exist based on different cultivation set up. Root systems may be grown in hydroponics, gellan gum or soil (Clark et al., 2011; Hargreaves et al., 2009), agar plates (Nagel et al., 2009; Slovak et al., 2014), even if artificial growth mediums are not always or often representative for field conditions (Figure 16).

This because soil mineral or water content are highly variable in the field and hard to precisely reproduce, both spatially and under controlled conditions (Cairns et al., 2011). Root systems enclosure comprised soil filled tubes (Nagel et al., 2009; Ytting et al., 2014), cartridges (Dresboll et al., 2013; Nagel et al., 2012), cultivation in regular soil (Bucksch et al., 2014; Burton et al., 2012). These allow either 2D or 3D visualization of the root systems. As for examples, two-dimension root observation is provided by growth pouches (Adu et al., 2014; Hund et al., 2009), growth between paper in rhizoslides (Le Marié et al., 2014), tubular rhizotrons in 4PMI

Figure 16. (a) Petri dish (Carion, J.F., ACCES, 2013): pea seeds, germinated 2 days before implantation on agar were soaked 10 seconds in alcohol at 70 ° C, then 10 seconds in bleach, finally rinsed in the sterile water. (b) Rhizoboxes (Courtois, B., Cirad, 2013): Plants were grown in a hydroponic system filled with glass beads of 1.5 mm diameter. (c) 3D root growth system (Randy T. Clark et al, University of Cornell, 2011): Plants grown in semi-sterile glass growth cylinders containing solid gellant gum media replete with nutrients.

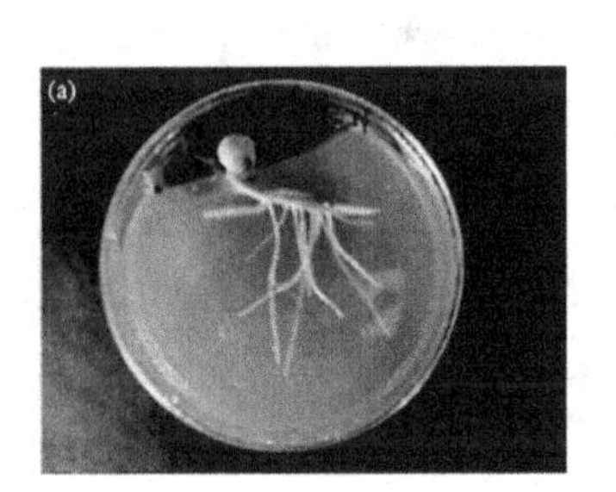

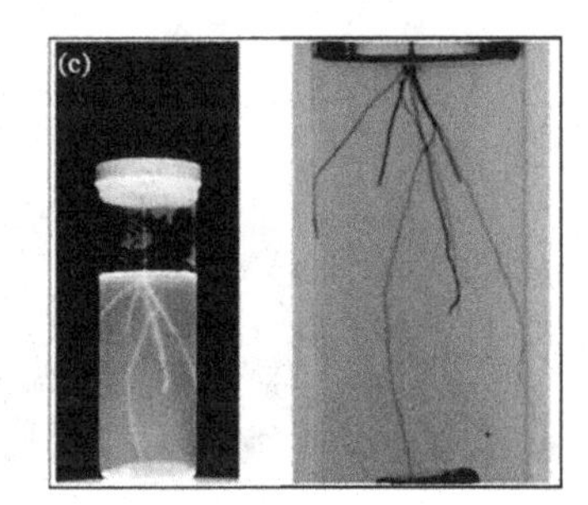

while 3D observation is also possible in agar or gellan gum (Clark et al., 2011; Fang et al., 2009; Iyer-Pascuzzi et al., 2010; Ribeiro et al., 2014; Topp et al., 2013) or hydroponics (Pace et al., 2014) or transparent soil (Downie et al., 2012). Gellan gum growth systems provide superior optical clarity to facilitate non-invasive three-dimensional (Fang et al., 2009) imaging and temporal studies of plant root systems while also allowing reproducible control of the rhizosphere (Clark et al., 2011). However, image-based methods appear to be more and more used in high throughput root phenotyping for measuring the size, architecture, and other structural root traits rely on cameras (Le Marié et al., 2014; Nagel et al., 2009) or scanners (Adu et al., 2014; Slovak et al., 2014; Salon et al., 2014).

The 4PMI Phenotyping platform of INRA Dijon allows daily imaging of shoot plants with dedicated imaging cabins including VIS, NIR, and Fluorescence for GFP localization and expression. Within HTP (High Throughput Platform) 4PMI, very innovative 3D rhizotrons called RhizoTubeHD and the associated imaging cabin (Rhizocab) can characterize automatically at high resolution (down to 7 microns) whole roots systems (Figure 17).

RhizoTubeHD is flexible enough to allow growing plants in several developmental stages and 2D/3D growth and image acquisition allows very high-throughput. Our system avoids disentangling roots from soil pots to organize them into 2D conformation and avoids root damage. The rhizocab uses laser scanning for 2D/3D digitizing of root systems over time and have been engineered to fit within the Scanalyzer3D plant conveying set up, allowing root characterization of genotypes

Figure 17. The principle of RhizoTubeHD at INRA Dijon

overall RSA (root length, number of lateral roots, size and location of nodules) (Figure 18).

While the variety of imaging systems dedicated to root systems analysis is reaching a mature stage, the new bottelneck in the phenotyping chain now tends to move toward image analysis (Minervini et al., 2015). If the description of root systems has now converged toward a unified language based on graph (Lobet et al., 2015), there is a huge diversity of image analysis software for root systems image processing (Lobet et al., 2013) and the selection of best approaches remains to be done. Till now, biologists interested in root phenotypic traits measurement used free or not commercial software which were the only non-manual solutions available, for

Figure 18. The principle of rhizocab (left), the scanner used (middle) and finally an example of acquired root image (right)

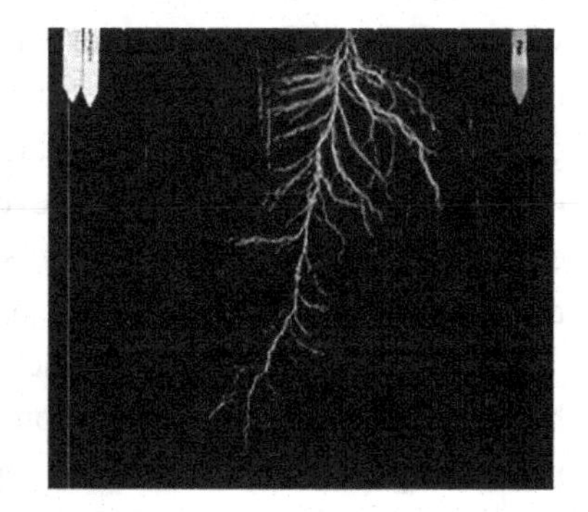

example GiaRoots, DART (Le Bot et al., 2010), EZ-Rhizo, RootNav, RootReader2D, RootSystemAnalyser, WinRHIZO (Arsenault et al., 1995), SmartRoot (ImageJ) (Lobet et al., 2011), RootTrace. Their main drawbacks are that the results on the architecture are not optimum, its use is not intuitive, with manual interactions by the operator, the necessity to calibrate the image before the measurement, the tough and long-time preparation, the problem of contrast optimization, the non-determination of all the root parameters (such as for example the local diameter).

Experiments and Results

All the drawbacks previously presented for the main software can be avoided by developing specific image processing allowing to take into account all the environmental conditions, and providing the main parameters of a root system. A fast and automatic image processing technique developed through the 4PMI platform of INRA Dijon is presented now, to determine the following parameters appearing essential for biologists: length and diameter of the primary root, number of secondary roots and nodules on the primary root, location and size of the nodules on the primary root and on the secondary roots, length and diameter of the secondary root, etc.

The image processing procedure consists first to automatically calibrate the images acquired, allowing the extraction of root parameters from the root images. Standard segmentation technique is applied providing a binary image, considering the biological characteristics of the plant. Several image segmentation methods have been proposed such as the Region approach, like Region growing or Split & Merge, the Frontier approach, like the Canny filter, or the classification or Thresholding technique, like the Histogram or the use of Support Vector Machine (SVM). In our case, a simple binarization method based on an automatic real-time thresholding has been used because of unsatisfied results obtained with Canny filter, and high computing time of the SVM method. The method developed is based on the optimization of the Otsu's method, since it appears as the most efficient for our application and it is an entirely automatic one. However, the problem of root crossings (Benoit et al., 2013; Benoit et al., 2014) is not really addressed here and we propose to improve the detection of the whole RSA using inpainting methods based on anisotropic diffusion with partial differential equations. This technique gives good performances to separate roots when they cross with rather large angle (typically larger than 3°). However, the situation of almost parallel roots is very common. This may cause artefacts in the root-system analysis and thus remains, to the best of our knowledge, to be addressed.

The main crucial root parameter being the length of the primary root, from the binary image, the skeletonization technique is applied as it is well used in image processing for a great number of applications such as pattern recognition, solid modelling for shape conception, organization of clusters. The standard skeletonization method based on distance transforms has been applied in our images. It consists in searching all the centers of a structural element when touching simultaneously all the sides of an object. Figure 19 provides an example of skeleton result obtained on one of our root image.

The previous algorithms and methodology have been tested on a database of 105 root images with different kinds of tap roots and growth stages. To calculate the length of the primary root, the number of pixels of primary root is extracted from the skeleton. However, the skeletons obtained present some errors due to the presence of nodules and lateral roots which implies to use an interpolation technique to optimize the determination of the "real" skeleton. Figure 20 presents the length of the primary root before and after the interpolation operation, compared to manual results.

Figure 19. The original image (a) and a zoom on a ROI (Region Of Interest). (b) the skeleton obtained

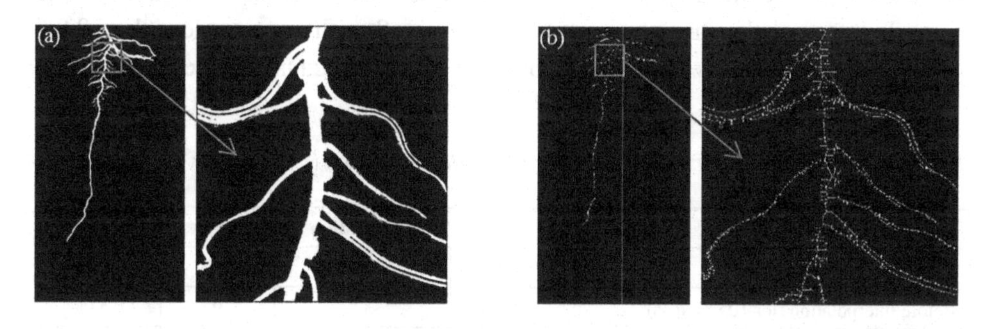

Figure 20. Comparison of manual and image processing results for primary root length determination

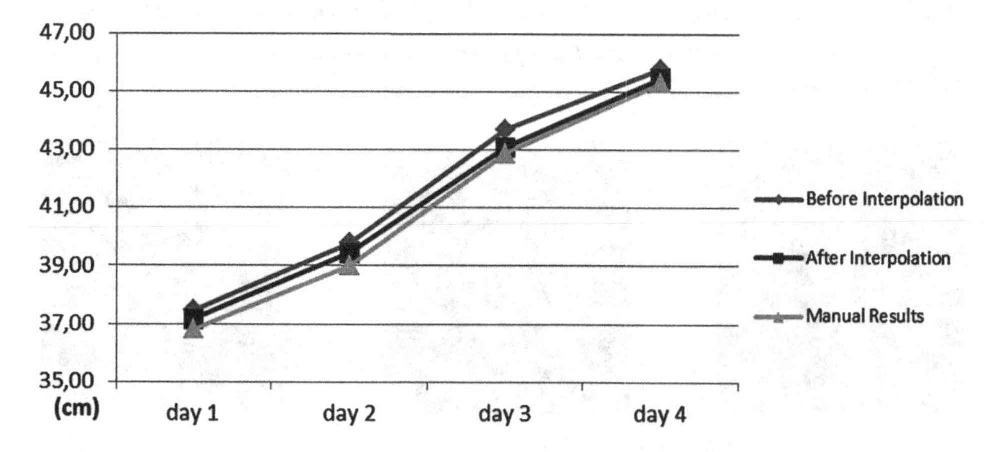

The Figurc 20 shows the interest to do the interpolation: the accuracy of measurements for the primary root length was increased, up to almost 1% correction on the total primary root length. For the 105 images tested, the error rate on the primary root length determination is presented in the Table 1.

Table 1 shows that the correction by interpolation method gives improvement: average error before interpolation is 0.51% for the 105 images, and is 0.39% after. This corresponds to values varying between 0.3 and 1.0 cm for the length; this gap is sufficiently high for our further research since in the root detection, root begins with a width of about 2 mm.

The second parameter discussed here concerns the determination of the local diameter of the primary root automatically and accurately which is not proposed with the standard software. On Figure 21a, the white pixels obtained by reconstruction of the primary root are superimposed with the original image (Figure 21b), which represents exactly the original primary root (Figure 21c).

From the number of white pixels for each line of the binary image, the local diameter of the primary root is obtained very accurately. It means that our method provides the correlation between the diameter and the location in the image. The local diameter of the primary root was determined with an error less than 0.9% compared to manual evaluation, on only 15 images because of the difficulty in

Table 1. Error rate of primary root length detection for 105 images

	Correlation			Average
Error span	0~0.5%	0.6%~0.1%	+1%	
Before Interpolation (for 105 images)	63	33	9	0.51%
After Interpolation (for 105 images)	82	21	2	0.39%

Figure 21. (a) Reconstructed image of a primary root; (b) superimposition of the original root on the reconstructed image; (c) representation of the primary root detected on the whole original image

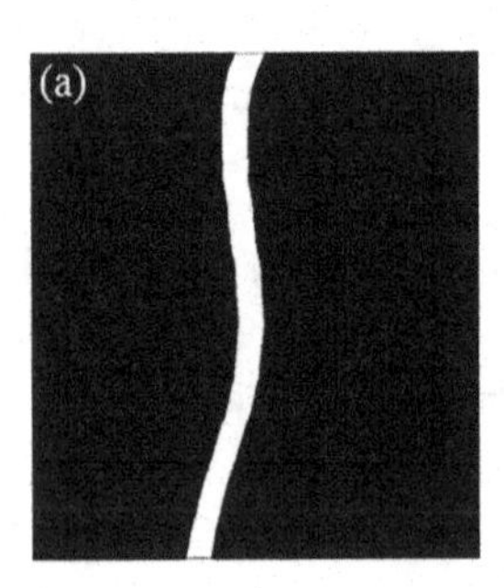

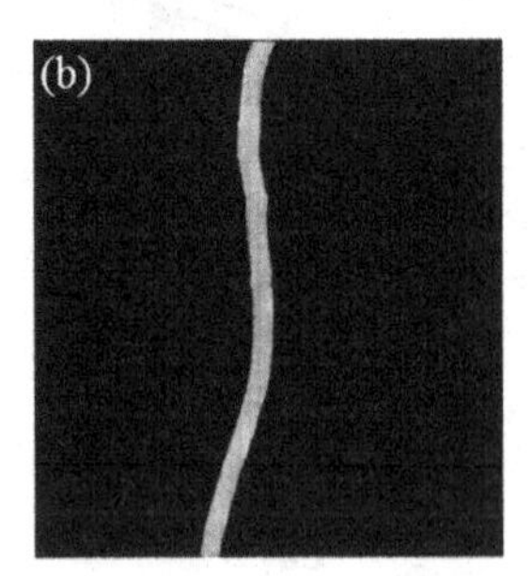

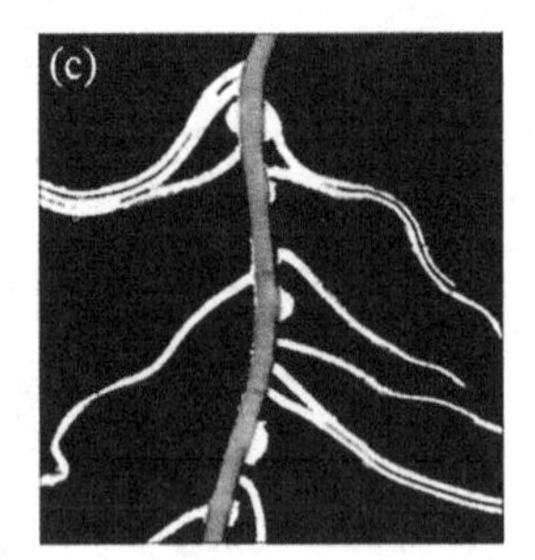

determining the local diameter manually. Compared with several commercial tools or software available on the market to process rhizotron images, one of our advantages is also the computation time since our technic needs only 1min30 for a 144Mo image, with completely automatic method.

FUTURE RESEARCH DIRECTIONS

We discussed in this chapter the use of 3D technologies and associated 3D image processing in agriculture for two different precision farming applications.

The first one deals with the use of Structure from motion of Shape from focus techniques for crop characterization. The evolution and/or improvement of the previous techniques are currently done in the research laboratories. Perspectives of the previous research works are tied not only on the improvement of the reconstruction process but also on their use for other specific agronomical applications. Concerning the improvement of the existing systems, the objectives are to overcome the limits of the reconstruction method, such as the small field of view when a low field of view is required, and the calculation time tied to the measurement of the sharpness applied for each pixel of an image. Since the field of view is not sufficiently extended, one solution could be the mutualization of acquisition systems. Another solution is based on the evolution of the lenses allowing a decreasing of the focus length with a high aperture diameter to reduce the depth of field, like the lenses proposed by Thalès Angénieux in France or Voigtlander in Germany. For the decreasing of the calculation time, the solution is to use parallelized algorithm implementation, like GPGPU (General-purpose Processing on Graphics Processing Units). This will finally allow combination of 3D information by SfM or SFF directly inside phenotyping platform.

Concerning the phenotyping applications, improvements and research directions could be done in two different ways. The first will be the optimization of the root characteristics determination by developing new image processing methods and the second will be to use other 2D/3D imaging techniques in comparison with those already existing. For example, other methods such as 2D neutron radiography and tomography (Leitner et al., 2014) are only low-throughput hence less suitable for breeding purposes. 3D-imaging methods for in situ root systems visualization grown in gel can be imaged by laser scanners (Fang et al., 2009) while in soil X-ray computed tomography (CT) (Mooney et al., 2012) or magnetic resonance imaging (MRI) (Rascher et al., 2011) are used. These two techniques have recently been compared (Metzner et al., 2015) for root phenotyping. While MRI is fully noninvasive, the effect of X-ray dose on the development of plant at least at mature stage after the very early imbibition and germination stages show no disturbance (Zappala et al., 2013) at the dose levels which allow good segmentation of the root-systems (Mairhofer et al., 2013). CT gives anatomical information while MRI can give more chemical related information such as water or lipid content. Functional information in 3D comes with the coupling of CT or MRI with positons emission tomography (PET). This method, although very low throughput, expensive and low resolution gives access to invaluable information for physiological studies such as the fluxes of nutrients in root systems (Garbout et al., 2012). The importance of the choice of the soil for phenotyping experiments is also an open problem since the actual observed biological scene is the root system in a given soil and the resulting soil-water transport (Amin et al., 1996).

Such other applications could be found some advantages in the use of 3D techniques such as for example the characterization of the centrifugal spreading process. Over the past 60 years the use of mineral fertilizers allowed farmers to drastically increase their crop yields, to the detriment of environment protection. The majority of farmers use a broadcast spinner spreader, also known as centrifugal spreader, because of their large working width, low cost, robustness and spreading efficiency. Several factors affect the fertilizer distribution in the field such as the spreader settings and the physical properties of the fertilizers. This distribution or spreading pattern should correspond to the crop's needs as closely as possible (Van Grinsven et al., 2012). To be able to spread the exact amount of fertilizer on the right place in the field, correct spreader settings are determined by performing a calibration test taking into account both machine and fertilizer properties. In most cases the fertilizer particles are collected in standardized trays and then weighted. Because this is a long and fastidious method, several alternative techniques have been developed to characterize the spreading process (ejection parameters such as direction, high speed and size of the fertilizer granules) more efficiently and calibrate spreaders (Cointault et al., 2002). However, because it is based on a 2D imaging system, the method can only be applied for particles moving parallel to the image sensor, although in practice, fertilizer particles are generally ejected with a vertical angle. Hence a 3D approach would increase the accuracy of the motion estimation for particles leaving flat disks and, more important, also enable correct particle motion estimation for inclined spreaders or spreaders with conical disks. A 3D imaging technique using a high speed binocular stereovision system has been proposed in combination with corresponding image processing algorithms for accurate determination of the parameters of particles leaving the spinning disks of centrifugal fertilizer spreaders (Hijazi et al., 2011). Validation of the stereo-matching algorithm using a virtual 3D stereovision simulator indicated an error of less than 2 pixels for 90% of the particles (Hijazi et al., 2014). The setup was validated using a cylindrical spread pattern of an experimental spreader.

CONCLUSION

3D imaging methods, both concerning acquisition and processing, have been developed since more than thirty years for a lot of applications. In the agricultural context, three-dimensional images can be acquired by Lidar or Radar, photogrammetry, stereovision, shape/structure from focus techniques … The dedicated applications cover all the sectors of agricultural from plant phenotyping in the field to vineyard biomass characterization or soil mapping. Since all these applications cannot be exposed in details we focused this chapter on two main applications dealing with crop characterization and root phenotyping.

The 3D acquisition systems developed or adapted for the previous applications are based on laser-scanning or on shape from focus or structure from motion. However, only 3D image acquisition is not sufficient to determine plant or process parameters. Indeed, they are coupled to 3D image processing using depth of map, point matching techniques or motion estimation with high accuracy for the parameter estimations. The drawbacks of 3D imaging such as complex data reconstruction, cost of the systems and specific illumination sometimes required especially for some laser scanning instruments are however alleviated by the following advantages such as the high accuracy of the visualization and the high resolution of the shoot. Perspectives of these works are not only dedicated to the improvement of reconstruction process but also for other agronomical applications previously briefly described. 3D imagery systems are thus more and more relevant for the agricultural domain.

REFERENCES

Adu, M. O., Chatot, A., Wiesel, L., Bennett, M. J., Broadley, M. R., White, P. J., & Dupuy, L. X. (2014). A scanner system for high-resolution quantification of variation in root growth dynamics of *Brassica rapa* genotypes. *Journal of Experimental Botany*, *65*(8), 2039–2048. doi:10.1093/jxb/eru048 PMID:24604732

Amin, M. H. G., Richards, K. S., Chorley, R. J., Gibbs, S. J., Carpenter, T. A., & Hall, L. D. (1996). Studies of soil-water transport by MRI. *Magnetic Resonance Imaging*, *14*(7), 879–882. doi:10.1016/S0730-725X(96)00171-3 PMID:8970099

Arsenault, J. L., Pouleur, S., Messier, C., & Guay, R. (1995). WinRHIZO, a root-measuring system with a unique overlap correction method. *HortScience*, *30*, 906.

Bengough, A. G., McKenzie, B. M., Hallett, P. D., & Valentine, T. A. (2011). Root elongation, water stress, and mechanical impedance: A review of limiting stresses and beneficial root tip traits. *Journal of Experimental Botany*, *62*(1), 59–68. doi:10.1093/jxb/erq350 PMID:21118824

Benoit, L., Belin, É., Dürr, C., Chapeau-Blondeau, F., Demilly, D., Ducournau, S., & Rousseau, D. (2015). Computer vision under inactinic light for hypocotyl–radicle separation with a generic gravitropism-based criterion. *Computers and Electronics in Agriculture*, *111*, 12–17. doi:10.1016/j.compag.2014.12.001

Benoit, L., Rousseau, D., Belin, E., Demilly, D., & Chapeau-Blondeau, F. (2014). Simulation of image acquisition in machine vision dedicated to seedling elongation to validate image processing root segmentation algorithms. *Computers and Electronics in Agriculture*, *114*, 84–92. doi:10.1016/j.compag.2014.04.001

Benoit, L., Rousseau, D., Belin, E., Demilly, D., Ducournau, S., Chapeau-Blondeau, F., & Dürr, C. (2013). *Locally oriented anisotropic image diffusion: application to phenotyping of seedlings.* Paper presented at Special session GEODIFF, 8th International Joint Conference on Computer Vision, Imaging and Computer Graphics Theory and Applications (VISAPP 2013), Barcelona, Spain.

Bernardini, F., Mittleman, J., Rushmeier, H., Silva, C., & Taubin, G. (1999). The ball-pivoting algorithm for surface reconstruction. *Visualization and Computer Graphics. IEEE Transactions on*, *5*(4), 349–359.

Billiot, B. (2013). *Conception d'un dispositif d'acquisition d'images agronomiques 3D en extérieur et développement des traitements associés pour la détection et la reconnaissance de plantes et de maladies.* (Ph-D Thesis). University of Burgundy.

Billiot, B., Cointault, F., Journaux, L., Simon, J. C., & Gouton, P. (2013). 3D image acquisition system based on Shape from Focus technique. *Sensors (Basel, Switzerland)*, *13*(4), 5040–5053. doi:10.3390/s130405040 PMID:23591964

Bucksch, A., Burridge, J., York, L. M., Das, A., Nord, E., Weitz, J. S., & Lynch, J. P. (2014). Image-based high-throughput field phenotyping of crop roots. *Plant Physiology*, *166*(2), 470–486. doi:10.1104/pp.114.243519 PMID:25187526

Burton, A. L., Williams, M., Lynch, J. P., & Brown, K. M. (2012). RootScan: Software for high-throughput analysis of root anatomical traits. *Plant and Soil*, *357*(1-2), 189–203. doi:10.1007/s11104-012-1138-2

Cairns, J. E., Impa, S. M., O'Toole, J. C., Jagadish, S. V. K., & Price, A. H. (2011). Influence of the soil physical environment on rice (*Oryza sativa* L.) response to drought stress and its implications for drought research. *Field Crops Research*, *121*(3), 303–310. doi:10.1016/j.fcr.2011.01.012

Chéné, Y., Rousseau, D., Lucidarme, P., Bertheloot, J., Caffier, V., Morel, P., & Chapeau-Blondeau, F. et al. (2012). On the use of depth camera for 3D phenotyping of entire plants. *Computers and Electronics in Agriculture*, *82*, 122–127. doi:10.1016/j.compag.2011.12.007

Clark, R. T., MacCurdy, R. B., Jung, J. K., Shaff, J. E., McCouch, S. R., Aneshansley, D. J., & Kochian, L. V. (2011). Three-Dimensional Root Phenotyping with a Novel Imaging and Software Platform. *Plant Physiology*, *156*(2), 2455–2465. doi:10.1104/pp.110.169102 PMID:21454799

Cointault, F., Guérin, D., Guillemin, J. P., & Chopinet, B. (2008). In-Field Wheat ears Counting Using Color-Texture Image Analysis. *New Zealand Journal of Crop and Horticultural Science*, *36*, 117–130. doi:10.1080/01140670809510227

Cointault, F., Paindavoine, M., & Sarrazin, P. (2002). Fast imaging system for particle projection analysis: application to fertilizer centrifugal spreading. Journal Measurement Science and Technology, 13, 1087-1093.

Downie, H., Holden, N., Otten, W., Spiers, A. J., Valentine, T. A., & Dupuy, L. X. (2012). Transparent soil for imaging the rhizosphere. *PLoS ONE*, 7. PMID:22984484

Dresboll, D. B., Thorup-Kristensen, K., McKenzie, B. M., Dupuy, L. X., & Bengough, A. G. (2013). Timelapse scanning reveals spatial variation in tomato (*Solanum lycopersicum* L.) root elongation rates during partial waterlogging. *Plant and Soil*, *369*(1-2), 467–477. doi:10.1007/s11104-013-1592-5

Fang, S., Yan, X., & Liao, H. (2009). 3D reconstruction and dynamic modeling of root architecture in situ and its application to crop phosphorus research. *The Plant Journal*, *60*(6), 1096–1108. doi:10.1111/j.1365-313X.2009.04009.x PMID:19709387

Fiorani, F., & Schurr, U. (2013). Future Scenarios for Plant Phenotyping. *Annual Review of Plant Biology*, *64*(1), 267–291. doi:10.1146/annurev-arplant-050312-120137 PMID:23451789

Garbout, A., Munkholm, L. J., Hansen, S. B., Petersen, B. M., Munk, O. L., & Pajor, R. (2012). The use of PET/CT scanning technique for 3D visualization and quantification of real-time soil/plant interactions. *Plant and Soil*, *352*(1-2), 113–127. doi:10.1007/s11104-011-0983-8

Gauthier, J.-P., Bornard, G., & Silbermann, M. (1991). Motions and pattern analysis: Harmonic analysis on motion groups and their homogeneous spaces. *Systems, Man and Cybernetics. IEEE Transactions on, 21*(1), 159–172.

Han, S., Bujoreanu, D., Cointault, F., & Rousseau, D. (2015). *Graph-based denoising of skeletonized root-systems*. Paper presented at the 4th edition of the International Workshop on Image Analysis Methods for the Plant Sciences (IAMPS), Louvain-la-Neuve, Belgium.

Hargreaves, C., Gregory, P., & Bengough, A. G. (2009). Measuring root traits in barley (*Hordeum vulgare* ssp. *vulgare* and ssp. *spontaneum*) seedlings using gel chambers, soil sacs and x-ray microtomography. *Plant and Soil, 316*(1-2), 285–297. doi:10.1007/s11104-008-9780-4

Hartley, R. (2013). *Multiple View Geometry in Computer Vision.* Cambridge University Press.

Hijazi, B., Cool, S., Cointault, F., Vangeyte, J., Mertens, K., Paindavoine, M., & Pieters, J. (2014). High speed stereovision setup for position and motion estimation of fertilizer particles leaving a centrifugal spreader. *Sensors (Basel, Switzerland), 14*(11), 21466–21482. doi:10.3390/s141121466 PMID:25401688

Hijazi, B., Vangeyte, J., Cointault, F., Dubois, J., Coudert, S., Paindavoine, M., & Pieters, J. (2011). Two-steps cross-correlation based algorithm for motion estimation applied to fertilizer granule motion during centrifugal spreading. *Optical Engineering (Redondo Beach, Calif.), 50*. doi:10.1117/1.3582859

Hund, A., Ruta, N., & Liedgens, M. (2009). Rooting depth and water use efficiency of tropical maize inbred lines, differing in drought tolerance. *Plant and Soil, 318*(1-2), 311–325. doi:10.1007/s11104-008-9843-6

Iyer-Pascuzzi, A. S., Symonova, O., Mileyko, Y., Hao, Y., Belcher, H., Harer, J., & Benfey, P. N. et al. (2010). Imaging and analysis platform for automatic phenotyping and trait ranking of plant root systems. *Plant Physiology, 152*(3), 1148–1157. doi:10.1104/pp.109.150748 PMID:20107024

Jacquemoud, S., Verhoef, W., Baret, F., Bacour, C., Zarco-Tejada, P. J., Asner, G. P., & Ustin, S. L. et al. (2009). PROSPECT + SAIL models: A review of use for vegetation characterization. *Remote Sensing of Environment, 113*, 56–66. doi:10.1016/j.rse.2008.01.026

Jay, S., Rabatel, G., Hadoux, X., Moura, D., & Gorretta, N. (2015). In-field crop row phenotyping from 3D modeling performed using Structure from Motion. *Computers and Electronics in Agriculture, 110*, 70–77. doi:10.1016/j.compag.2014.09.021

Le Bot, J., Serra, V., Fabre, J., Draye, X., Adamowicz, S., & Pages, L. (2010). DART: A software to analyse root system architecture and development from captured images. *Plant and Soil*, *326*(1-2), 261–273. doi:10.1007/s11104-009-0005-2

Le Marié, C., Kirchgessner, N., Marschall, D., Walter, A., & Hund, A. (2014). Rhizoslides: Paper-based growth system for non-destructive, high throughput phenotyping of root development by means of image analysis. *Plant Methods*, *10*(13). PMID:25093035

Leitner, D., Felderer, B., Vontobel, P., & Schnepf, A. (2014). Recovering root system traits using image analysis exemplified by two-dimensional neutron radiography images of lupine. *Plant Physiology*, *164*(1), 24–35. doi:10.1104/pp.113.227892 PMID:24218493

Lobet, G., Draye, X., & Périlleux, C. (2013). An online database for plant image analysis software tools. *Plant Methods*, *9*(38). PMID:24107223

Lobet, G., Pagès, L., & Draye, X. (2011). A Novel Image Analysis Toolbox Enabling Quantitative Analysis of Root System Architecture. *Plant Physiology*, *157*(1), 29–39. doi:10.1104/pp.111.179895 PMID:21771915

Lobet, G., Poun, M., Diener, J., Prada, C., Draye, X., Godin, C., & Schnepf, A. et al. (2015). Root System Markup Language: Toward a unified root architecture description language. *Plant Physiology*, *167*(3), 617–627. doi:10.1104/pp.114.253625 PMID:25614065

Lowe, D. G. (1999). Object recognition from local scale-invariant features. *Proceedings of the International Conference on Computer Vision.*

Mairhofer, S., Zappala, S., Tracy, S., Sturrock, C., Bennett, M., Mooney, S., & Pridmore, T. (2013). Recovering complete plant root system architectures from soil via X-ray mu-Computed Tomography. *Plant Methods*, *9*(8). PMID:23514198

Malamy, J. E. (2005). Intrinsic and environmental response pathways that regulate root system architecture. *Plant, Cell & Environment*, *28*(1), 67–77. doi:10.1111/j.1365-3040.2005.01306.x PMID:16021787

Manschadi, A. M., Kaul, H. P., Vollmann, J., Eitzinger, J., & Wenzel, W. (2014). Reprint of 'Developing phosphorus-efficient crop varieties. An interdisciplinary research framework'. *Field Crops Research*, *165*, 49–60. doi:10.1016/j.fcr.2014.06.027

Metzner, R., Eggert, A., van Dusschoten, D., Pflugfelder, D., Gerth, S., Schurr, U., & Jahnke, S. (2015). Direct comparison of MRI and X-ray CT technologies for 3D imaging of root systems in soil: Potential and challenges for root trait quantification. *Plant Methods*, *11*(1), 17. doi:10.1186/s13007-015-0060-z PMID:25774207

MicMac. (n.d.). IGN. Retrieved from http://logiciels.ign.fr/?-Micmac,3-

Minervini, M., Scharr, H., & Tsaftaris, S. A. (2015). Image analysis: The new bottleneck in plant phenotyping. *IEEE Signal Processing Magazine*, *32*(4), 126–131. doi:10.1109/MSP.2015.2405111

Mooney, S. J., Pridmore, T. P., Helliwell, J., & Bennett, M. J. (2012). Developing X-ray Computed Tomography to non-invasively image 3-D root systems architecture in soil. *Plant and Soil*, *352*(1-2), 1–22. doi:10.1007/s11104-011-1039-9

Nagel, K. A., Kastenholz, B., Jahnke, S., van Dusschoten, D., Aach, T., Mühlich, M., & Schurr, U. et al. (2009). Temperature responses of roots: Impact on growth, root system architecture and implications for phenotyping. *Functional Plant Biology*, *36*(11), 947–959. doi:10.1071/FP09184

Nagel, K. A., Putz, A., Gilmer, F., Heinz, K., Fischbach, A., Pfeifer, J., & Schurr, U. et al. (2012). GROWSCREEN-Rhizo is a novel phenotyping robot enabling simultaneous measurements of root and shoot growth for plants grown in soil-filled rhizotrons. *Functional Plant Biology*, *39*(11), 891–904. doi:10.1071/FP12023

Pace, J., Lee, N., Naik, H. S., Ganapathysubramanian, B., & Lubberstedt, T. (2014). Analysis of maize (*Zea mays* L.) seedling roots with the highthroughput image analysis tool ARIA (Automatic Root Image Analysis). *PLoS ONE*, 9.

Rascher, U., Blossfeld, S., Fiorani, F., Jahnke, S., Jansen, M., Kuhn, A. J., & Schurr, U. et al. (2011). Non-invasive approaches for phenotyping of enhanced performance traits in bean. *Functional Plant Biology*, *38*(12), 968–983. doi:10.1071/FP11164

Ribeiro, K. M., Barreto, B., Pasqual, M., White, P. J., Braga, R. A., & Dupuy, L. X. (2014). Continuous, high-resolution biospeckle imaging reveals a discrete zone of activity at the root apex that responds to contact with obstacles. *Annals of Botany*, *113*(3), 555–563. doi:10.1093/aob/mct271 PMID:24284818

Salon, C., Jeudy, C., Bernard, C., Cointault, F., Han, S., Lamboeuf, M., & Baussard, C. (2014). *Nodulated root imaging within the high throughput plant phenotyping platform (PPHD, INRA, Dijon)*. Paper presented in “Institute Days of MPI-MP”. Potsdam-Golm, Allemagne.

Santos, T. T., & de Oliveira, A. (2012). *Image-based 3D digitizing for plant architecture analysis and phenotyping.* Paper presented at Workshop on Industry Applications (WGARI) in SIBGRAPI 2012 (XXV Conference on Graphics, Patterns and Images), Ouro Preto, MG, Brazil.

Slovak, R., Goschl, C., Su, X. X., Shimotani, K., Shiina, T., & Busch, W. (2014). A scalable open-source pipeline for large-scale root phenotyping of *Arabidopsis. The Plant Cell, 26*(6), 2390–2403. doi:10.1105/tpc.114.124032 PMID:24920330

Tilman, D., Balzer, C., Hill, J., & Befort, B. L. (2011). Global food demand and the sustainable intensification of agriculture. *Proceedings of the National Academy of Sciences of the United States of America, 108*(50), 20260–20264. doi:10.1073/pnas.1116437108 PMID:22106295

Topp, C. N., Iyer-Pascuzzi, A. S., Anderson, J. T., Lee, C.-R., Zurek, P. R., Symonova, O., & Benfey, P. N. et al. (2013). 3D phenotyping and quantitative trait locus mapping identify core regions of the rice genome controlling root architecture. *Proceedings of the National Academy of Sciences of the United States of America, 110*(18), E1695–E1704. doi:10.1073/pnas.1304354110 PMID:23580618

Van Grinsven, H. J., Ten Berge, H. F., Dalgaard, T., Fraters, B., Durand, P., Hart, A., & Willems, W. J. et al. (2012). Management, regulation and environmental impacts of nitrogen fertilization in northwestern Europe under the Nitrates Directive; a benchmark study. *Biogeoscience, 9*(12), 5143–5160. doi:10.5194/bg-9-5143-2012

VisTex. (2002). *Vision texture database.* Maintained by the Vision and Modeling group at the MIT Media Lab. Retrieved from http://vismod.media.mit.edu/vismod/imagery/VisionTexture/

Voisin, A. S., Salon, C., Jeudy, C., & Warembourg, F. R. (2003). Root and nodule growth in Pisum sativum L. in relation to photosynthesis. Analysis using 13C labelling. *Annals of Botany, 92*(4), 557–563. doi:10.1093/aob/mcg174 PMID:14507741

Wasson, A. P., Richards, R. A., Chatrath, R., Misra, S. C., Prasad, S. V. S., Rebetzke, G. J., & Watt, M. et al. (2012). Traits and selection strategies to improve root systems and water uptake in water-limited wheat crops. *Journal of Experimental Botany, 63*(9), 3485–3498. doi:10.1093/jxb/ers111 PMID:22553286

Ytting, N. K., Andersen, S. B., & Thorup-Kristensen, K. (2014). Using tube rhizotrons to measure variation in depth penetration rate among modern North-European winter wheat (*Triticum aestivum* L.) cultivars. *Euphytica, 199*(1-2), 233–245. doi:10.1007/s10681-014-1163-8

Zappala, S., Helliwell, J., Tracy, S., Mairhofer, S., Sturrock, C. J., Pridmore, T., & Mooney, S. et al. (2013). Effects of X-ray dose on rhizosphere studies using X-ray computed tomography. *PLoS ONE*, *8*(6), e67250. doi:10.1371/journal.pone.0067250 PMID:23840640

KEY TERMS AND DEFINITIONS

Phenotyping: Determination of the whole observable characteristics of an object.

Precision Agriculture: Concept of management of agricultural fields based on intra-parcellar variabilities.

Rhizotron: Transparent structure designed to grow plants, comprising a transparent side from where underground plant parts such as roots can be observed non invasively.

Shape from Focus: Method for depth recovering by means of taking several images with a monocular camera whose extrinsic or intrinsic settings are changed for every frame.

Sharpness Operator: Imaging operator for SFF technique allowing optimizing sharp pixel selection for each image of a sequence.

Stereovision: Method which consists to take images under different point of views to determine the dimensions, shapes or locations of objects.

Structure from Motion: Range imaging technique; it refers to the process of estimating three-dimensional structures from two-dimensional image sequences which may be coupled with local motion signals.

Chapter 9
Analysis of New Opotoelectronic Device for Detection of Heavy Metals in Corroded Soils:
Design a Novel Optoelectronic Devices

Gustavo Lopez Badilla
Universidad Politecnica de Baja California, Mexico

Juan Manuel Terrazas Gaynor
CETYS University, Mexico

ABSTRACT

This study was made to design and an analysis the purpose of novel optoelectronic sensor, with three steps, being the first the use of MatLab software to simulate with a mathematical equation, the operation of the sensor as the electrical conductivity. Other step, was the evaluations with Wein2k using the Density of States (DOS) Theory, as a computational system of the physicochemical, to analyze different materials used, to fabricate the novel sensor, being the material proposed the Gallium Nitride (GaN), in according to the electrical conductivity, and was obtained, to fabricate this type of material with the Chemical Vapor Deposition (CVD) and then was made some evaluations to validate this material. Also was made a microanalysis with the Scanning Electron Microscopy (SEM) to evaluate the metallic surfaces of the electrical connections at micro scale.

DOI: 10.4018/978-1-5225-0632-4.ch009

INTRODUCTION

This study was made to evaluate the design and analysis of operation of a new optoelectronic sensor, with the objective to detect heavy metals (HM) in damaged soils. The main function of the new sensor is detecting the HM with a quick response and the minimum percentage of error, and with an optoelectronic device at low cost and high efficiency. The new sensor can be improved, and have a percentage of zero error. In this analysis three steps mentioned below were developed. The first step was an analysis of a mathematical simulation to evaluate the electrical operation of the new sensor with the MatLab software to obtain a mathematical equation that developed the simulations. Other step was the evaluation with the Wein2k software, using the theory of density of states (DOS), to simulate the physical-chemistry properties of the materials utilized of test, in the new sensor and evaluate its efficiency. One of two activitics of this step was the analysis of the materials used to fabricate the new sensor, being the best material proposed in this study the gallium nitride (GaN) for its electrical properties, as an excellent electrical conductor. Other activity of this step was the test of electrical operation with the fabrication of the new sensor with the chemical vapor deposition (CVD) technique, and with the elaboration of some evaluations of the electrical conductivity to its validated. The third step was a microanalysis of a small concentration of a corroded soil, with small particles of HM as copper, silver, steel and tin at micro scale. This analysis was made with the Scanning Electron Microscopy (SEM) and with the new sensor operating, to evaluate the electrical conductivity and time of response in the detection of the HM, mentioned above. Once evaluated and validated the new sensor, was proposed to use it for quick recovery of corroded soils. The principal heavy metals evaluated is because some industrial companies as the electronics industry in the Mexicali city, in some times dump waste metal in soils annexed to its buildings, where these metallic particles can be dragged up almost one kilometer of its source and generates a damages to soils, and are converted to infertile soils. These heavy metals analyzed, are the most commonly used in the electronics industry, and this type of companies represents the 80% of the industrial plants in this city. This study was made from 2012 to 2014.

INTRODUCTION

One of factors used to design and evaluate the new sensor based in the chromatic systems, with rods and cones and tetrachromatic systems. This is the same function of the electronic optocouplers that are very used in industrial activities, scientific research, medicine activities and biology, and also in chemical and physical operations (Mendenhall, et al, 2007; O'Regan et al, 2015). An area which it has been of great interest in the last ten years is the microbiology by the presence of microorganisms in soils, because of heavy metals of chemical substances of the electronics industry. The action of pouring liquid or solid substances to soil generates a deterioration of soils. The damage of soils generates the growth and reproduction of microorganisms that results in a reduction in green areas and they cannot be used to any type of activities, in essential in the agricultural operations to produce food to the populations in the world. The reduction of soils to produce food causes that more populations in development countries suffers from hungry (N. G. et al, 2011; Peña et al, 2007). The recovery actions of soils are made by microorganisms that eliminate these heavy metals. To determine the HM is necessary an optocoupler very sensitive to put the necessary quantity of microorganisms to break down these heavy metals in soils corroded (Girit, et al, 2012; Ju, et al, 2010). The basic electronic devices as transistors, were utilized as macro devices as showed in Figure 1 as a first step to fabricate the new sensor. The use of optoelectronics devices to detect the microorganisms and monitoring which are detected the heavy metals that are the great interesting of ecologists to restore soils.

The Optoelectronics Technology

The optoelectronics is a modern technology that has made great strides in the 90's and much greater potential in the century XXI. It has been developed since the era of the 50's, where he was born as a necessity of military activities. Through many years of innovation and research, the optoelectronics has reached a point of development in which was decreased the costs and increased the quantity of functions of the electronic optocouplers, being an important technology that can be used in without number of operations, being one of them of importance today in the industry (Louie et al, 2011). With the passage of time, new methods appeared in this technology,

Figure 1. Electronic diagram of a microdevice used an electronic optocoupler

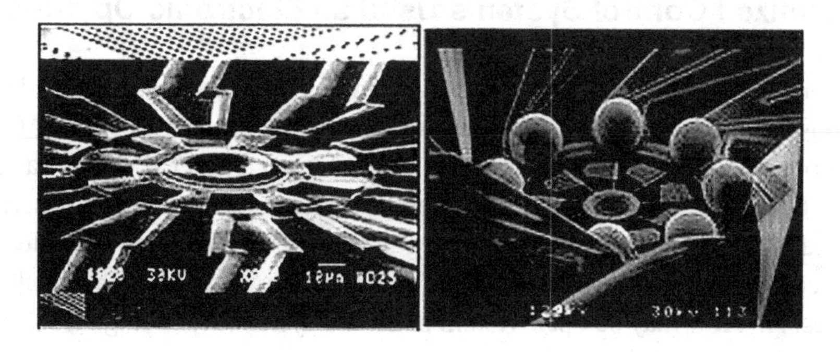

facilitating its usc in industrial companies. Today, exist countless optoelectronic devices and systems, which have been miniaturized and automatized to be used in industrial processes depending of the technology.

Objective of the Investigation

The main characteristic of the study is to propose a new system with optoelectronic technology to support in the detection of heavy metals in corroded soils, which generates deterioration and soils infertile in zones annexed to the electronics industry. This occurs normally in the Mexicali city, and to reverse the process of damage is necessary detected with the new optoelectronic device with high efficiency. This is achieved by inserting of microorganisms that consume the heavy metals (Chen et al, 2011). This will generate fertile soils and more potential in the cultivation processes, being an impacting in the social development, health and economy of the region of Mexicali, located in the northwest of Mexico. In the world exist many quantity of techniques to recover corroded soils, but the majorly are of agricultural techniques. The difference of this study with other studies is the use of the new sensor as electronic optocoupler and the microorganisms to break down the HM.

Optoelectronic Systems

The optoelectronic systems have increased the operations, and exist many electronic devices and systems with optoelectronic devices. For example, most of the Walkman feature a red light (LED) that tells us that the batteries are exhausted and must be replaced. The television screens use this type of electronic devices, and also oscilloscopes, TV, LCD screens and modern communications systems, to name a few. The optoelectronic devices are called opto isolators or optical coupling devices (An et al, 2013; Vogel et al, 2010). The optoelectronics systems have the function of detect any quantity of objects or particles as HM, which uses the light properties to function.

Automatized Control Systems Used as Electronic Optocoupler

An automatized control system (ACS) is equipment which operates with many functions to control any type of operations as the industrial activities. The ACSs are used as optoelectronic devices in the activities mentioned (Li et al, 2011; Yang et al, 2010). The ACS are used as macro and micro optoelectronic sensors that functions to check, adjust and operation of industrial regulations of each product, with a feedback equipment to regulate the variations of the output which are signal of the inputs of the activities. An ACS may be open system or closed system depending on if it makes a feedback system, using the same system output as input to make decisions based on past states of the system. For example, an open system may be one that receives the amount of light of the environment and this intensity can be controlled in any object, but uncontrolled of the source (Kim et al, 2010; Majumdar et al, 2008). By other way, the closed systems can control functions regulating the effect in the object and the source. It makes to the automation as a group of optoelectronic technology that electromechanical machines, controlled by computer systems, with autonomous or independent functions. The benefits of this are flexibility, lower cost, higher capacity and quality. An ACS with optoelectronic devices is an elemental sensor that performs the function of measuring the manufacturing process, while active an actuator which implements the control action on the industrial operations, and change the behavior of the system according the adequate regulations performs the actions of process control (Hecht et al, 2014; July al, 2009), manipulates the actuators and bases its decision on information received by the sensors. In addition, the micro optoelectronic devices are used in real life control systems to perform tasks that require powerful output, meaning that it is easier for robust systems. A diagram of control system is showed below that express the importance of optoelectronic sensors and actuators in a control system (Figure 2).

Optoelectronics Sensors Characteristics

The most important characteristic in an optocoupler is to transfer the electricity efficiently from photon particles from light. An efficiency of an optocoupler is measured through the current transfer ratio (CTR), which is the ratio between the change of current in the output side of the barrier and the change of current in the input side of the barrier of the light. Most work of the CTR system in an optocoupler between 10% and 50%. Generally, the electronic optocouplers are between a trans-

Figure 2. Block diagram of an optoelectronic device as a microcomponent

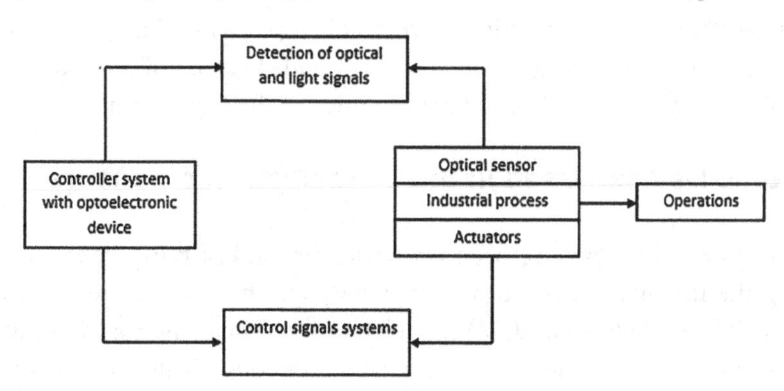

mitter and a rcceiver in an electrical circuit (Cline et al, 2008; De Franco, 2007). The optocouplers are often constructed with light emitting diodes (LEDs) as part of the side of the optical transmitter. A LED produces light when voltage is added to this, making it a perfect light source for an optocoupler. Incandescent lamps are also sometimes used in optocouplers, however, they are not as efficient as LEDs, as they distort the input signals and do not last long. The optocouplers have become essential in the daily live, by the upgrade of technology and the complexity of the digital components. Today, we live in a world where our activities depend on a multitude of technological devices based on the conduction of electricity, electronics and light, as TV, radio, cell phone, CD player and DVDs, optical reader supermarket, computers, etc., and we have become accustomed to these electronic devices (Kenneth et al, 2010). These type of equipment works with electronics components developed in the previous century, which in turn is based on small devices called diodes and transistors that uses electronic chips in all circuits (electronic cards), of the components mentioned.

In Figure 3 is observed a part of automatic pilot (AP) of an airplane used to detect the activation of the AP when missing a person (pilot and copilot) after 30 seconds. If this optoelectronic device has an electrical failure, the AP not will be active and generate a great disaster in an airplane. This is very important because can cause an emergency in a flight, being a great concern of the aerospace industry of the world, by the very strict regulations of this type of industries (Mars et al, 2007; Higgins et al, 2011). By other way, the diodes and transistors are the second most fabricated electronic devices with the 28%, after the resistors, inductors and capacitors with 39%. Silicon is the chemical element most used in the photoelectronic devices, and then copper, silver and; principally. Since about 20 years ago, scientists are studying and developing new electronic devices based in gallium with carbon particles. This is a technology where chemical compounds containing carbon atoms in their structures are used. In addition to conducting electricity, this new technology requires to its operation, with manipulated and emitted light, hence the name of optoelectronics. This scientific and technological field of optoelectronics is a best cause of the revolution in the electronic topic, as well as in our lives as an area of intense research and development worldwide because of its great academic, economic, energy and social impact as variables of the engineering and society.

Electronic Devices Used in the Optoelectronic Sensors

These electronic devices have been evolved over the last thirty years, as well as changing the raw material used in companies, which manufacture these type of electronic devices (Peña et al, 2007). The electronic components forming equipment and electronic systems are designed with the basics components, like the diode and

Figure 3. Optoelectronic sensor of an aerospace system of automatic pilot

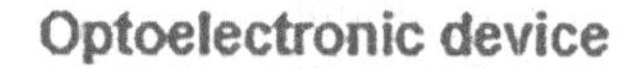

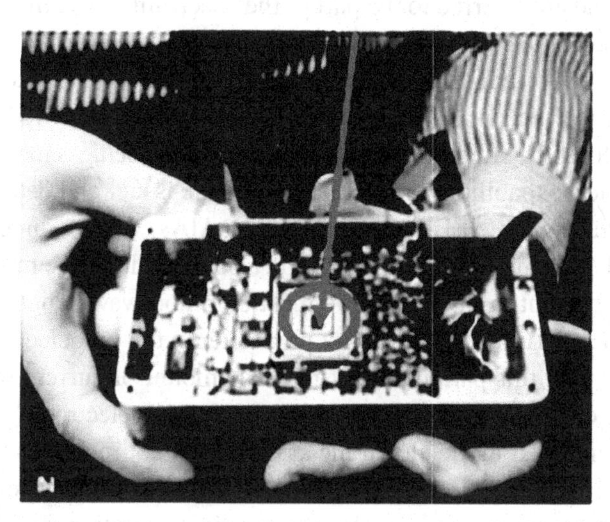

transistor as elemental activators and resistance, coil and capacitor as elemental components that store electrical energy. But thanks to the advances in the technology in the late 80s to the present, were incorporated new devices to develop operations as the light radiation, which has utilized principally in industries, areas of medicine, aerospace and marine (Chen et al, 2011). These devices revolutionized the area of electronics systems and equipment to be manufactured with LEDs and detection operations other electronic components, which not develop these functions and it was difficult to detect objects and particles to micro and macroscopic level. The optoelectronic devices are developed mostly with semiconductor, according to the manufacturing materials, being the most common of copper, silver, gold, nickel, silicon, tin. In the last 15 years are made the optoelectronic devices with materials of arsenic, gallium and nobio essentially to improve the physical and chemical properties and optimal performance of these devices (Vogel et al, 2010). These chemicals improve the properties of the interaction of light with matter and thus are better to detect objects and particles in the macro and micro operations that cannot be detected at naked eye. Light levels represent different radiation intensity, and based on this the required the chemical elements are used. Infrared rays are of great importance in the detection of objects and particles in industrial activities and security systems. Also, are applied optoelectronic devices with the aspect mentioned above, and it can be noted that from the standpoint of basic knowledge and the properties of semiconductor materials, the experiments carried out show that illuminating adequate and controlled semiconductors have helped greatly to the understanding of thc properties thereof. So both from a fundamental point of view applied, these chemical elements are of growing importance in the world of electronics. Usually, the term light, and are referred to the part of the spectrum of electromagnetic radiation to which the human eye is sensitive. By encapsulating an emitter and an optical detector, the optical relationship is always established and this makes the nature of use will be entirely electronic. This eliminates the need have knowledge of optics for the user. Therefore, for effective implementation alone needless meet some electrical characteristics, capabilities and limitations of the issuer and detector. The most common materials for opto LED are GaAs and GaAlAs. When a direct current, pass through to the LED, photons are emitted, and also the voltage is emitted power and depends of the current in the LED. The most common wave lengths are 660, 850 and 940 nanometers (nm). An important parameter in the optoelectronic devices is its efficiency, where this parameter defines the amount of current required by the LED to gct the desired current in the output of the optoelectronic device. One of electronic designs used in the electronic systems and equipment is the Darlington transistor configuration. This electronic configuration can obtain a relation between the input and the output, being a simple mathematical operation, dividing the output current between the input current required. In the case of the Schmitt trigger output and the triac device its efficiency is defined by the amount of current required for the transmitter to operate. Another important parameter is the voltage in the insulation process in the specialized optocouplers, which is 7500 Volts for 1 second.

Types of Optoelectronic Devices

The main optoelectronic devices used are the semiconductors, where are several types of elements optocoupled (Brian C. O'Regan et al, 2015). The semiconductors more used are the phototransistor and photodarlington, photoSCR and phototriac. They function in according to the theory of semiconductor optoelectronic devices based on silicon (Si) or Germanium (Ge). Several types of optocouplers that have a difference from each other, which depend on the devices output to be inserted into the component. Accordingly exist the following main types:

1. **Phototransistor:** It comprises an optocoupler with an output stage formed by a transistor BJT (Bipolar Junction Transistor).
2. **Phototriac:** Consists of an optocoupler with an output stage formed by a triac in the output.
3. **Phototriac Pass to Zero:** Is an optocoupler whose output stage is crossing a triac-zero. The internal circuit of the zero crossing triac switches only in the zero crossing of the current AC.

4. **Optotiristor:** Is designed for applications where isolation between a logic signal and the network needed.

The following is the general block diagram for the connection of a digital system to a power amplifier using an optocoupler, showed in Figure 4.

High Technology with Optoelectronic Devices

When a digital system interconnected to any high power system exists, is necessary an optocoupler coupled to ensure the electrical insulation (Hecht, D. S. et al, 2014). This protection avoids the risks in the optocoupler device to solve problems of electrical safety, costly damage to digital control systems and damage to the production process of the operation equipment. The optocoupler is a device easy to use, with a wide variety of engaging and very low cost. Therefore, it would be unforgivable not to use it when is necessary be controlled. As for the calculation of the load or handling device current in the power stage it is always strongly recommended to use the safety criterion of 30% over the maximum rates indicated by the manufacturer. It is the only way to avoid headaches, sometimes irreversible, in managing power devices. The optocouplers who have as source an incandescent lamp as a photosensitive element and a photoresistor, are often are called photorheostats. These devices are similar to other with a led source and output as a photosensitive element being a device (photodiode or phototransistor and phototiristor). An important property of the optocouplers is the construction of their high galvanic isolation between input and output. Among the most materials used are the plastic devices, and are already coming to the global marketing, being some type of plastic, where are utilized in electronic components called organic light-emitting diodes (OLEDs for short) for optical displays. These devices are similar to the screens of cell phones, calculators, watches, computer monitors, TVs and many more liquid crystal devices (LCDs) technology. Moreover, perhaps within a few years, other devices that may already be present commercially are organic solar cells (OPV) (Mendenhall et al, 2007) related with the OLED. In some activities where is applied the light as a physical variable, are produced systems that capture as the sun light and transform it into electricity.

Figure 4. Block diagram of an electronic system using an optocoupler

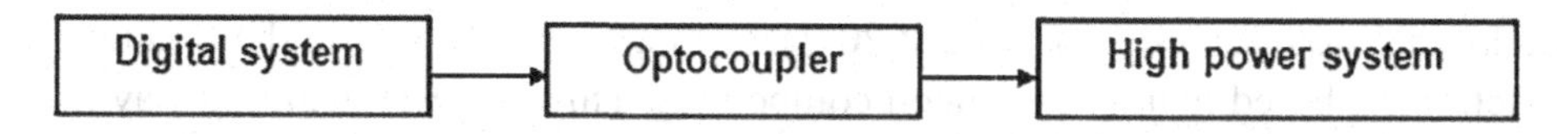

A fundamental importance in research and development of such equipment are the materials used. Scientists from the area of physical sciences, chemistry, optics, materials science and engineering, investigates about new materials with different and better features for a multitude of applications (Li et al, 2011).

Studies with Optoelectronic Devices

Of these new materials, molecules and polymers with large chains of molecules as organic semiconductors are in an interdisciplinary area. In the 70s, the scientists began to study the properties of organic semiconductor materials but it was up in the 80s when chemists and people in the area of materials science managed to design and synthesize new molecules and polymers with improved properties. These novel compounds have exceptional characteristics, such as easy processing, low cost, mechanical flexibility and room temperature deposition on a variety of substrates (which is essential for electronic devices plastics) (De Franco, 2007). Moreover, these organic materials can be designed by molecular engineering in a virtually infinite variety of ways to optimize one or more of electrical, mechanical and optical properties, that is, it is something like what they do architects and engineers: it can design and build a house with many materials and structures as internal and external sides, and with varied furnishings for comfort and decoration. For these reasons, it is expected that in the near future, can be use of a wide variety of technological devices that use plastic materials so that, gradually, the traditional electronics based mainly on silicon, is giving way to the new plastic technology (July al, 2009). The photonic devices are referring to an electronic-based photon and not a flow of electrons represented in the optoelectronic topic, as similar to the electronic devices that work with light and electrons, based on driving electricity, as well as the manipulation of light and based on organic and polymeric as plastic materials. This have been under intense research over the last 20 years and, gradually, and in the traditional electronics, based mainly on silicon compounds. This material is giving way to a new technology of organic materials as plastics. Silicon is one of the nonmetallic chemical elements and the most abundant in the world, and has been used for several decades in the manufacture of circuit integrated. The photonic devices are referring to an electronic-based photon and not a flow of electrons represented in the optoelectronic topic, as similar to the electronic devices that work with light and electrons, based on driving electricity, as well as the manipulation of light and based on organic and polymeric as plastic materials. This have been under intense research over the last 20 years and, gradually, and in the traditional electronics, based mainly on silicon compounds. This material is giving way to a new technology of organic materials as plastics. Silicon is one of the nonmetallic chemical element and the most abundant in the world, and has been used for several decades in the manufacture of circuit integrated Mars et al, 2007).

Density of State (DOS) Analysis

The distribution of a fixed amount of energy depends of a number of identical particles, also of the density of energy states available and the probability of it occupied a given state. The probability of an electron occupies a particular state of energy is given by the distribution function, but if there are more energy states available in a range of energy given, then this will give greater weight to the probability of the internal energy (Nirmalraj et al, 2013; Boland, 2012). The distribution of energy between identical particles depends in part on how many states and is available on a given energy range. This density of states as a function of energy, have the number of states per unit volume in a range of energy. The term statistical weight, is sometimes used synonymously, particularly in situations where the available states are discrete. Physical limitations on the particles, determining the shape function of the density of states. For electrons in a metal, the density of states comes from the wave nature of the electrons in the characteristic type of particle in a box (Parr et al, 2004; Chigo et al, 2004).

SEM Analysis in the Optoelectronic Devices

The Scanning Electron Microscope or SEM (Scanning Electron Microscopy), uses a beam of electrons instead of a beam of light to form an enlarged image of the surface of an object. It is an instrument that allows observation and surface characterization of inorganic and organic solids. It has a large depth of field, which allows them to focus at the same time a large part of the sample (Fiolhais et al, 2003; López-Badilla, 2008). The scanning electron microscope is equipped with various sensors, among which may be mentioned: the secondary electron detector for high resolution images SEI (Secondary Electron Image), a backscattered electron detector which allows the imaging composition and topography of the BEI (Backscattered Electron Image) surface, and energy dispersive detector EDS (Energy Dispersive Spectrometer) that can collect the X-rays generated by the sample and perform various semi quantitative and distribution of elements in surface analysis. It can be used in studies of the morphological aspects of microscopic areas of different materials with which scientific researchers and private companies work, in addition to processing and analyzing the images obtained. The major uses are the high resolution SEM (1μm), and also the large depth of field that gives three-dimensional look to images and simple sample preparation. The sample preparation is relatively simple to determine the main features as solid sample to detect the thin films of the conductive material as electronic system with optoelectronic component. These processes are made with small pieces of conductive samples in low vacuum chamber. Computer applications are varied, ranging from the petrochemical industry and metallurgy to forensics. The analysis with the SEM analysis (López et al, 2010), is very used in the electronics devices to evaluate the type of materials used in the optoelectronic systems, to determine its physical properties. With this technique can know the chemical elements of the materials used in the optoelectronic devices and some features as the length or wide of the materials to determine the adequate function of these type of electronic components. An example of an SEM analysis is showed in Figure 5, where is observed a micro optoelectronic device (MOED) damaged when was exposed in indoor of an electronics industry located in Mexicali in the

Figure 5. SEM analysis of a damage of a MOED exposed in an acidic environment in different seasons the electronics industry in Mexicali: (a) and (b) in summer and (c) and (d) in winter (2014)

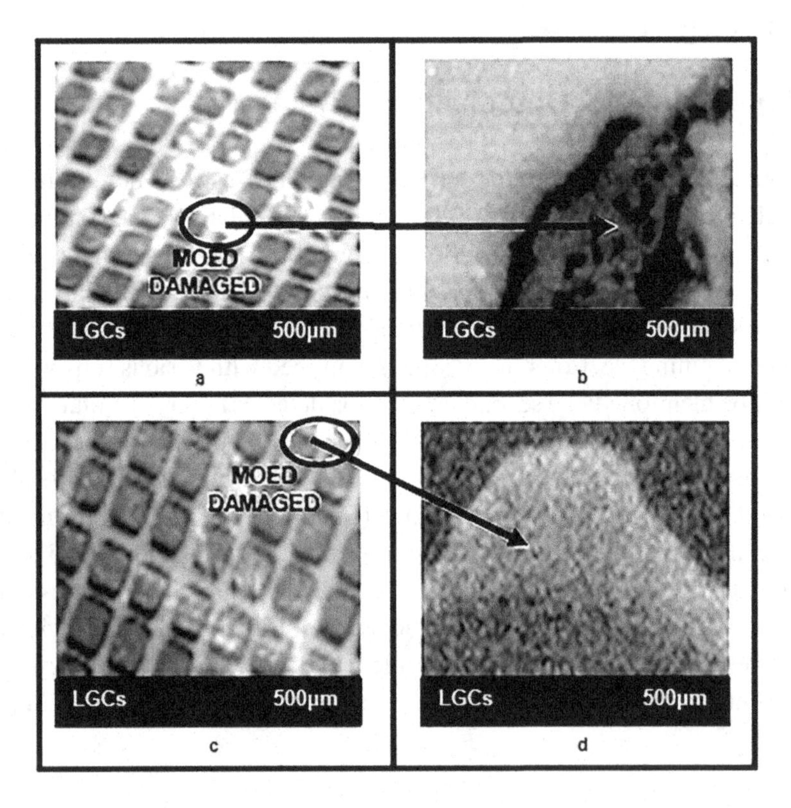

northwest of Mexico in the border with California State of the United States. As showed in Figure 5, the MOED the metallic connections of a silicon wafer are used in an industrial plant where manufactured microelectronic devices were. This deterioration of the MOED is presented principally when in the Mexicali city are indices of relative humidity (RH) and temperature higher than 80% and 25 °C, which are common in this region of Mexico essentially in some days of July and August in summer and December and January.

Types of Chromatic Signals

According to the structure that exists at the level of the ocular retina, both in animas and in man, there are two types of specialized cells in the photoreception: cones (showing systems) and canes (photoreceptors) which contain pigments producing chemical energy when exposed to light (July et al, 2009). This energy is transmitted through the optical path to the visual cortex to be performed. The photoreceptors have different functions, and each of the pigments are different. The cones are those with pigments that are selectively sensitive to different wavelengths of each color (red, green and blue are the primary colors). Each of these pigments absorbs a wavelength range having a peak absorption (maximum absorption) and it is particular. These are mixed or overlap between them, and generates different palettes. The full stimulation of all the cones gives the sensation of white (Vogel et al, 2010). Depending on the number of visual pigments possessing, its vision is classified as:

1. **Monochrome System:** Contains one type of cone and cane. This system detects only the black and white colors.
2. **Dichromatic System:** Contains two types of cones and cane. This system detects only the red and green colors.
3. **Trichromatic System:** Contains three types of cones and cane. This system detect the red, green and blue colors and is the normal vision of humans.
4. **Tetrachromatic System:** Contains four or more cones and cane. This system detect the red, green, blue, black and white colors and can detect the ultraviolet light.

Heavy Metals

The heavy metals may include light elements such as carbon and may exclude some of the heavier metals. Heavy metals are free and naturally in some ecosystems and can vary in concentration. However, there are a number of elements in some form can represent a serious environmental problem and is common to refer to them by the generic term heavy metal (López et al, 2007). At present, there are anthropogenic sources of heavy metals, such as pollution, which has been introduced into ecosystems. For examples, fuels derived from waste (not organic) usually provide these metals, so you should consider heavy metals when wastes are used as fuel. The best known toxic heavy metals are mercury, lead, cadmium and arsenic, and on rare occasions, a non-metal such as selenium. Sometimes also it is spoken of contamination by heavy metals including other lighter toxic elements such as beryllium or aluminum (López et al, 2011).

Relationship with Living Organisms

Living organisms require different amounts of heavy metals. Small amounts of iron, cobalt, copper, manganese, molybdenum, and zinc are required by humans. Excessive amounts can damage our bodies (Berresheim et al, 2005). Other heavy metals such as mercury, plutonium, and lead are toxic metals that have no vital or beneficial effect to the body and its accumulation in time and in the body of animals can cause serious illness. Some items that are normally toxic to some organisms, under some conditions may be beneficial. For example, vanadium, tungsten, including cadmium.

Heavy Metals and Pollution

The heavy metals may include light elements such as carbon and may exclude some of the heavier metals. Heavy metals are free in the atmosphere and naturally in some ecosystems and can vary in concentration. However, there are a number of elements that in some form can represent a serious environmental problem and is common to refer to them by the generic term heavy metal (López et al, 2007). At present, there are anthropogenic sources of heavy metals, such as pollution, which has been introduced into ecosystems, being principally of the electronics industry. For examples, fuels derived from waste (not organic) usually provide these metals, so you should consider heavy metals when wastes are used as fuel. The best known toxic heavy metals are mercury, lead, cadmium and arsenic, and on rare occasions, a non-metal such as selenium. Sometimes also it is spoken of contamination by heavy metals including other lighter toxic elements such as beryllium or aluminum. From the electronics industry are the copper, silver, steel and tin (López et al, 2011).

Relationship with Living Organisms

Living organisms require different amounts of heavy metals. Small amounts of iron, cobalt, copper, manganese, molybdenum, and zinc are required by humans. Excessive amounts can damage our bodies (Berresheim et al, 2005). Other heavy metals such as mercury, plutonium, and lead are toxic metals that have no vital or beneficial effect to the body and its accumulation in time and in the body of animals can cause serious illness. Some items that are normally toxic to some organisms, under some conditions may be beneficial. For example, vanadium, tungsten, including cadmium.

Heavy Metals and Pollution

The best way for controlling the concentrations of heavy metals in gaseous streams is diverse. Some of them are dangerous to the environment and health, for example mercury (Hg), cadmium (Cd), leads (Pb) and chromium (Cr). Others cause corrosion, such as zinc or lead, or are harmful in other ways (for example arsenic may contaminate catalysts) (De Hayes et al, 2001). The heavy metal contamination can arise from many sources, but more commonly of the electronics industry (Villa et al, 2009). By precipitation of these compounds or ion exchange into soil and sludge, heavy metals can be located and be deposited. Unlike organic pollutants, heavy metals do not decay and presented other challenges to overcome them. Currently, plants (phytoremediation) and microorganisms are used to remove heavy metals such as mercury. Certain plants exhibiting hyper accumulation can be used to remove these metals soil concentration by biomaterial. In some tailings impoundments vegetation which is then incinerated to recover heavy metals is used. One of the biggest problems associated with the onset of heavy metals is the potential for bioaccumulation and bio magnification causing increased exposure of these metals to a body which could be alone in the environment. There are other types of heavy metals that are used in the electronics and metallic industry, whose density is at least five times greater than that of water. They have direct application in many processes of production of goods and services. The most important are: Arsenic (As), (Cd), cobalt (Co), copper (Cu), nickel (Ni), tin (Sn) and zinc (Zn). Toxic metals are those whose concentration in the atmosphere can damage the health of people. The terms heavy metals and toxic metals are used as synonyms but only some of them belong to both groups. Some metals are essential in low concentrations, as part of enzyme systems, such as cobalt, zinc, molybdenum, or iron as part of hemoglobin. Its absence causes diseases and their excess can be poison (Harte et al, 2008). Technological development, massive and indiscriminate consumption and production of mainly urban waste, has led to the presence of many metals in significant quantities in soils, causing numerous effects on the health and balance of ecosystems. They incorporated with food or breathe particles and accumulate in the body, reaching toxicity limits. If the incorporation is slow chronic poisoning, damages the tissues or organs in which accumulate occur. For many years the lead oxide was used as a white pigment used in paint (now replaced by titanium oxide) and in many countries is still used tetraethyl lead as an additive in gasoline. Mercury is used pure or in the form of amalgams. Its use in dentistry and some batteries frequently. Cadmium alloys used and also in various batteries. What makes them toxic or become heavy metal these metals are generally not heavy metals in food essential features, but the concentrations in which they can occur and the molecular density. These metals change their natural biological structure by chemical processes, such as oxidation, industrial processes, etc. (Fergusson, 2009).

Corrosion Process

Corrosion is defined as the deterioration of a metallic material as a result of electrochemical etching by their environment, and can be understood as the general tendency of material to find it is most stable or lower internal energy form. Provided that the corrosion is caused by an electrochemical reaction (oxidation), the rate at which occurs will depend to some extent on the temperature, the salinity of the fluid in contact with the metal and the properties of the metals in question. The corrosion process is natural and spontaneous. The best known factors are the chemical alterations of metals from the air, as rust iron and steel or the formation of green patina on copper and its alloys (bronze, brass).

METHODOLOGY

This study was made to support investigations in environmental and microbiology areas being an important analysis to recover corroded soils, which are evaluated in all regions of the world. With this study can improve some methods and techniques to be detected microorganisms with optoelectronic devices that have exactitude in its detection processes. The investigation was made in three steps, where was evaluated some optoelectronic devices to the evaluation process that mention in the next stages:

1. Was made an evaluation of simulations with the MatLab software to determine a mathematical equation to design and fabricate a new sensor as optoelectronic device.
2. Was made evaluations with some materials with the DOS theory, to determine the best to the new sensor and were fabricate with the CVD. To the electrical conductivity evaluations were use some devices with macro and micro components annexed to electronic systems, using internal BJT (Bipolar Junction Transistor), FET (Field Effect Transistor), tiristors and relays. Also, was made an analysis of corroded soils of zones where were pulled HM in areas annexed to industrial plants, principally of the electronics industry in the Mexicali city was made to analyze the optoelectronic devices necessary to this study.

3. A microscopy evaluation was made to determine the type of materials used in the optoelectronic devices and systems and the electronic diagrams in the micro components to know the operation and the type of the optoelectronic devices and systems necessary to this study.

RESULTS

The evaluation of the optoelectronic devices and systems as macro and micro components in the microbiology area to detect some type of HM to recover corroded soils, is necessary to determine the type of this macro and microelectronic devices and systems. This is very important to have the adequate results in the necessary times and convert these contaminated soils in fertile soils as green areas or zones to cultivate. For this reasons, the investigations with optoelectronics devices in the microbiology is important to be calculated the quantity or percentage of microorganisms recognized as the support actions to break down metallic materials or the microorganisms that damage very fast the soils.

Evaluation of Operation of Optoelectronic Device (Infrared)

Exist diverse methodologies to design and make some testing and probe the operation the operation of optoelectronic devices principally the infrared systems to determine the type of applications in activities as industrial, medical, aerospace, space, and marine, and in security operations. One of the design and probe in a simulation process is the Wein2k with Linux, where are observed the characteristics of the chemical elements that forms the materials to the infrared systems. This program has complex operations, but the results are very reliable, because the 95% of the analysis of chemical elements of materials and the mathematical simulation of the functionality shows the same process when the infrared systems is fabricated and in the operation activities. An example of this, is the evaluation of a new material used to an infrared with the program Wein2k, generating some evaluations in according the Fermi Energy that is a factor in the circulation of the electrical current (analyzing the flow of the electrons in the band gap).

The Figure 6 shows the information mentioned, representing the energy in electron volts (eV), from negative and positive values and showing the level of the Fermi Energy. In the positive values, the optoelectronic device will be an adequate function and will be more time of life in their operations. When the energy is same of the Fermi Energy, before pass to the negative values, the optoelectronic device will have a time of response very fast, to detect any object or particles. This analysis is very important to determine the use of these devices and depends of the activity

Figure 6. Classification map of metals discharged of industries and types of soils in Mexicali (2014)

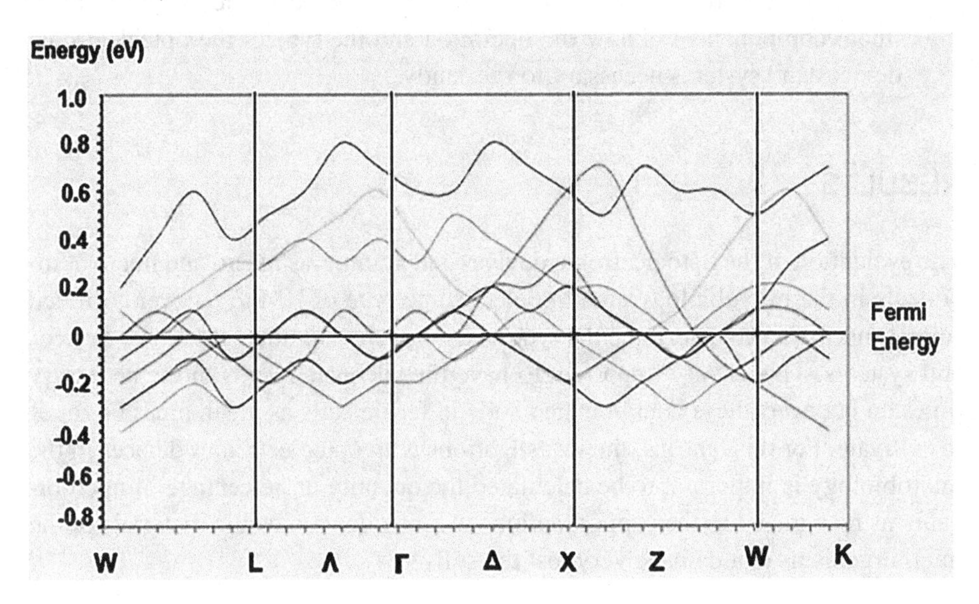

is utilized the chemical elements which forms the materials. The different smoothed lines are of the evaluations in groups of seven different times of response time to detect the objects very fast o very slow, depending of the function to do. The letters and symbols indicates the different times of response, and its illustration is for the representation of the Wein 2k program.

Analyses of Discharged Metals

One of the contaminants that can affect the soil, are metals, can be found in the form of minerals to be part of the crust, and originates corroded soils. During the collection, use and industrialization as well as their processes of smelting, refining and recycling; large quantities of waste, with negative effects in the form of particles that can reach larger, contaminating them areas are generated. A metallic element, such as mercury, nickel, zinc and cadmium, are environmental concern because they do not degrade over time. A metal having specific gravity of 5 or greater is considered a heavy metal, generally are toxic to organisms in relatively low concentration, and tend to accumulate in the food chain. Although many are necessary nutrients, they are sometimes magnified being toxic to soils. Heavy metals and metalloids, are one of the most important groups of pollutants, affecting all agencies to biological chemical level, even at microscopic levels. From soil nutrients from our food are taken, is a source of production of plant and animal biomass: providing material resources and minerals of economic value, serves as a filter, buffer and storage for water, it is home to millions of microorganisms in biogeochemical cycles happen. So, the soil is considered the basis for life and our civilization, as it is the physical support for the development and construction of all economic and recreational sectors. An example of this is showed in the map of Mexicali that represents the processes of the discharge of metals from industries, in special of the electronic industries, pulled in soils as raw material or finished products fabricated with important defects to be declined. This information is showed in Figure 7, indicating in four zones where the main places of the industries in the Mexicali city are.

The map of the Figure 7 represented by the four zones in industrial parks, located the zone 1 in the west of the city, where are around 15 companies of the electronics industry. The zone 2 are located in the northeast of Mexicali where are around 25 industries of the same type of activities mentioned and the zone 3 is around to the east with 15 industries. The major zone with industries is the zone 4 with around 45 industries. The other 20 industries resting are located in some places of Mexicali. The major zone with more trash pulled is the zone 4. Government authorities of ecology departments are working with managers to reduce it in all zones of Mexicali to decrement the soils damaged by the materials with heavy

Figure 7. Classification map of metals discharged of industries and types of soils in Mexicali (2014)

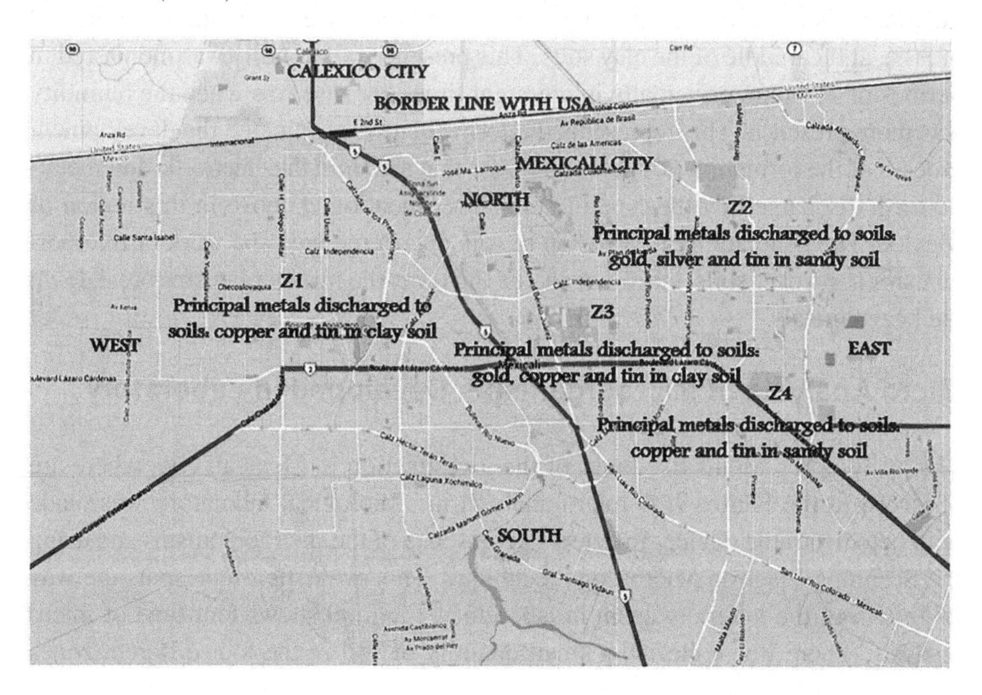

metals discharge by the industries. The four zones will be divided in two types of soils: clay and sandy, because this city have both soils. This was observed with the detection of an optoelectronic device as infrared, with special conditions to detect the heavy metals and other specialized infrared to detect the microorganisms that damage soils. The infrared devices were designed and fabricated to this investigation by the researchers of the study mentioned above.

Micro Evaluation of Microorganism Growth in Soils

The microorganisms (MO) in the two type of soils evaluated is showed in Figure 8, where was observed the difference of formation in clay and sandy soils.

In clay soils was illustrated the formation process of the microorganism, where the light color of particles are the microorganisms formed in some periods of times and the dark color represents the particles of soils in this case of the clay soil showed in Figures 8a and 8b and the sandy soils observed in Figure 8c an 8d. This was detected with the specialized optoelectronic devices designed and fabricated in our educational institution. The microanalysis was at scale of 100X to increase the size of the particles and the microorganisms evaluated in both type of soils. This evaluation was made with the technique of Scanning Electron Microscopy with a test of the both soils mentioned. In the clay soil were presented humidity with a 60% of the presence of humidity and the microorganism formed were in small groups and some MO were developed in separated locations. In change in the sandy soils the microorganism was formed in big groups and the presence of humidity was around of 30%, at the middle of the clay soils. This presence of humidity was monitored in periods of 12 hours principally in the night times because was when the humidity was more presence. This represents the forms that was evaluated the development process of the formation of microorganisms that are from the electronic and metallic trash discharged by the type of companies mentioned above in this region of Mexico. The generation of corrosion in soils interfered with the work of microorganisms to recover soils, but when passed the time, the microorganisms breakdown the heavy metals.

Micro Analysis of Microorganisms Developed in Laboratory

The analysis represents the zones of the investigation in Mexicali city, where are observed in the Figure 9. A micro analysis in a biological laboratory was made with optoelectronic devices to detect the presence of the microorganisms and using the Scanning Electron Microscopy technique. This evaluation represents the way to growth of the microorganism in laboratory. Figure 9 shows four type of micro analysis, where was collected a small quantity of soil in the four different zones

Figure 8. Microanalysis of MO generated by waste in different soils in a zone 1 (a, b) and zone 2 (c, d) (2014)

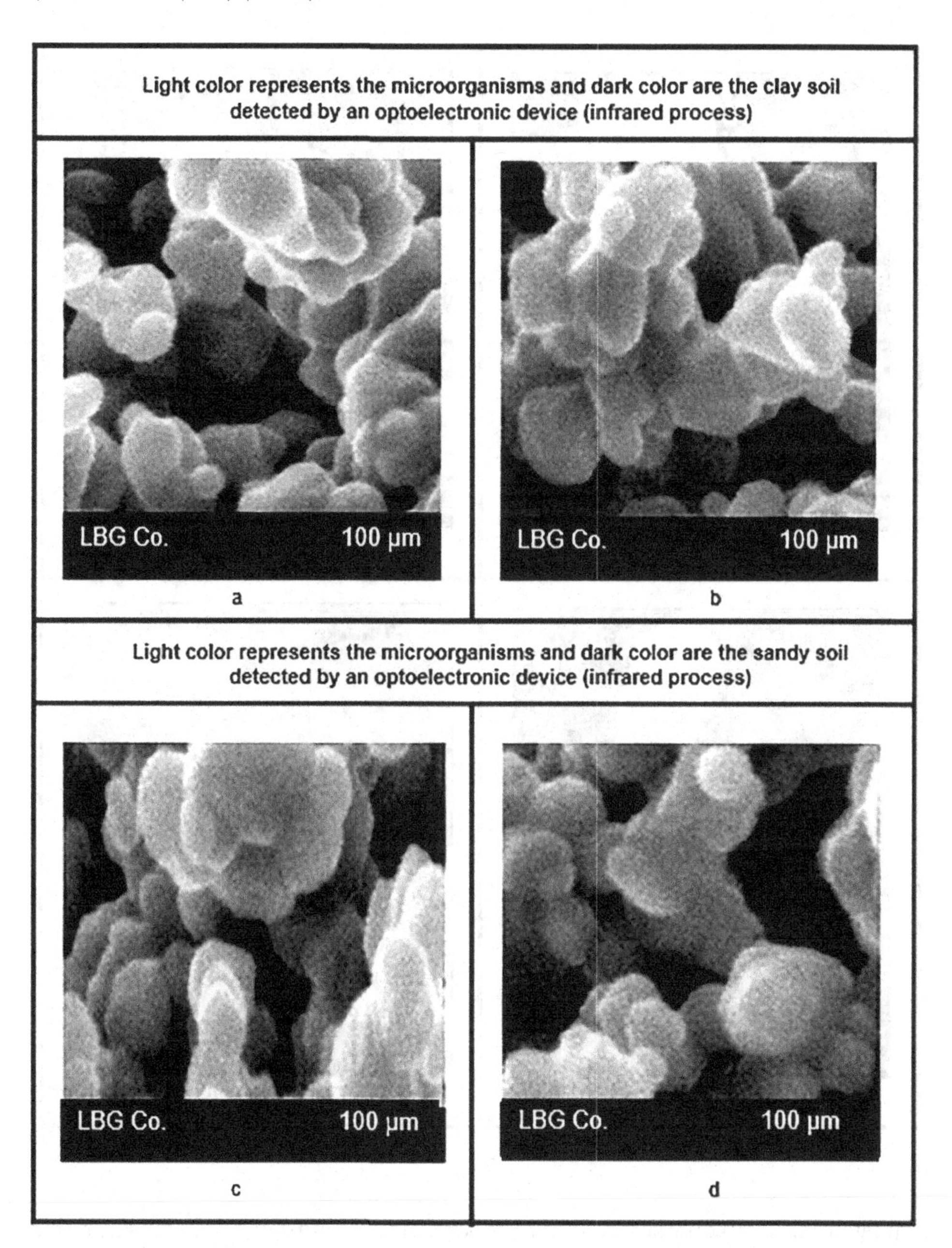

Figure 9. Microanalysis of MO developed in clay soil of zone 1 (a) and zone 3 (c) and sandy soil in zone 2 (b) and zone 4 (d) in Mexicali (2014)

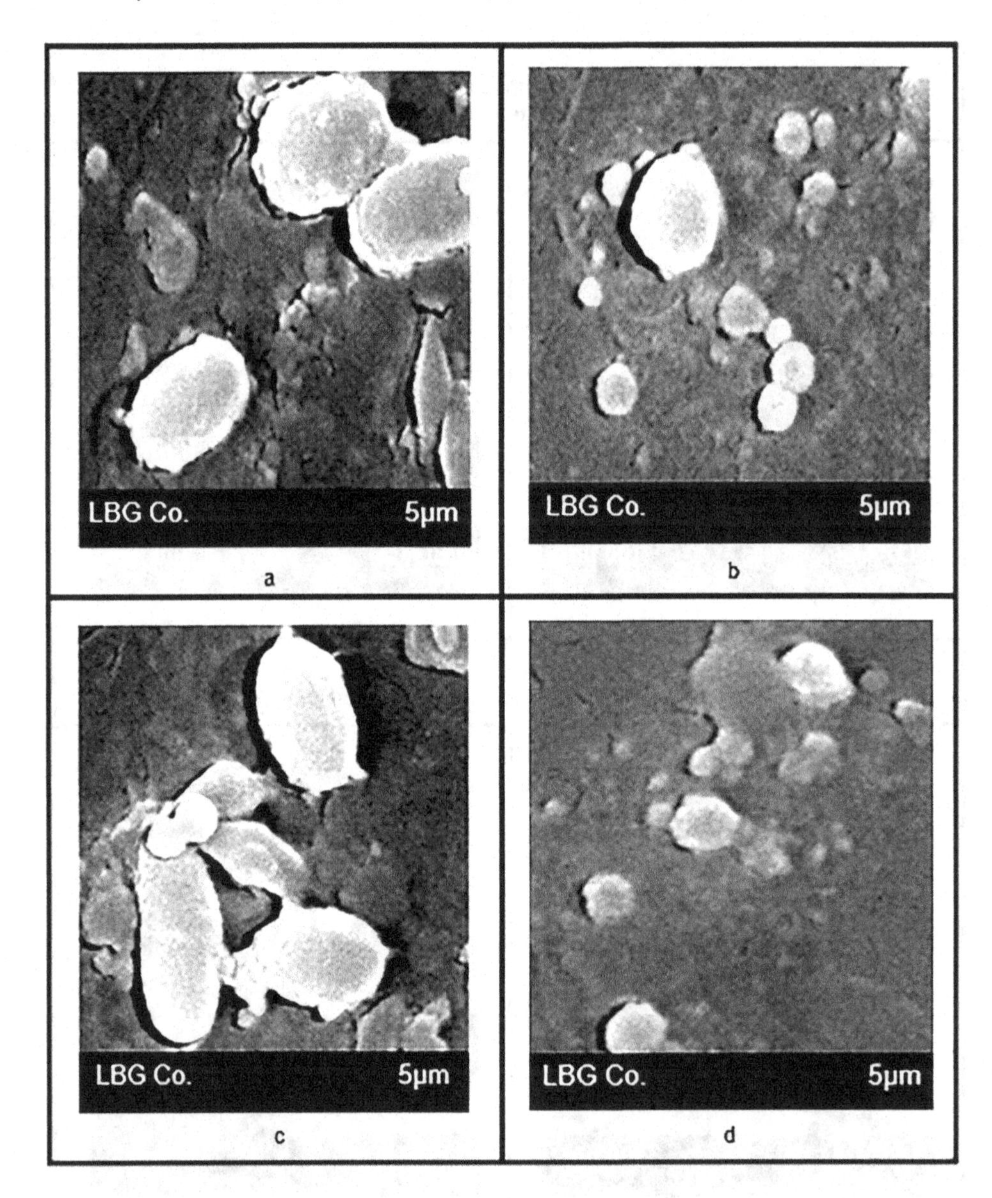

evaluated and used some specialized methods to design and development of a growth of microorganism necessary to break down heavy metals based in the characteristics that put out of it body from the gastrointestinal and respiratory systems of microorganisms with sulfur try to destroy very fast the heavy metals. The different groups of microorganisms involved heavily in the biological treatment of liquid and solid waste. These treatments may be of aerobic type, anaerobic or mixed. Many of them are useful in bioremediation of soil and water. Contribute to soil fertility, organic matter degraded remains of dead animals and plants, as well as the participation of some bacteria in the fixation of atmospheric nitrogen. It can be used in the design and construction of biosensors that can be used in the determination of analyses of environmental interest. A number of microorganisms are useful in biotechnology in obtaining a variety of products of interest and production of single cell protein source.

The Figure 9a and 9c shows the growth of microorganism in the test of clay soils and Figures 9b and 9d, shows the microorganisms in sandy soils and were illustrated big microorganisms in separated locations in clay zones and small microorganisms with joint cologness in sandy soils. The microanalysis was at 5X of increment of size with the SEM technique. These microorganisms were pulled in the zones evaluated to control the process of the microorganism presented in this type of soils to avoid the deterioration of the soils.

Micro Analysis of Soils with Optoelectronic Devices

As mention in the last section, was made the analysis in the zone 2, for the great quantity of waste generated in this location of the Mexicali city, where are showed the deteriorated soils (Figure 10).

The evaluation indicates the forms of deterioration in clay and sandy soils in summer and winter seasons in 2014, where was observed the major deterioration in clay soils and the minor damage in sandy soils. This was detected by the specialized optoelectronic devices fabricated by the personal of the investigation to monitoring the deterioration of soils. The function of the optoelectronic device detects the not presence of particles or objects, and where are holes send a signal of a not uniform of the soil, indicating the damage of soils. In the Figure 9 was observed the holes formed after the pulled of the trash and damage made by the HM. In the clay soils were formed big holes and in the sandy soils were formed small holes. This difference is for the type of material that in clay soils have the particles more joint and in the sandy soils have the particle more separated. The SEM technique was applied to determine with a major precision the results.

Evaluation of DOS Method

The variation of the electron density cause changes in electrical conductivity properties of the materials used in the electronics industry. Local Density Approximations or (DLA) is an approximation for the exchange-correlation of the energy functional in density functional theory (DFT) that depend only on the value of the electron density at each point in space (and not, for example, derived from the orbital density).

Figure 10. Microanalysis of soils of clay in summer (a) and in winter (b) and sandy in summer (c) and sandy in winter (d) in Mexicali (2014) in zone 2 of the map evaluation

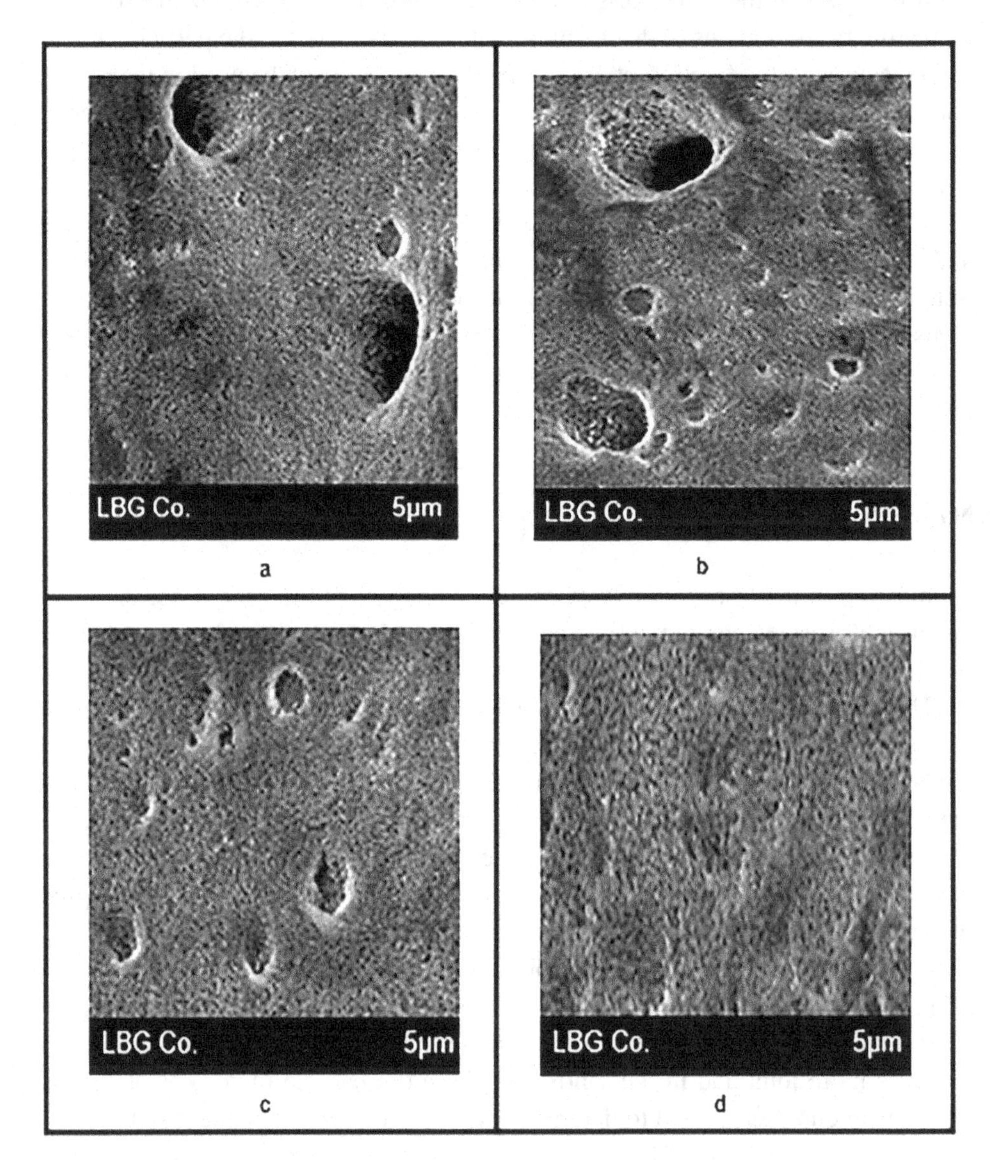

Many approaches can produce local energy of the orbital approaches. The successful approaches are those that have been derived from a model of homogeneous electron gas (HEG). In this regard, the DLA is generally synonymous with the functions HEG based approach, which are applied to realistic systems (molecules and solids). This can be monitored by thermic analysis as showed in Figures 11 and 12 of this study.

The Figures 11 and 12 represents two analyses of thermic evaluations about the movements of the molecules of the soils with microorganism that are in movement every time for look food for it. In both figures are used the optoelectronic devices as infrared to monitoring the presence of microorganisms in the both type of soils in 2014 in the Mexicali city. The Figure 11 indicates the major detection of the presence with red color of microorganisms detected by the infrared device, where in the yellow color indicates the medium term of the detection and the blue color the less detection of microorganisms with the specialized optoelectronic devices fabricated and used in this study with reliable results. In Figure 12 was observed a major detection with the optoelectronic device with more presence of red and yellow color with a lot zones of the graphic of the evaluation.

DISCUSSION OF THE RESULTS

The use of optoelectronic devices in studies of the ecology and microbiology is very important for the reliable results obtained. In the micro science exists some aspects

Figure 11. Analysis with optoelectronic device (infrared) of clay soil in the zone 1 of Mexicali (2014)

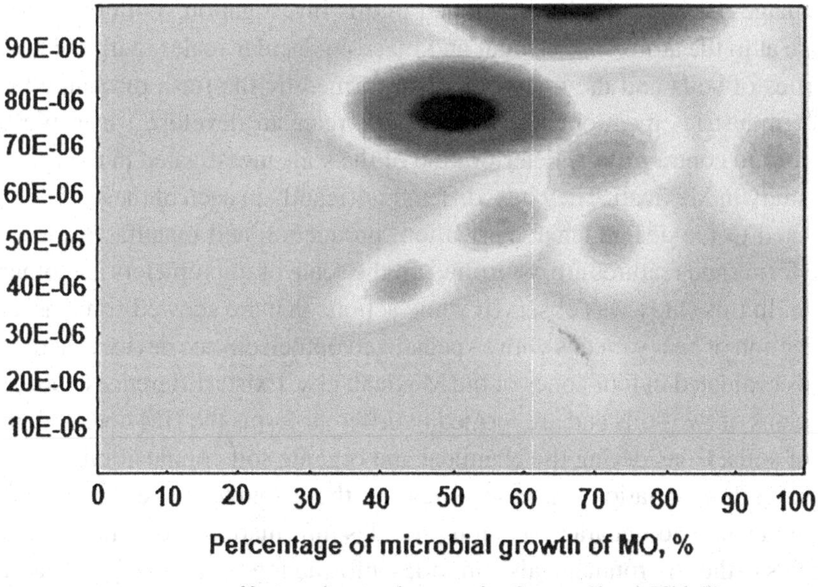

Figure 12. Analysis with optoelectronic device (infrared) of sandy soil in the zone 2 of Mexicali (2014)

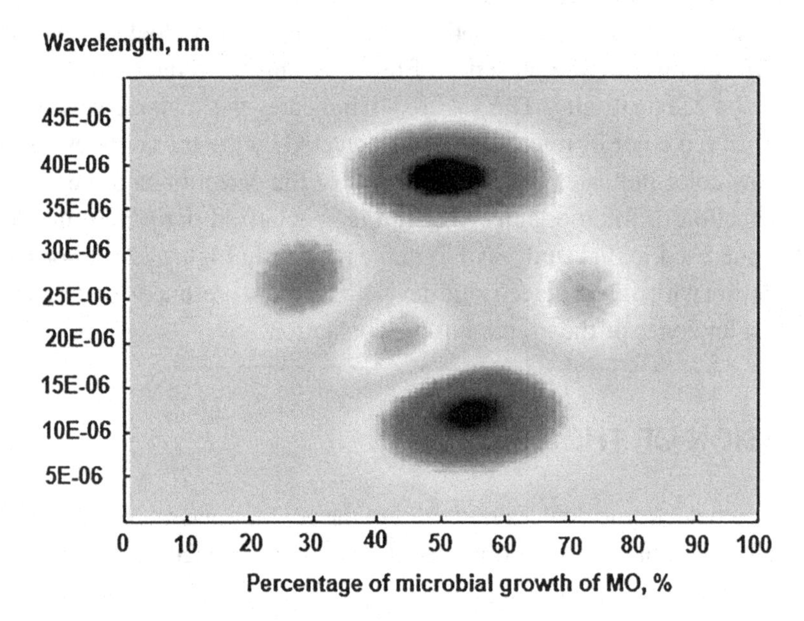

that not are observed at naked eye, and occurs some functions that are generated by the climatic and pollution factors, which in some times the human cannot control in this study. The presence of phenomena and manipulation of the material of soils and damage the type of soils mentioned in this investigation. With this study can evaluate at to the atomic, molecular and macromolecular scales, with the principal properties of soils and the microorganism formed by the trash discharged by the type of industries mentioned and the microorganism developed in a biological laboratory to contra rest the deterioration of the soils investigated in the four zones of the study in Mexicali. They may differ significantly in each of these scales. They are related to the design, characterization, production and manufacturing Device whose forms and features are controlled on the scale of micrometers and other type of scale. In this study was observed some actions that are showed some aspects of the detection of heavy metals with a specialized optoelectronic device that damaged the soils evaluated in four zones of the Mexicali city. Exists differences between the evaluations of two soils and are formed in different forms the HM detected, in both types of soils. Considering the chemical and organic soil composition, in addition to its natural degradation processes, many of these sources generate the different damage in soils. Soil characterization provides information about the influence of pollutants in the environment, also provides information on the soil quality and level of involvement; giving rise to the search for adequate means for their remediation.

CONCLUSION

The development of studies in ecology and microbiology has led to the need for be ever faster, with exactitude and more the use of powerful analytical tools to estimate more efficiently the microorganisms' growth in the course of some bioprocesses. In this regard, the ecology and microbiology with optoelectronic techniques are useful tools that meet these expectations and applied in life sciences. Some systems used to operate optoelectronic activities to obtain measurement with precision, and based on optical phenomena that occur when affecting a beam of light through a medium, are very important to determine very fast any microbiology process evaluated. The light is scattered due to the existence of suspended within the medium in analysis, and it are also involved in reducing the intensity of the resulting light beam particles in the detection of the microorganisms in soils to know the required process to recover the contaminated soils. The manner in which the sample suspended in the medium interferes with light is related to the size, shape, composition of particles in the suspension and the wavelength of incident light. The microbiological techniques can differ in the geometry of the light detector used. The progress of optical systems has led to its effective use for the production of information in different fields of measurement including biotechnological processes. Thus, the microbiological methods constitute an alternative in obtaining information and control systems for biology processes. In this study was determined the quantity and percentage of HM in the zones evaluated in the Mexicali city, where was observed near of companies of the electronics industry. This study is the first step to recover soils in this city, where the next stage will be the use of some electronic devices to growth and maintain green areas and cultivated zones. Heavy metals and metalloids, are one of the most important groups of pollutants, affecting all agencies to biological chemical level, even at microscopic levels.

ACKNOWLEDGMENT

Authors thanks to authorities of companies in the four zones of evaluation and the social areas of this city. Also authors thanks to companies that cooperate with this investigation with information about the waste putted in the soils next and generate microbiological processes.

REFERENCES

An, , J., Suk, J. W., Han, B., Borysiak, M., Cai, W., & Velamakanni, A. (2013). The use of solid state electronic devices in the microbiology. *Journal of Optocouplers Systems*, *6*(2), 89–101.

Berresheim, H., Wine, P. H., & Davies, D. D. (2005). *Sulfur dioxide in the atmosphere*. Ed. H.B. Singh. Van Nostran Rheingold.

Boland, J. J., & Coleman, J. N. (2012). Micro contamination in the soils used to agriculture activities. *Journal of Microbiology (Seoul, Korea).*

Chen, C., Park, C., Boudouris, B. W., & Horng, J., & Geng. (2011). The optoelectronics functions in the agriculture. *Journal of Agriculture and Soils*, *9*(4), 67–90.

Chigo, E., & Rivas-Silva, J. F. (2004). La aproximación LDA+U en la teoría DFT. *Revista Mexicana de Física*, *50*(2), 88–95.

Cline, D., Hofstetter, H. W., & Griffin, J. R. (2008). *Analysis of visual science* (4th ed.). Boston: Butterworth-Heinemann.

De Franco. (2007). Materials evaluation to obtain the better vision to optoelectronic sensors. McGraw Hill Ed.

De Hayes, D. H., Schaberg, P. G., & Strimbeck, G. R. (2001). Red Spruce Hardiness and Freezing Injury Susceptibility. In F. Bigras (Ed.), *Conifer Cold Hardiness*. Kluwer Academic Publishers. doi:10.1007/978-94-015-9650-3_18

Fergusson, J. F. (2009). The principal heavy elements. In *Environmental and impact and health effects*. Pergamum Press.

Fiolhais, C., & Nogueira, F. (2003). A Primer in Density Functional Theory. Springer.

Girit, B., Zettl, C., Crommie, A., & Segalman, M. F. (2012). The optoelectronic devices. An overview of the use to detect microorganisms. *Journal of Microbiology (Seoul, Korea)*, *8*(3), 23–45.

Harte, J., Holden, C., Scheneider, R., & Shirey, C. (2008). Principal toxics chemical elements as heavy metals. McGraw Hill.

Hecht, D. S., Hu, L., & Irvin, G. (2014). Beneficial organisms for agriculture and sustainable environment. *Journal of Environment and Technology*, *7*(4), 56–72.

Higgins, S., Lyons, T. M., & Doherty, P. E. (2011). Optoelectronics systems in the food industry. *Journal of Agriculture and Food Operations*, *13*(5), 99–121.

Ju, L., Geng, B., Horng, J., & Girit, C. (2010). The optoelectronics in the recovering of soils. *Journal of Electronics Industry*, *7*(3), 56–72.

July, T., & Deane, B. (2009). Color analysis of the optical vision. In Applied Optics (3rd ed.). New York: Wiley-Interscience.

Kim, P., Yoo, I., & Chung, H. (2010). Trichoderma fungus use to recovery soils detected by optoelectronics systems. In *Use of optoelectronics in the agriculture* (10th ed.). Elsevier Books.

Li, X., & Magnuson, C. W. (2011). Optoelectronic devices used in food industries to detect microorganisms. *Journal of Applied Microbiology*, *14*(7), 78–95.

López,, B.G., Valdez,, S.B., & Schorr,, W. M., Zlatev, R., Tiznado, V.H., Soto, H.G., & De la Cruz, W. (2011). AES in Corrosion of Electronic Devices in Arid in Marine Environments. *Anti-Corrosion Methods and Materials*, *4*(2), 134–149.

López, B. G., Valdez, S. B., Schorr, W. M., Tiznado, V. H., & Soto, H. G. (2010). *Influence of Climate Factors on Copper Corrosion in Electronic Equipment and Devices*. Anti-Corrosion Methods and Materials.

López, B.G., Valdez, S.B., Zlatev, K.R., Flores, P.J., Carrillo, B.M., & Schorr, W.M. (2007). *Corrosion of Metals at Indoor Conditions in the Electronics Manufacturing Industry*. Anti-Corrosion Methods and Materials.

López, G., Tiznado, H., Soto, G., De la Cruz, W., Valdez, B., Schorr, M., & Zlatev, R. (2010). *Corrosión de dispositivos electrónicos por contaminación atmosférica en interiores de plantas de ambientes áridos y marinos*. Nova Scientia. (in Spanish)

López-Badilla, G. (2008). *Caracterización de la corrosión en materiales metálicos de la industria electrónica en Mexicali*. PhD Thesis.

Majumdar, A. K., & Ricklin, J. C. (2008). *Free-Space Laser Communications, Principles and Advances*. New York: Springer. doi:10.1007/978-0-387-28677-8

Mars, U., & Johnson, G. (2007). Evolution, techniques, industrial operations with optoelectronic devices with chromatic systems. Mc Graw Hill Ed.

Mendenhall, J. A., Candell, L. M., Hopman, P. I., Zogbi, G., Boroson, D. M., Caplan, D. O., . . . Shoup, R. C. (2007). Design of an Optical Photon Counting Array Receiver System for Deep-Space Communications. *Proceedings of the IEEE, 5*(1), 2059-2069.

Morrinsense, & Lewis, D. (2010). Laser epithelial evaluated a chromatic systems. John Wiley and Sons Ed.

N. G., Jr., & Larson, D. R. (2009). Reference Ballistic Chronograph. *Optical Engineering Journal, 48*(4), 43602-43607.

O'Regan, B. C., Barnes, P. R. F., Li, X., Law, C., Palomares, E., & Marin-Beloqui, J. M. (2015). Optoelectronic Studies of Methyl ammonium Lead Iodide Perovskite Solar Cells with Mesoporous TiO_2: Separation of Electronic and Chemical Charge Storage, Understanding Two Recombination Lifetimes, and the Evolution of Band Offsets during J–V Hysteresis. *Journal of the American Chemical Society, 137*(15), 5087–5099. doi:10.1021/jacs.5b00761 PMID:25785843

Parr, R. G., & Weitao, Y. (2004). *Density-Functional Theory of Atoms and Molecules in metals*. Oxford, UK: Oxford University Press.

Peña, J. M. S., Marcos, C., Fernández, M. Y., & Zaera, R. (2007). Cost-Effective Optoelectronic System to Measure the Projectile Velocity in High-Velocity Impact Testing of Aircraft and Spacecraft Structural Elements. *Optical Engineering, 46*(5), 510141-510146.

Louie, S. G., & Wang, F. (2011). The electronic devices in the agricultural operations. *Journal of Agronomy Engineering, 11*(4), 67–84.

Nirmalraj, P. N., & Blau, W. J. (2013). The presence of microbial in soils by the satured pollution. *Journal of Environment and Industry, 8*(3), 32–47.

Villa, J. L., Sallán, J., Llombart, A., & Sanz, J. (2009). Design of a high frequency Inductively Coupled Power Transfer system for electric vehicle battery charge. *Applied Energy, 86*(3), 355–363.

Vogel, E. M., Voelkl, E., Colombo, L., & Ruoff, R. S. (2010). The optocouplers functions. *Journal of Electronics Systems, 12*(6), 11–34.

Yang, H., Heo, J., Park, S., Song, H. J., & Seo, D. H. (2010). Detection of microbial organisms in soils. In Use of optoelectronics in the agriculture (10th ed.). Elsevier Books.

Compilation of References

Abdul Aziz, Z. A., & Marzuki, A. (2010). Residual Folding Technique adopting switched capacitor residue amplifiers and folded cascode amplifier with novel pmos isolation for high speed pipelined ADC applications. *Proceedings of the 3rd AUN/SEED-Net Regional Conference in Electrical and Electronics Engineering:International Conference on System on Chip Design Challenges (ICoSoC 2010)* (pp. 14–17).

Abidi, M. A. (1992). *Data fusion in robotics and machine intelligence*. Academic Press Professional, Inc.

Adu, M. O., Chatot, A., Wiesel, L., Bennett, M. J., Broadley, M. R., White, P. J., & Dupuy, L. X. (2014). A scanner system for high-resolution quantification of variation in root growth dynamics of *Brassica rapa* genotypes. *Journal of Experimental Botany*, *65*(8), 2039–2048. doi:10.1093/jxb/eru048 PMID:24604732

Agranov, G., Mauritzson, R., Barna, S., Jiang, J., Dokoutchaev, A., Fan, X., & Li, X. (2007). Super Small, Sub 2μm Pixels for Novel CMOS Image Sensors. *Proceedings of the Extended Programme of the 2007 International Image Sensor Workshop* (pp. 307–310).

Aha, D. W., Kibler, D., & Albert, M. K. (1991). Instance-based learning algorithms. *Machine Learning*, *6*(1), 37–66. doi:10.1007/BF00153759

Alexander, C., Erenskjold Moeslund, J., Klith Bøcher, P., Arge, L., & Svenning, J.-C. (2013, July). Airborne laser scanner (LiDAR) proxies for understory light conditions. *Remote Sensing of Environment, 134*, 152–161. doi:10.1016/j.rse.2013.02.028

Amin, M. H. G., Richards, K. S., Chorley, R. J., Gibbs, S. J., Carpenter, T. A., & Hall, L. D. (1996). Studies of soil-water transport by MRI. *Magnetic Resonance Imaging*, *14*(7), 879–882. doi:10.1016/S0730-725X(96)00171-3 PMID:8970099

An, , J., Suk, J. W., Han, B., Borysiak, M., Cai, W., & Velamakanni, A. (2013). The use of solid state electronic devices in the microbiology. *Journal of Optocouplers Systems*, *6*(2), 89–101.

Anderson, S. A., & Tallapally, L. K. (1996). Hydrologic response of a steep tropical slope to heavy rainfall. 7th International Syposium on Landslides (pp. 1489-1498). Trondheim, Norway: Brookfield, Rotterdam.

Apostoloff, N., & Zisserman, A. (2007). Who are you? - real time person identification. *Proceedings of British Machine Vision Conference* (pp. 48.1-48.10). BMVA Press. doi:10.5244/C.21.48

Aptina. (2010). *An Objective Look at FSI and BSI.* Aptina White Paper.

Ardeshirpour, Y., Deen, M. J., Shirani, S., West, M. S., & Ls, O. N. (2004). 2-D CMOS based image sensor system for fluorescent detection. *Proceedings of the Canadian Conference on Electrical and Computer Engineering* (pp. 1441–1444). IEEE.

Arsenault, J. L., Pouleur, S., Messier, C., & Guay, R. (1995). WinRHIZO, a root-measuring system with a unique overlap correction method. *HortScience*, *30*, 906.

Baker, R. J. (2010). *CMOS: Circuit Design, Layout, and Simulation* (3rd ed.). Wiley-IEEE Press. doi:10.1002/9780470891179

Bandyopadhyay, A., Lee, J., Robucci, R., & Hasler, P. (2006). MATIA: A Programmable 80 μW/frame CMOS Block Matrix Transformation Imager Architecture. *IEEE Journal of Solid-State Circuits*, *41*(3), 663–672. doi:10.1109/JSSC.2005.864115

Banerjee, T. P., & Das, S. (2012). Multi-sensor data fusion using support vector machine for motor fault detection. *Information Sciences*, *217*, 96–107. doi:10.1016/j.ins.2012.06.016

Baron, I. R. S. (2010). State of the Art of Landslide Monitoring in Europe: Preliminary results of the Safeland Questionnaire. In R. S. Baron (Ed.), Landslide Monitoring Technologies & Early Warning Systems (pp. 15-21). Vienna: Geological Survey of Austria.

Barr, K. (2007). *ASIC Design in the Silicon Sandbox: A Complete Guide to Building Mixed-Signal Integrated Circuits.* McGraw Hill Professional.

Basaca-Preciado, L. C.-Q.-L.-B., Sergiyenko, O. Y., Rodríguez-Quinonez, J. C., García, X., Tyrsa, V. V., Rivas-Lopez, M., & Starostenko, O. et al. (2014). Optical 3D laser measurement system for navigation of autonomous mobile robot. *Optics and Lasers in Engineering*, *54*, 159–169. doi:10.1016/j.optlaseng.2013.08.005

Bass, M. (1995). Handbook of Optics, Vol. II- Devices, Measurements and Properties. New York: McGraw-Hill, Inc.

Beiser, L. (1995). Fundamental architecture of optical scanning systems. *Applied Optics*, *31*(34), 7307–7317. doi:10.1364/AO.34.007307 PMID:21060601

Bengough, A. G., McKenzie, B. M., Hallett, P. D., & Valentine, T. A. (2011). Root elongation, water stress, and mechanical impedance: A review of limiting stresses and beneficial root tip traits. *Journal of Experimental Botany*, *62*(1), 59–68. doi:10.1093/jxb/erq350 PMID:21118824

Benoit, L., Rousseau, D., Belin, E., Demilly, D., Ducournau, S., Chapeau-Blondeau, F., & Dürr, C. (2013). *Locally oriented anisotropic image diffusion: application to phenotyping of seedlings.* Paper presented at Special session GEODIFF, 8th International Joint Conference on Computer Vision, Imaging and Computer Graphics Theory and Applications (VISAPP 2013), Barcelona, Spain.

Benoit, L., Belin, É., Dürr, C., Chapeau-Blondeau, F., Demilly, D., Ducournau, S., & Rousseau, D. (2015). Computer vision under inactinic light for hypocotyl–radicle separation with a generic gravitropism-based criterion. *Computers and Electronics in Agriculture, 111*, 12–17. doi:10.1016/j.compag.2014.12.001

Benoit, L., Rousseau, D., Belin, E., Demilly, D., & Chapeau-Blondeau, F. (2014). Simulation of image acquisition in machine vision dedicated to seedling elongation to validate image processing root segmentation algorithms. *Computers and Electronics in Agriculture, 114*, 84–92. doi:10.1016/j.compag.2014.04.001

Bergman, M. (2014). *Knowledge-based Artificial Intelligence.* AI3 Web page. Retrieved from http://www.mkbergman.com/1816/knowledge-based-artificial-intelligence

Bernardini, F., Mittleman, J., Rushmeier, H., Silva, C., & Taubin, G. (1999). The ball-pivoting algorithm for surface reconstruction. *Visualization and Computer Graphics. IEEE Transactions on, 5*(4), 349–359.

Berresheim, H., Wine, P. H., & Davies, D. D. (2005). *Sulfur dioxide in the atmosphere.* Ed. H.B. Singh. Van Nostran Rheingold.

Bertrand, O., Queval, R., & Maitre, H. (1982). Shape interpolation using Fourier descriptors with application to animation graphics. *Signal Processing, 4*(1), 53–58. doi:10.1016/0165-1684(82)90039-1

Besl, P., & McKay, N. (1992, February). A method for registration of 3-D shapes. In *Pattern Analysis and Machine Intelligence* (pp. 239-256).

Beucher, S., & Lantuejoul, C. (1979). Use of Watershed in contour detection. *Proceedings of theInternational Workshop on Image Processing: Real-time Edge and Motion Detection/Estimation.*

Beucher, S., & Meyer, F. (1993). *The Morphological Approach to Segmentation: The Watershed Transformation.* Palo Alto, CA: E. R. Dougherty.

Bigas, M., Cabruja, E., Forest, J., & Salvi, J. (2006). Review of CMOS image sensors. *Microelectronics Journal, 37*(5), 433–451. doi:10.1016/j.mejo.2005.07.002

Billiot, B. (2013). *Conception d'un dispositif d'acquisition d'images agronomiques 3D en extérieur et développement des traitements associés pour la détection et la reconnaissance de plantes et de maladies.* (Ph-D Thesis). University of Burgundy.

Billiot, B., Cointault, F., Journaux, L., Simon, J. C., & Gouton, P. (2013). 3D image acquisition system based on Shape from Focus technique. *Sensors (Basel, Switzerland), 13*(4), 5040–5053. doi:10.3390/s130405040 PMID:23591964

Boissier, O., Bordini, R. H., Hubner, J. F., Ricci, A., & Santi, A. (2013). Multi-agent Oriented Programming with JaCaMo. *Science of Computer Programming, 78*(6), 747–761. doi:10.1016/j.scico.2011.10.004

Boland, J. J., & Coleman, J. N. (2012). Micro contamination in the soils used to agriculture activities. *Journal of Microbiology (Seoul, Korea)*.

Bordini, R. H., Hubner, J. F., & Wooldridge, M. (2007). *Programming Multi-Agent Systems in AgentSpeak Using Jason*. Chichester, UK: John Wiley & Sons.

Brockherde, W., Bussmann, A., Nitta, C., Hosticka, B. J., & Wertheimer, R. (2004). High-Sensitivity, High-Dynamic Range 768 x 576 Pixel CMOS Image Sensor. Proceedings of the ESSCIRC (pp. 411–414).

Bucksch, A., Burridge, J., York, L. M., Das, A., Nord, E., Weitz, J. S., & Lynch, J. P. (2014). Image-based high-throughput field phenotyping of crop roots. *Plant Physiology*, *166*(2), 470–486. doi:10.1104/pp.114.243519 PMID:25187526

Burton, A. L., Williams, M., Lynch, J. P., & Brown, K. M. (2012). RootScan: Software for high-throughput analysis of root anatomical traits. *Plant and Soil*, *357*(1-2), 189–203. doi:10.1007/s11104-012-1138-2

Cairns, J. E., Impa, S. M., O'Toole, J. C., Jagadish, S. V. K., & Price, A. H. (2011). Influence of the soil physical environment on rice (*Oryza sativa* L.) response to drought stress and its implications for drought research. *Field Crops Research*, *121*(3), 303–310. doi:10.1016/j.fcr.2011.01.012

Camenzind, H. (2005). *Designing Analog Chips*. Academic Press.

Canny, J. (1986). A Computational Approach to Edge Detection. *IEEE Transactions on Pattern Analysis and Machine Intelligence*, *PAMI-8*(6), 679–698. doi:10.1109/TPAMI.1986.4767851 PMID:21869365

Castanedo, F. (2013). A review of data fusion techniques. *The Scientific World Journal*, 1–19. PMID:24288502

Cevik, I., Huang, X., Yu, H., Yan, M., & Ay, S. (2015). An Ultra-Low Power CMOS Image Sensor with On-Chip Energy Harvesting and Power Management Capability. *Sensors (Basel, Switzerland)*, *15*(3), 5531–5554. doi:10.3390/s150305531 PMID:25756863

Chae, Y., Cheon, J., Lim, S., Kwon, M., Yoo, K., Jung, W., & Han, G. (2011). A 2.1 M Pixels, 120 Frame/s CMOS Image Sensor ADC Architecture with Column-Parallel Delta Sigma ADC Architecture. *IEEE Journal of Solid-State Circuits*, *46*(1), 236–247. doi:10.1109/JSSC.2010.2085910

Chen, C., Park, C., Boudouris, B. W., & Horng, J., & Geng. (2011). The optoelectronics functions in the agriculture. *Journal of Agriculture and Soils*, *9*(4), 67–90.

Chéné, Y., Rousseau, D., Lucidarme, P., Bertheloot, J., Caffier, V., Morel, P., & Chapeau-Blondeau, F. et al. (2012). On the use of depth camera for 3D phenotyping of entire plants. *Computers and Electronics in Agriculture*, *82*, 122–127. doi:10.1016/j.compag.2011.12.007

Chigo, E., & Rivas-Silva, J. F. (2004). La aproximación LDA+U en la teoría DFT. *Revista Mexicana de Física*, *50*(2), 88–95.

Cho, H.-J., & Park, T.-H. (2008). Template matching method for SMD inspection using discrete wavelet transform. *Proceedings of the2008 SICE Annual Conference* (pp. 3198-3201). Tokyo: IEEE.

Cho, T., & Gray, P. R. (1995). A 10b 20MSamples/s 35mW Pipeline A/D Converter. *IEEE Journal of Solid-State Circuits, 30*(3), 166–172. doi:10.1109/4.364429

Clark, R. T., MacCurdy, R. B., Jung, J. K., Shaff, J. E., McCouch, S. R., Aneshansley, D. J., & Kochian, L. V. (2011). Three-Dimensional Root Phenotyping with a Novel Imaging and Software Platform. *Plant Physiology, 156*(2), 2455–2465. doi:10.1104/pp.110.169102 PMID:21454799

Cline, D., Hofstetter, H. W., & Griffin, J. R. (2008). *Analysis of visual science* (4th ed.). Boston: Butterworth-Heinemann.

Cointault, F., Paindavoine, M., & Sarrazin, P. (2002). Fast imaging system for particle projection analysis: application to fertilizer centrifugal spreading. Journal Measurement Science and Technology, 13, 1087-1093.

Cointault, F., Guérin, D., Guillemin, J. P., & Chopinet, B. (2008). In-Field Wheat ears Counting Using Color-Texture Image Analysis. *New Zealand Journal of Crop and Horticultural Science, 36*, 117–130. doi:10.1080/01140670809510227

Cokes, D., & Lewis, P. (1969). *Stochastic analysis of chains of events*. Academic Press.

Colombo, D., Colosimo, B., & Previtali, B. (2013, January). Comparison of methods for data analysis in the remote monitoring of remote laser welding. *Optics and Lasers in Engineering, 51*(1), 34–46. doi:10.1016/j.optlaseng.2012.07.022

Corke, P. (2013). *Robotics, Vision and Control*. Berlin: Springer.

Cruz, C., Sucar, L. E., & Morales, E. F. (2008). Real–time face recognition for human–robot interaction.*Proceedings of 8th IEEE International Conference on Automatic Face & Gesture Recognition* (pp.1-6). IEEE Xplore Digital Library.

Cruz-Perez, C., Starostenko, O., Alarcon-Aquino, V., & Rodriguez-Asomoza, J. (2015). Automatic image annotation for description of urban and outdoor scenes. In T. Sobh & K. Elleithy (Eds.), *Innovations and advances in computing, informatics, systems sciences, networking and engineering* (vol. 313, pp.139-147). Retrieved from http://link.springer.com/book/10.1007%2F978-3-319-06773-5

Cunningham, I. A., & Shaw, R. (1999). Signal-to-noise optimization of medical imaging sytems. *JOSA A, 16*(3), 621–632. doi:10.1364/JOSAA.16.000621

Dagenais, M. a. (1995). *Integrated optoelectronics*. Academic Press.

Damelin, S. B. (2012). *The mathematics of signal processing*. Cambridge University Press.

Dasarathy, B. V. (1997). Sensor fusion potential exploitation-innovative architectures and illustrative applications. *Proceedings of the IEEE, 1*(85), 24–38. doi:10.1109/5.554206

Databases. (2015). *Face recognition home page*. Retrieved from http://www.face-rec.org/databases/

David, Y. (1999). *Digital pixel cmos image sensors*. Stanford University.

De Franco. (2007). Materials evaluation to obtain the better vision to optoelectronic sensors. McGraw Hill Ed.

De Hayes, D. H., Schaberg, P. G., & Strimbeck, G. R. (2001). Red Spruce Hardiness and Freezing Injury Susceptibility. In F. Bigras (Ed.), *Conifer Cold Hardiness*. Kluwer Academic Publishers. doi:10.1007/978-94-015-9650-3_18

Deguchi, J., Tachibana, F., Morimoto, M., Chiba, M., Miyaba, T., & Tanaka, H. K. T. (2013). A 187.5μVrms -Read-Noise 51mW 1.4Mpixel CMOS Image Sensor with PMOSCAP Column CDS and 10b Self-Differential Offset-Cancelled Pipeline SAR-ADC. Proceedings of the ISSCC 2013 (pp. 494–496)

Deshmukh, A. C., Gaikwad, P. R., & Patil, M. P. (2015). A comparative study of feature detection methods. *International Journal of Electrical and Electronic Engineering & Telecommunications.*, *4*(2), 1–7. Retrieved from http://ijeetc.com/ijeetcadmin/upload/IJEETC_5524b6e0c6527.pdf

Downie, H., Holden, N., Otten, W., Spiers, A. J., Valentine, T. A., & Dupuy, L. X. (2012). Transparent soil for imaging the rhizosphere. *PLoS ONE*, 7. PMID:22984484

Dresboll, D. B., Thorup-Kristensen, K., McKenzie, B. M., Dupuy, L. X., & Bengough, A. G. (2013). Timelapse scanning reveals spatial variation in tomato (*Solanum lycopersicum* L.) root elongation rates during partial waterlogging. *Plant and Soil*, *369*(1-2), 467–477. doi:10.1007/s11104-013-1592-5

El Gamal, A. (2002). Trends in CMOS Image Sensor Technology and Design. In *Electron Devices Meeting*, (pp. 805–808). doi:10.1109/IEDM.2002.1175960

Elfes, A. (1992). Multi-source spatial data fusion using Bayesian reasoning. *Data fusion in robotics and machine intelligence*, 137-163.

Erlank, A. O., & Steyn, W. H. (2014). Arcminute Attitude Estimation for CubeSats with a Novel Nano Star Tracker. *Proceedings of the19th World Congress The international Federation of Automatic Control* (pp. 9679–9684). Cape Town: IFAC. doi:10.3182/20140824-6-ZA-1003.00267

Faceli, K., de Carvalho, A. C. P. L. F., & Rezende, S. O. (2004). Combining intelligent techniques for sensor fusion. *Applied Intelligence*, *3*(20), 199–213. doi:10.1023/B:APIN.0000021413.05467.20

Fang, S., Yan, X., & Liao, H. (2009). 3D reconstruction and dynamic modeling of root architecture in situ and its application to crop phosphorus research. *The Plant Journal*, *60*(6), 1096–1108. doi:10.1111/j.1365-313X.2009.04009.x PMID:19709387

Fang, W., Huang, X., Zhang, F., & Li, D. (2015, February). Intensity Correction of Terrestrial Laser Scanning Data by Estimating Laser Transmission Function. *IEEE Transactions on Geoscience and Remote Sensing*, *53*(2), 942–951. doi:10.1109/TGRS.2014.2330852

Faro Focus. (2016). *Faro Focus 3D stationary laser scanner* [Online Image]. Retrieved from http://www.faro.com

Faro Freestyle. (2016). *Faro Hand-held Scanner Freestyle* [Online Image]. Retrieved from http://www.faro.com

Farrell, J. E., Xiao, F., Catrysse, P. B., & Wandell, B. A. (2004). A simulation tool for evaluating digital camera image quality. In *Electronic Imaging 2004* (pp. 124–131). International Society for Optics and Photonics.

Fedoseev, V. I. (2011). Priem prostranstvenno-vremennyh signalov v optiko-jelektronnyh sistemah (puassonovskaja model'). *Universitetskaja kniga, 232.*

Feng, Z. (2014). *Méthode de simulation rapide de capteur d'image CMOS prenant en compte les paramètres d'extensibilité et de variabilité.* Ecole Centrale de Lyon.

Fergusson, J. F. (2009). The principal heavy elements. In *Environmental and impact and health effects*. Pergamum Press.

Fiolhais, C., & Nogueira, F. (2003). A Primer in Density Functional Theory. Springer.

Fiorani, F., & Schurr, U. (2013). Future Scenarios for Plant Phenotyping. *Annual Review of Plant Biology*, *64*(1), 267–291. doi:10.1146/annurev-arplant-050312-120137 PMID:23451789

Fischer, R. (2003). Elektrische Maschinen (12th ed.). Carl Hanser Verlag GmbH & Co. KG.

Flores Fuentes, W., Rivas López, M., Srgiyenko, O., González Navarro, F. F., Rivera Castillo, J., & Hernández Balbuena, D. (2014). Machine Vision Supported by Artificial Intelligence Applied to Rotary Mirror Scanners. *Proceedings of the 2014 IEEE 23th International Symposium on Industrial Electronics* (pp. 1949-1954). Istambul, Turkey: ISIE. doi:10.1109/ISIE.2014.6864914

Flores-Fuentes, W. L.-N.-C.-B.-Q., Rivas-Lopez, M., Sergiyenko, O., Gonzalez-Navarro, F. F., Rivera-Castillo, J., Hernandez-Balbuena, D., & Rodríguez-Quiñonez, J. C. (2014a). Combined application of power spectrum centroid and support vector machines for measurement improvement in optical scanning systems. *Signal Processing*, *98*, 37–51. doi:10.1016/j.sigpro.2013.11.008

Flores-Fuentes, W. L.-Q.-B.-C., Rivas-Lopez, M., Sergiyenko, O., Rodriguez-Quinonez, J. C., Hernandez-Balbuena, D., & Rivera-Castillo, J. (2014b). Energy Center Detection in Light Scanning Sensors for Structural Health Monitoring Accuracy Enhancement. *Sensors Journal, IEEE*, *7*(14), 2355–2361. doi:10.1109/JSEN.2014.2310224

Flores-Pulido, L., Starostenko, O., Contreras-Gómez, R., & Alvarez-Ochoa, L. (2008). Wavelets families and similarity metrics analysis in VIR system design. *WSEAS Transactions on Information Science and Applications Journal.*, *5*(4), 436–448.

Forsyth, D. A., & Ponce, J. (2002). *Computer Vision: A modern Approach.* Prentice Hall Professional Technical Reference.

Fouda, Y. M. (2015). A robust template matching algorithm based on reducing dimensions. *Journal of Signal and Information Processing.*, *6*(02), 109–122. doi:10.4236/jsip.2015.62011

Fowler, B., Balicki, J., How, D., & Godfrey, M. (2001). Low FPN High Gain Capacitive Transimpedance Amplifier for Low Noise CMOS Image Sensors. *Proceedings of the SPIE*(Vol. 4306). doi:10.1117/12.426991

Franca, J. G. (2005). A 3D scanning system based on laser triangulation and variable field of view. *Image Processing,2005. ICIP 2005. IEEE International Conference on*. IEEE.

Frank, U., Giese, H., Klein, F., Oberschelp, O., Schmidt, A., Schulz, B., et al. (2004). *Selbstoptimierende Systeme des Maschinenbaus - Definitionen und Konzepte* (Vol. 155). Padeborn: W. V. Westfalia Druck.

Fu, Q., Zhan, W., Lin, Q., & Wu, N. (2009). A Novel CMOS Color Pixel for Vision Chips. *Proceedings of the IEEE Sensors 2009 Conference*. doi:10.1109/ICSENS.2009.5398508

Fukuda, M. (1999). *Optical semiconductor devices*. John Wiley and Sons.

Gal'jardi, R., & Karp, S. (1978). Opticheskaja svjaz. Moscow: Svjaz.

Gao, W., Cao, B., Shan, S., & Chen, X. (2008). The CAS-PEAL large-scale Chinese face database and baseline evaluations. *IEEE Transactions on Systems, Man, and Cybernetics. Part A, Systems and Humans*, *38*(1), 149–161. doi:10.1109/TSMCA.2007.909557

Garbout, A., Munkholm, L. J., Hansen, S. B., Petersen, B. M., Munk, O. L., & Pajor, R. (2012). The use of PET/CT scanning technique for 3D visualization and quantification of real-time soil/plant interactions. *Plant and Soil*, *352*(1-2), 113–127. doi:10.1007/s11104-011-0983-8

Garcia-Amaro, E., Nuño-Maganda, M., & Morales-Sandoval, M. (2012). Evaluation of machine learning techniques for face detection and recognition. *Proceedings of 22nd International Conference on Electrical Communications and Computers* (pp. 213–218). IEEE Xplore Digital Library. doi:10.1109/CONIELECOMP.2012.6189911

Garcia-Cruz, X., Sergiyenko, O., Tyrsa, V., Rivas-Lopez, M., Hernandez-Balbuena, D., Rodríguez-Quinoñez, J., & Mercorelli, P. (2014, March). Optimization of 3D laser scanning speed by use of combined variable step. *Optics and Lasers in Engineering*, *54*, 141–151. doi:10.1016/j.optlaseng.2013.08.011

Garcia-Talegon, J., Calabres, S., Fernandez-Lozano, J., Inigo, A., Herrero-Fernandez, H., Arias-Perez, B., & Gonzalez-Aguilera, D. (2015). Assessing Pathologies On Villamayor Stone (Salamanca, Spain) By Terrestrial Laser Scanner Intensity Data. *Remote Sensing and Spatial Information Sciences, XL-5/W4*, 445-451.

Gascon-Moreno, J., Ortiz-Garcia, E., Salcedo-Sanz, S., Paniagua-Tineo, A., Saavedra-Moreno, B., & Portilla-Figueras, J. (2011). Multi-parametric Gaussian Kernel Function Optimization for ε-SVMr Using a Genetic Algorithm. In J. Cabestany, I. Rojas, & G. Joya (Eds.), *Advamces in Computational Intelligence* (pp. 113–120). Berlin: Springer. doi:10.1007/978-3-642-21498-1_15

Gauthier, J.-P., Bornard, G., & Silbermann, M. (1991). Motions and pattern analysis: Harmonic analysis on motion groups and their homogeneous spaces. *Systems, Man and Cybernetics. IEEE Transactions on*, *21*(1), 159–172.

Gil, R. J., & Martin.Bautista, M. J. (2012). A novel integrated knowledge support system based on ontology learning: Model specification and a case study. *Knowledge-Based Systems*, *36*, 340–352. doi:10.1016/j.knosys.2012.07.007

Ginhac, D., Dubois, J., Heyrman, B., & Paindavoine, M. (2009). A high speed programmable focal-plane SIMD vision chip. *Analog Integrated Circuits and Signal Processing*, *65*(3), 389–398. doi:10.1007/s10470-009-9325-7

Girit, B., Zettl, C., Crommie, A., & Segalman, M. F. (2012). The optoelectronic devices. An overview of the use to detect microorganisms. *Journal of Microbiology (Seoul, Korea)*, *8*(3), 23–45.

Gonzalez-Aguilera, D. L.-G. (2010). *Camera and laser robust integration in engineering and architecture applications*. INTECH Open Access Publisher. doi:10.5772/9959

Gonzalez, R. C., Woods, R. E., & Eddins, S. L. (2009). *Digital Image processing using MATLAB* (3rd ed.). Gatesmark Pub.

Goy, J., Courtois, B., Karam, J. M., & Pressecq, F. (2001). Design of an APS CMOS Image Sensor for Low Light Level Applications Using Standard CMOS Technology. *Analog Integrated Circuits and Signal Processing*, *29*(1/2), 95–104. doi:10.1023/A:1011286415014

Granger, E., Radtke, P., & Gorodnichy, D. (2014). *Survey of academic research and prototypes for face recognition in video*. CBSA Science and Engineering Directorate, Division Report, 25. Retrieved from http://pubs.drdc-rddc.gc.ca/BASIS/pcandid/www/engpub/DDW?W%3DSYSNUM=800522

Granlund, G. H. (1972). Fourier Preprocessing for Hand Print Character Recognition. *IEEE Transactions on Computers*, *C-21*(2), 195–201. doi:10.1109/TC.1972.5008926

Gray, P. R., Hurst, P. J., Lewis, S. H., & Meyer, R. (2010). *Analysis and Design of Analog Integrated Circuits (5th ed.)*. Wiley.

Grayscale Panoramic Image. (n.d.). Retrieved from http://www.iff.fraunhofer.de

Gunn, S. R. (1998). *Support vector machines for classification and regression*. University of Southhampton.

Hall, D. L., & Llinas, J. (1997). An introduction to multisensor data fusion. *Proceedings of the IEEE*, *1*(85), 6–23. doi:10.1109/5.554205

Hamami, S., Fleshel, L., Yadid-pecht, O., & Driver, R. (2004). CMOS Aps Imager Employing 3.3V 12 bit 6.3 ms/s pipelined ADC. *Proceedings of the 2004 International Symposium onCircuits and Systems ISCAS'04*.

Han, S., Bujoreanu, D., Cointault, F., & Rousseau, D. (2015). *Graph-based denoising of skeletonized root-systems*. Paper presented at the 4th edition of the International Workshop on Image Analysis Methods for the Plant Sciences (IAMPS), Louvain-la-Neuve, Belgium.

Hargreaves, C., Gregory, P., & Bengough, A. G. (2009). Measuring root traits in barley (*Hordeum vulgare* ssp. *vulgare* and ssp. *spontaneum*) seedlings using gel chambers, soil sacs and x-ray microtomography. *Plant and Soil, 316*(1-2), 285–297. doi:10.1007/s11104-008-9780-4

Harte, J., Holden, C., Scheneider, R., & Shirey, C. (2008). Principal toxics chemical elements as heavy metals. McGraw Hill.

Hartley, R. (2013). *Multiple View Geometry in Computer Vision.* Cambridge University Press.

Hecht, D. S., Hu, L., & Irvin, G. (2014). Beneficial organisms for agriculture and sustainable environment. *Journal of Environment and Technology*, *7*(4), 56–72.

Hidayat, E., Fajrian, N. A., Muda, N. A., Huoy, C. Y., & Ahmad, S. (2011). A comparative study of feature extraction using PCA and LDA for face recognition.*Proceedings of 7th International Conference on Information Assurance and Security* (pp. 354-359). doi:10.1109/ISIAS.2011.6122779

Higgins, S., Lyons, T. M., & Doherty, P. E. (2011). Optoelectronics systems in the food industry. *Journal of Agriculture and Food Operations*, *13*(5), 99–121.

Hijazi, B., Cool, S., Cointault, F., Vangeyte, J., Mertens, K., Paindavoine, M., & Pieters, J. (2014). High speed stereovision setup for position and motion estimation of fertilizer particles leaving a centrifugal spreader. *Sensors (Basel, Switzerland)*, *14*(11), 21466–21482. doi:10.3390/s141121466 PMID:25401688

Hijazi, B., Vangeyte, J., Cointault, F., Dubois, J., Coudert, S., Paindavoine, M., & Pieters, J. (2011). Two-steps cross-correlation based algorithm for motion estimation applied to fertilizer granule motion during centrifugal spreading. *Optical Engineering (Redondo Beach, Calif.)*, *50*. doi:10.1117/1.3582859

Hinks, T., Carr, H., Gharibi, H., & Laefer, D. (2015, June). Visualisation of urban airborne laser scanning data with occlusion images. *ISPRS Journal of Photogrammetry and Remote Sensing*, *104*, 77–87. doi:10.1016/j.isprsjprs.2015.01.014

Hiremath, S., van der Heijden, G., van Evert, F., Stein, A., & ter Braak, C. (2014, January). Laser range finder model for autonomous navigation of a robot in a maize field using a particle filter. *Computers and Electronics in Agriculture*, *100*, 41–50. doi:10.1016/j.compag.2013.10.005

Hitomi, K. (1996). *Manufacturing Systems Engineering: A Unified Approach to Manufacturing Technology, Production Management, and Industrial Economics* (2nd ed.). London: Taylor & Francis.

Hong-Seok, P., & Chintal, S. (2015). Development of High Speed and High Accuracy 3D Dental Intra Oral Scanner. *Proceedings of the25th DAAAM International Symposium on Intelligent Manufacturing and Automation.* Elsevier. doi:10.1016/j.proeng.2015.01.481

Hoon Sohn, C. F. (2002). *A Review of Structural Health Monitoring Literature 1996-2001.* Retrieved from www.lanl.gov.projects/damage

Hou, B. (2011). Charge-coupled devices combined with centroid algorithm for laser beam deviation measurements compared to a position-sensitive device. *Optical Engineering (Redondo Beach, Calif.)*, *50*(3), 033603–033603. doi:10.1117/1.3554379

Huang, Z., Li, W., Wang, J., & Zhang, T. (2015). Face recognition based on pixel-level and feature-level fusion of the top-level's wavelet sub-bands. *Information Fusion*, *22*, 95–104. doi:10.1016/j.inffus.2014.06.001

Hug, C., & Wehr, A. (1997). Detecting And Identifying Topographic Objects In Imaging Laser Altimeter Data. *IAPRS*, *32*, 19-26.

Hund, A., Ruta, N., & Liedgens, M. (2009). Rooting depth and water use efficiency of tropical maize inbred lines, differing in drought tolerance. *Plant and Soil*, *318*(1-2), 311–325. doi:10.1007/s11104-008-9843-6

Hurwitz, J., Smith, S., & Murray, A. A. (2001). *A Miniature Imaging Module for Mobile Applications*. ISSCC. doi:10.1109/ISSCC.2001.912559

Hwang, S. H. (2012). *CMOS image sensor: current status and future perspectives*. Retrieved from http://www.techonline.com

Innocent, M. (n. d.). *General introduction to CMOS image sensors*. Academic Press.

IPC, Association Connecting Electronics Industries. (2014, February 1). *Industry Data: Trends New Orders*. Retrieved from http://www.ipc.org/3.0_Industry/3.1_Industry_Data/2014/Trends-New-Orders-0214.pdf

Iyer-Pascuzzi, A. S., Symonova, O., Mileyko, Y., Hao, Y., Belcher, H., Harer, J., & Benfey, P. N. et al. (2010). Imaging and analysis platform for automatic phenotyping and trait ranking of plant root systems. *Plant Physiology*, *152*(3), 1148–1157. doi:10.1104/pp.109.150748 PMID:20107024

Jacquemoud, S., Verhoef, W., Baret, F., Bacour, C., Zarco-Tejada, P. J., Asner, G. P., & Ustin, S. L. et al. (2009). PROSPECT + SAIL models: A review of use for vegetation characterization. *Remote Sensing of Environment*, *113*, 56–66. doi:10.1016/j.rse.2008.01.026

Janesick, J. R. (2001). *Scientific charge-coupled devices*. SPIE Press.

Jay, S., Rabatel, G., Hadoux, X., Moura, D., & Gorretta, N. (2015). In-field crop row phenotyping from 3D modeling performed using Structure from Motion. *Computers and Electronics in Agriculture*, *110*, 70–77. doi:10.1016/j.compag.2014.09.021

Johnson, D. H. (2013). *Statistical signal processing*. Retrieved from http://www.ece.rice.edu/~dhj/courses/elec531/notes. pdf

Ju, L., Geng, B., Horng, J., & Girit, C. (2010). The optoelectronics in the recovering of soils. *Journal of Electronics Industry*, *7*(3), 56–72.

July, T., & Deane, B. (2009). Color analysis of the optical vision. In Applied Optics (3rd ed.). New York: Wiley-Interscience.

Kamencay, P., Hudec, R., Benco, M., & Zachariasova, M. (2014). 2D-3D face recognition method based on a modified cca-pca algorithm. *International Journal of Advanced Robotic Systems*, *11*(36), 1–8. doi:10.5772/58251

Kennedy, W. P. (2012). *The Basics of triangulation sensors.* Retrieved from http://archives.sensorsmag.com/articles/0598/tri0598/main.shtml

Kerle, N., Stump, A., & Malet, J.-P. (2010). Object oriented and cognitive methods for multidata event-based landslide detection and monitoring. *Landslide Monitoring Technologies & Early Warning Systems.*

Kim, P., Yoo, I., & Chung, H. (2010). Trichoderma fungus use to recovery soils detected by optoelectronics systems. In *Use of optoelectronics in the agriculture* (10th ed.). Elsevier Books.

Klein, L. A. (2003). Sensor and data fusion: a tool for information assessment and decision making. Bellingham, WA: SPIE Press.

Kleinfelder, S., Lim, S., Liu, X., & El Gamal, A. (2001). A 10 000 Frames/s CMOS Digital Pixel Sensor. *IEEE Journal of Solid-State Circuits*, *36*(12), 2049–2059. doi:10.1109/4.972156

Kohavi, R. (1995). A Study of Cross-Validation and Bootstrap for Accuracy Estimation and Model Selection.*Proceedings of the 14th international joint conference on Artificial intelligence.*

Krippner, P., & Beer, D. (2004, April 1). AOI testing positions in comparison. In *Viscom Vision Technology*. Hanover, Germany: Viscom AG.

Kumar, P., McElhinney, C., Lewis, P., & McCarthy, T. (2013, November). An automated algorithm for extracting road edges from terrestrial mobile LiDAR data. *ISPRS Journal of Photogrammetry and Remote Sensing*, *85*, 44–55. doi:10.1016/j.isprsjprs.2013.08.003

Labayrade, R., Royere, C., Gruyer, D., & Aubert, D. (2005). Cooperative fusion for multi-obstacles detection with use of stereovision and laser scanner. *Autonomous Robots*, *2*(19), 117–140. doi:10.1007/s10514-005-0611-7

Langfelder, G., Longoni, A., & Zaraga, F. (2009). Further developments on a novel Color sensitive CMOS detector.*Proc. of SPIE*(Vol. 7356). doi:10.1117/12.822291

Le Bot, J., Serra, V., Fabre, J., Draye, X., Adamowicz, S., & Pages, L. (2010). DART: A software to analyse root system architecture and development from captured images. *Plant and Soil*, *326*(1-2), 261–273. doi:10.1007/s11104-009-0005-2

Le Marié, C., Kirchgessner, N., Marschall, D., Walter, A., & Hund, A. (2014). Rhizoslides: Paper-based growth system for non-destructive, high throughput phenotyping of root development by means of image analysis. *Plant Methods*, *10*(13). PMID:25093035

Leitner, D., Felderer, B., Vontobel, P., & Schnepf, A. (2014). Recovering root system traits using image analysis exemplified by two-dimensional neutron radiography images of lupine. *Plant Physiology*, *164*(1), 24–35. doi:10.1104/pp.113.227892 PMID:24218493

Lienhart, R., & Maydt, J. (2002). An extended set of Haar-like features for rapid object detection. *Proceedings of IEEE International Conference on Image Processing* (Vol. 1, pp. 900–903). Retrieved from http://www.lienhart.de/Prof._Dr._Rainer_Lienhart/Source_Code_files/ICIP2002.pdf

Li, F. a. (2006). *CCD image sensors in deep-ultraviolet: degradation behavior and damage mechanisms*. Springer Science and Business Media.

Li, F. M. (2003). Degradation behavior and damage mechanisms of CCD image sensor with deep-UV laser radiation. *Electron Devices. IEEE Transactions on, 12*(51), 2229–2236.

Li, L., Ma, G., & Du, X. (2013). Edge detection in potential-field data by enhanced mathematical morphology filter. *Pure and Applied Geophysics, 4*(179), 645–653. doi:10.1007/s00024-012-0545-x

Lim, K., Lee, J. C., Panotopulos, G., & Heilbing, R. (2006). Illumination and Color Management in Solid state lighting. *Proceedings of the 41th IAS Annual Meeting of theIEEE Industry Applications Conference*. doi:10.1109/IAS.2006.256908

Lindgren, L., Melander, J., Johansson, R., & Möller, B. (2005). A Multiresolution 100-GOPS 4-Gpixels/s Programmable Smart Vision Sensor for Multisense Imaging. *IEEE Journal of Solid-State Circuits, 40*(6), 1350–1359. doi:10.1109/JSSC.2005.848029

Lindner, L., Sergiyenko, O., Rodríguez-Quinoñez, J., Tyrsa, V., Mercorelli, P., Fuentes-Flores, W., . . . Nieto-Hipolito, J. (2015). Continuous 3D scanning mode using servomotors instead of stepping motors in dynamic laser triangulation. *Industrial Electronics (ISIE), 2015 IEEE 24th International Symposium on* (pp. 944-949). Buzios: IEEE.

Lindner, L., Sergiyenko, O., Tyrsa, V., & Mercorelli, P. (2014, June 01-04). An approach for dynamic triangulation using servomotors. *Proceedings of the 2014 IEEE 23rd International Symposium on Industrial Electronics (ISIE)* (pp. 1926-1931). Istanbul: IEEE.

Lin, S.-c., & Su, C.-h. (2006). A Visual Inspection System for Surface Mounted Devices on Printed Circuit Board.*IEEE Conference on Cybernetics and Intelligent Systems* (pp. 1-4). Bangkok: IEEE. doi:10.1109/ICCIS.2006.252237

Litex Electricals Pvt. Ltd . (2015). Retrieved from http://www.litexelectricals.com/halogen_lamp_characteristics.asp

Liu, X., Tan, X., & Chen, S. (2012). Eyes Closeness detection using appearance based methods. In Z. Shi, D. Leake, & S. Vadera (Eds.), *IFIP advances in information and communication technology* (Vol. 385, pp. 398–408). Springer Berlin Heidelberg. doi:10.1007/978-3-642-32891-6_49

Li, X., & Magnuson, C. W. (2011). Optoelectronic devices used in food industries to detect microorganisms. *Journal of Applied Microbiology, 14*(7), 78–95.

Li, X., Zhao, H., Liu, Y., Jiang, H., & Bian, Y. (2014). Laser scanning based three dimensional measurement of vegetation canopy structure. *Optics and Lasers in Engineering, 54*, 152–158. doi:10.1016/j.optlaseng.2013.08.010

Li, Y., Qiao, Y., & Ruichek, Y. (2015). Multiframe-Based High Dynamic Range Monocular Vision System for Advanced Driver. *IEEE Sensors Journal*, *15*(10), 5433–5441. doi:10.1109/JSEN.2015.2441653

Lobet, G., Draye, X., & Périlleux, C. (2013). An online database for plant image analysis software tools. *Plant Methods*, *9*(38). PMID:24107223

Lobet, G., Pagès, L., & Draye, X. (2011). A Novel Image Analysis Toolbox Enabling Quantitative Analysis of Root System Architecture. *Plant Physiology*, *157*(1), 29–39. doi:10.1104/pp.111.179895 PMID:21771915

Lobet, G., Poun, M., Diener, J., Prada, C., Draye, X., Godin, C., & Schnepf, A. et al. (2015). Root System Markup Language: Toward a unified root architecture description language. *Plant Physiology*, *167*(3), 617–627. doi:10.1104/pp.114.253625 PMID:25614065

Loinaz, M. J., Singh, K. J., Blanksby, A. J., Member, S., Inglis, D. A., Azadet, K., & Ackland, B. D. (1998). A 200-mW, 3.3-V, CMOS color Camera IC producing 352x288 24-b Video at 30 Frames/s. *IEEE Journal of Solid-State Circuits*, *33*(12), 2092–2103. doi:10.1109/4.735552

López, B.G., Valdez, S.B., Zlatev, K.R., Flores, P.J., Carrillo, B.M., & Schorr, W.M. (2007). *Corrosion of Metals at Indoor Conditions in the Electronics Manufacturing Industry*. Anti-Corrosion Methods and Materials.

López,, B.G., Valdez,, S.B., & Schorr,, W. M., Zlatev, R., Tiznado, V.H., Soto, H.G., & De la Cruz, W. (2011). AES in Corrosion of Electronic Devices in Arid in Marine Environments. *Anti-Corrosion Methods and Materials*, *4*(2), 134–149.

López, B. G., Valdez, S. B., Schorr, W. M., Tiznado, V. H., & Soto, H. G. (2010). *Influence of Climate Factors on Copper Corrosion in Electronic Equipment and Devices*. Anti-Corrosion Methods and Materials.

López-Badilla, G. (2008). *Caracterización de la corrosión en materiales metálicos de la industria electrónica en Mexicali*. PhD Thesis.

López, G., Tiznado, H., Soto, G., De la Cruz, W., Valdez, B., Schorr, M., & Zlatev, R. (2010). *Corrosión de dispositivos electrónicos por contaminación atmosférica en interiores de plantas de ambientes áridos y marinos*. Nova Scientia. (in Spanish)

Lorenser, D. A. (2003). Towards wafer-scale integration of high repetition rate passively mode-locked surface-emitting semiconductor lasers. *Applied Physics. B, Lasers and Optics*, *8*(79), 927–932.

Louie, S. G., & Wang, F. (2011). The electronic devices in the agricultural operations. *Journal of Agronomy Engineering*, *11*(4), 67–84.

Lowe, D. (2004). Distinctive image features from scale-invariant keypoints. *International Journal of Computer Vision, 60*(2), 91–110. Retrieved rom http://link.springer.com/article/10.1023/B:VISI.0000029664.99615.94

Lowe, D. G. (1999). Object recognition from local scale-invariant features.*Proceedings of the International Conference on Computer Vision.*

Lu, G.-F., Zou, J., & Wang, Y. (2012). Incremental learning of complete linear discriminant analysis for face recognition. *Knowledge-Based Systems, 31*, 19–27. doi:10.1016/j.knosys.2012.01.016

Lulé, T., Benthien, S., Keller, H., Mütze, F., Rieve, P., Seibel, K., & Böhm, M. (2000). Sensitivity of CMOS Based Imagers and Scaling Perspectives. *IEEE Transactions on Electron Devices, 47*(11), 2110–2122. doi:10.1109/16.877173

Luo, B., Gao, Y., Sun, Z., & Zhao, S. (2013). SMT Components Model Inspection Based on Characters Image Matching and Verification.*IEEE International Conference on Green Computing and Communications and IEEE Internet of Things and IEEE Cyber, Physical and Social Computing*. doi:10.1109/GreenCom-iThings-CPSCom.2013.252

Luo, R. C.-C., & Yih, . (2002). Multisensor fusion and integration: Approaches, applications, and future research directions. *Sensors Journal, IEEE, 2*(2), 107–119. doi:10.1109/JSEN.2002.1000251

Lytyuga, A. (2009). Mathematical model of signals in television systems with low-orbit space objects observation in daytime. *Collected Works of Kharkiv University of Air Force, 4*(22), 41–46.

Lyu, T., Yao, S., Nie, K., & Xu, J. (2014). A 12-bit high-speed column-parallel two-step single-slope analog-to-digital converter (ADC) for CMOS image sensors. *Sensors (Basel, Switzerland), 14*(11), 21603–21625. doi:10.3390/s141121603 PMID:25407903

Ma, G., Zhao, B., & Fan, Y. (2013, May). Non-diffracting beam based probe technology for measuring coordinates of hidden parts. *Optics and Lasers in Engineering, 51*(5), 585–591. doi:10.1016/j.optlaseng.2012.12.011

Mahmood, A., & Khan, S. (2012). Correlation coefficient based fast template matching through partial elimination. *IEEE Transactions on Image Processing, 21*(4), 2099–2108. doi:10.1109/TIP.2011.2171696 PMID:21997266

Mahmoodi, A., & Joseph, D. (2008). Pixel-Level Delta-Sigma ADC with Optimized Area and Power for Vertically-Integrated Image Sensors. *Proceedings of the 51st Midwest Symposium on Circuits and Systems MWSCAS '08* (pp. 41–44). doi:10.1109/MWSCAS.2008.4616731

Mairhofer, S., Zappala, S., Tracy, S., Sturrock, C., Bennett, M., Mooney, S., & Pridmore, T. (2013). Recovering complete plant root system architectures from soil via X-ray mu-Computed Tomography. *Plant Methods, 9*(8). PMID:23514198

Majumdar, A. K., & Ricklin, J. C. (2008). *Free-Space Laser Communications, Principles and Advances*. New York: Springer. doi:10.1007/978-0-387-28677-8

Malamy, J. E. (2005). Intrinsic and environmental response pathways that regulate root system architecture. *Plant, Cell & Environment, 28*(1), 67–77. doi:10.1111/j.1365-3040.2005.01306.x PMID:16021787

Manschadi, A. M., Kaul, H. P., Vollmann, J., Eitzinger, J., & Wenzel, W. (2014). Reprint of 'Developing phosphorus-efficient crop varieties. An interdisciplinary research framework'. *Field Crops Research*, *165*, 49–60. doi:10.1016/j.fcr.2014.06.027

Mars, U., & Johnson, G. (2007). Evolution, techniques, industrial operations with optoelectronic devices with chromatic systems. Mc Graw Hill Ed.

Marston, R. M. (1999). *Optoelectronics circuits manual*. Butterworth-Heinemann.

Marzuki, A., Pang, K. L., & Lim, L. (2007). *Method and Apparatus for Integrating a Quantity of Light*. U.S. Patent 2007/0235632 A1.

Marzuki, A., Abdul Aziz, Z. A., & Abd Manaf, A. (2011). *A Review of CMOS Analog Circuits for Image Sensing Application. Proceedings of the 2011 IEEE international conference on Imaging systems and techniques*. Penang: IEEE.

Massari, N., Gottardi, M., Gonzo, L., Stoppa, D., & Simoni, A. (2005). A CMOS Image Sensor with Programmable Pixel-Level Analog Processing. *IEEE Transactions on Neural Networks*, *16*(6), 1673–1684. doi:10.1109/TNN.2005.854369 PMID:16342506

McClure, W. F. (2003). 204 years of near infrared technology: 1800-2003. *Journal of Near Infrared Spectroscopy*, *6*(11), 487–518. doi:10.1255/jnirs.399

Melexis. (n. d.). *MLX75412BD, Avocet HDR Image sensors*. Retrieved from www.melexis.com

Mello, A. R., & Stemmer, M. R. (2015). Inspecting Surface Mounted Devices Using K Nearest Neighbor and Multilayer Perceptron.*International Symposium on Industrial Electronics*. doi:10.1109/ISIE.2015.7281599

Mendenhall, J. A., Candell, L. M., Hopman, P. I., Zogbi, G., Boroson, D. M., Caplan, D. O., . . . Shoup, R. C. (2007). Design of an Optical Photon Counting Array Receiver System for Deep-Space Communications. *Proceedings of the IEEE, 5*(1), 2059-2069.

Metzner, R., Eggert, A., van Dusschoten, D., Pflugfelder, D., Gerth, S., Schurr, U., & Jahnke, S. (2015). Direct comparison of MRI and X-ray CT technologies for 3D imaging of root systems in soil: Potential and challenges for root trait quantification. *Plant Methods*, *11*(1), 17. doi:10.1186/s13007-015-0060-z PMID:25774207

Meva, D. T., & Kumbharana, C. K. (2014). Study of different trends and techniques in face recognition. *International Journal of Computers and Applications*, *96*(8), 1–4. doi:10.5120/16811-6548

Meynants, G. (2010). *Global Shutter Image Sensors for Machine*. ISE.

Michoud, , M. J.-H. (2012). New classification of landslide-inducing anthropogenic activities. *Geophysical Research Abstracts*.

MicMac. (n.d.). IGN. Retrieved from http://logiciels.ign.fr/?-Micmac,3-

Minervini, M., Scharr, H., & Tsaftaris, S. A. (2015). Image analysis: The new bottleneck in plant phenotyping. *IEEE Signal Processing Magazine*, *32*(4), 126–131. doi:10.1109/MSP.2015.2405111

Ming-Chih, L., Su-Chin, C., Yu-Shen, L., & Cheng-Pei, T. (2015, March). *Machine Vision in the realization of Landslide monitoring applications*. Retrieved from INTERPRAEVENT: http://www.interpraevent.at/?tpl=startseite.php&menu=41

Moeini, A., Moeini, H., & Faez, K. (2015). Unrestricted pose-invariant face recognition by sparse dictionary matrix. *Image and Vision Computing*, *36*, 9–22. doi:10.1016/j.imavis.2015.01.007

Mooney, S. J., Pridmore, T. P., Helliwell, J., & Bennett, M. J. (2012). Developing X-ray Computed Tomography to non-invasively image 3-D root systems architecture in soil. *Plant and Soil*, *352*(1-2), 1–22. doi:10.1007/s11104-011-1039-9

Morandi, P., Breman, F., Doumalin, P., Germaneau, A., & Dupre, J. (2014, July). New Optical Scanning Tomography using a rotating slicing for time-resolved measurements of 3D full field displacements in structures. *Optics and Lasers in Engineering*, *58*, 85–92. doi:10.1016/j.optlaseng.2014.02.007

Morrinsense, & Lewis, D. (2010). Laser epithelial evaluated a chromatic systems. John Wiley and Sons Ed.

Motorola. (2016). *Motorola LI4278 Barcode Reader* [Online Image]. Retrieved from https://portal.motorolasolutions.com

N. G., Jr., & Larson, D. R. (2009). Reference Ballistic Chronograph. *Optical Engineering Journal*, *48*(4), 43602-43607.

Nagel, K. A., Kastenholz, B., Jahnke, S., van Dusschoten, D., Aach, T., Mühlich, M., & Schurr, U. et al. (2009). Temperature responses of roots: Impact on growth, root system architecture and implications for phenotyping. *Functional Plant Biology*, *36*(11), 947–959. doi:10.1071/FP09184

Nagel, K. A., Putz, A., Gilmer, F., Heinz, K., Fischbach, A., Pfeifer, J., & Schurr, U. et al. (2012). GROWSCREEN-Rhizo is a novel phenotyping robot enabling simultaneous measurements of root and shoot growth for plants grown in soil-filled rhizotrons. *Functional Plant Biology*, *39*(11), 891–904. doi:10.1071/FP12023

Nappi, M., Riccio, D., & Wechsler, H. (2015). Robust face recognition after plastic surgery using region-based approaches. *Pattern Recognition*, *48*(4), 1261–1276. doi:10.1016/j.patcog.2014.10.004

Nikitin, V. M., Fomin, V. N., Borisenkov, A. I., Nikolaev, A. I., & Borisenkov, I. L. (2008). *Adaptive noise protection of optical-electronic information systems*. Belgorod.

Nirmalraj, P. N., & Blau, W. J. (2013). The presence of microbial in soils by the satured pollution. *Journal of Environment and Industry*, *8*(3), 32–47.

NOAA. (2015). Retrieved from http://oceanservice.noaa.gov/facts/lidar.html

Norsworthy, S. R., Schreier, R., & Temes, G. C. (Eds.), (1997). *Delta-sigma data converters: theory, design, and simulation* (Vol. 97). New York: IEEE press.

Nowak, R. D., & Kolaczyk, E. D. (2000). A statistical multiscale framework for Poisson inverse problems. *Information Theory. IEEE Transactions on*, *46*(5), 1811–1825.

O'Regan, B. C., Barnes, P. R. F., Li, X., Law, C., Palomares, E., & Marin-Beloqui, J. M. (2015). Optoelectronic Studies of Methyl ammonium Lead Iodide Perovskite Solar Cells with Mesoporous TiO_2: Separation of Electronic and Chemical Charge Storage, Understanding Two Recombination Lifetimes, and the Evolution of Band Offsets during J–V Hysteresis. *Journal of the American Chemical Society*, *137*(15), 5087–5099. doi:10.1021/jacs.5b00761 PMID:25785843

Omnivision. (n. d.). *OV7740, 1/5" CMOS VGA (640 x 480) CameraChip™ sensor*. Retrieved from http://www.ovt.com

Optic, E. (2014). *Imaging Electronics 101: Understanding Camera Sensors for Machine Vision Applications*. Retrieved from http://www.edmundoptics.com/technical-resources-center/imaging/understanding-camera-sensors-for-machine-vision-applications/

Oresjo, S. (2009). *Results from 2007 Industry Defect Level and Test Effectiveness Studies*. APEX and Designers Summit.

ORL database. (2015). *Four face databases in Matlab format: Algorithms*. Retrieved from http://www.cad.zju.edu.cn/home/dengcai/Data/FaceData.html

Otsu, N. (1979). A Threshold Selection Method from Gray-Level Histograms. *IEEE Transactions on Systems, Man, and Cybernetics*, 62–66.

Pace, J., Lee, N., Naik, H. S., Ganapathysubramanian, B., & Lubberstedt, T. (2014). Analysis of maize (*Zea mays* L.) seedling roots with the highthroughput image analysis tool ARIA (Automatic Root Image Analysis). *PLoS ONE*, 9.

Parmar, D., & Mehta, B. (2013). Face recognition methods & applications. *International Journal on Computer Technology & Applications*, *4*(1), 84–86.

Parr, R. G., & Weitao, Y. (2004). *Density-Functional Theory of Atoms and Molecules in metals*. Oxford, UK: Oxford University Press.

Passeri, D., Placidi, P., Verducci, L., Pignatel, G. U., Ciampolini, P., Matrella, G., & Marras, A., and G. M. B. (2002). Active Pixel Sensor Architectures in Standard CMOS Technology for Charged-Particle Detection Technology analysis.*Proceedings of PIXEL 2002 International Workshop on Semiconductor Pixel Detectors for Particles and X-rays*.

Peña, J. M. S., Marcos, C., Fernández, M. Y., & Zaera, R. (2007). Cost-Effective Optoelectronic System to Measure the Projectile Velocity in High-Velocity Impact Testing of Aircraft and Spacecraft Structural Elements. *Optical Engineering*, *46*(5), 510141-510146.

Pfeifer, T., Schmitt, R., Pavim, A., Stemmer, M., Roloff, M., & Schneider, C. (2010). Cognitive Production Metrology: A new concept for flexibly attending the inspection requirements of small series production.*Proceedings of the 36th International MATADOR Conference*. doi:10.1007/978-1-84996-432-6_81

Phillips, P. J. (2015). *The FERET database and evaluation procedure for face recognition algorithms, FERET documents*. Retrieved from http://www.nist.gov/itl/iad/ig/feret-docs.cfm

Piprek, J. (2003). *Semiconductor optoelectronic devices: introduction to physics and simulation*. Academic Press.

Plaza, J. P. (2009). Multi-Channel Morphological Profiles for Classification of Hyperspectral Images Using Support Vector Machines. *Sensors (Basel, Switzerland)*, 197. PMID:22389595

Poplin, D. (2006). An Automatic Flicker Detection Method for Embedded Camera Systems. *IEEE Transactions on Consumer Electronics*, *52*(2), 308–311. doi:10.1109/TCE.2006.1649642

Quaglia, A., & Epifano, C. M. (2012). *Face recognition: methods, applications and technology*. Commack, NY: Nova Science Publishers, Inc.

Raducanu, B., Vitria, J., & Leonardis, A. (2010). Online pattern recognition and machine learning techniques for computer-vision: Theory and applications. *Image and Vision Computing*, *28*(7), 1063–1064. doi:10.1016/j.imavis.2010.03.007

Rascher, U., Blossfeld, S., Fiorani, F., Jahnke, S., Jansen, M., Kuhn, A. J., & Schurr, U. et al. (2011). Non-invasive approaches for phenotyping of enhanced performance traits in bean. *Functional Plant Biology*, *38*(12), 968–983. doi:10.1071/FP11164

Raylase. (2016). *Raylase Miniscan II* [Online Image]. Retrieved from http://www.raylase.de

Rentschler, K., & Moser, M. (1996). Geotechniques of claystone hillslopes-properties of weathered claystone and formation of sliding surfaces. *7th International Symposium on Landslides* (pp. 1571-1578). Tronheim, Norway: Brookfield, Rotterdam.

Ribeiro, K. M., Barreto, B., Pasqual, M., White, P. J., Braga, R. A., & Dupuy, L. X. (2014). Continuous, high-resolution biospeckle imaging reveals a discrete zone of activity at the root apex that responds to contact with obstacles. *Annals of Botany*, *113*(3), 555–563. doi:10.1093/aob/mct271 PMID:24284818

Riegl. (2016). *Riegl VMX 450 mobile laser scanner* [Online Image]. Retrieved from http://www.riegl.com

Rivas López, M., Flores Fuentes, W., Rivera Castillo, J., Sergiyenko, O., & Hernández Balbuena, D. (2013). A Method and Electronic Device to Detect the Optoelectronic Scanning Signal Energy Centre. In S. L. Pyshkin, & J. M. Ballato (Eds.), Optoelectronics (pp. 389-417). Crathia: Intech.

Rivas-Lopez, M. C.-F.-Q.-B.-B. (2014). Scanning for light detection and Energy Centre Localization Methods assesment in vision systems for SHM. *2014 IEEE 23rd International Symposium on* (pp. 1955-1960). Industrial Electronics (ISIE).

Rivera-Rubio, J., Alexiou, I. I., & Bharath, A. A. (2015). Appearance-based indoor localization: A comparison of patch descriptor performance. *Pattern Recognition Letters*. Retrieved from http://www.sciencedirect.com/science/article/pii/S0167865515000744

Rodriguez-Quiñonez, J. B.-L.-F.-P. (2014). Improve 3D laser scanner measurements accuracy using a FFBP neural network with Widrow-Hoff weight/bias learning function. *Opto-Electronics Review*, *4*(22), 224–235.

Rodriguez-Quiñonez, J. C., Sergiyenko, O. Y., Preciado, L. C. B., Tyrsa, V. V., Gurko, A. G., Podrygalo, M. A., & Balbuena, D. H. et al. (2014). Optical monitoring of scoliosis by 3D medical laser scanner. *Optics and Lasers in Engineering*, *54*, 175–186. doi:10.1016/j.optlaseng.2013.07.026

Rodríguez-Quinoñez, J., Sergiyenko, O., Gonzalez-Navarro, F., Basaca-Preciado, L., & Tyrsa, V. (2013, February). Surface recognition improvement in 3D medical laser scanner using Levenberg–Marquardt method. *Signal Processing*, *93*(2), 378–386. doi:10.1016/j.sigpro.2012.07.001

Roemer, G., & Bechthold, P. (2014). Electro-optic and Acousto-optic Laser Beam Scanners. *Proceedings of the8th International Conference on Laser Assisted Net Shape Engineering*. Elsevier.

Rogalski, A. (2002). Infrared detectors: An overview. *Infrared Physics & Technology*, *3*(43), 187–210. doi:10.1016/S1350-4495(02)00140-8

Roloff, M. L. (2014, September 14). Uma nova Abordagem para a Implementação de um Sistema Multiagente para a Configuração e o Monitoramento da Produção de Pequenas Spéries. *Uma nova Abordagem para a Implementação de um Sistema Multiagente para a Configuração e o Monitoramento da Produção de Pequenas Spéries*. Florianópolis, SC, Brazil: Universidade Federal de Santa Catarina.

Roth, P. M., & Winter, M. (2008). *Survey of appearance-based methods for object recognition.* Technical Report ICG–TR–01/08 Graz. Retrieved from http://machinelearning.wustl.edu/uploads/Main/appearance_based_methods.pdf

Salon, C., Jeudy, C., Bernard, C., Cointault, F., Han, S., Lamboeuf, M., & Baussard, C. (2014). *Nodulated root imaging within the high throughput plant phenotyping platform (PPHD, INRA, Dijon)*. Paper presented in "Institute Days of MPI-MP". Potsdam-Golm, Allemagne.

Samarasinghe, S. (2006). *Neural Networks for Applied Sciences and Engineering: From Fundamentals to Complex Pattern Recognition* (1st ed.). Auerbach Publications.

Sanchez-Garcia, I. (2004). *Model for automatic recognition of shapes based on Fourier descriptors* [Bachelor Thesis]. University de las Americas Puebla, Mexico.

Santos, T. T., & de Oliveira, A. (2012). *Image-based 3D digitizing for plant architecture analysis and phenotyping*. Paper presented at Workshop on Industry Applications (WGARI) in SIBGRAPI 2012 (XXV Conference on Graphics, Patterns and Images), Ouro Preto, MG, Brazil.

Saul, L. K., Jaakkola, T., & Jordan, M. I. (1996). Mean field theory for sigmoid belief networks. *Journal of Artificial Intelligence Research*, 61–76.

Savvaidis, P. D. (2003). Existing Landslide Monitoring Systems aand Techniques. School of Rural and Surveying Engineering, The Aristotle University of Thessaloniki.

Scanlab. (2016). *Scanlab dynAXIS S* [Online Image]. Retrieved from http://www.scanlab.de

Scheffer, D., Dierickx, B., & Meynants, G. (1997). Random addressable 2048x2048 active pixel image sensor. *IEEE Transactions on Electron Devices*, *44*(10), 1716–1720. doi:10.1109/16.628827

Schulze, M. (2008). Elektrische Servoantriebe: Baugruppen mechatronischer Systeme. Carl Hanser Verlag GmbH & Co. KG.

Seitz, P. (2011). Fundamentals of noise in optoelectronics. In *Single-photon imaging* (pp. 1–25). Springer Berlin Heidelberg. doi:10.1007/978-3-642-18443-7_1

Sergiyenko, O., Tyrsa, V., Basaca-Preciado, L., Rodríguez-Quinoñez, J., Hernandez, W., Nieto-Hipolito, J., & Starostenko, O. (2011). Electromechanical 3D Optoelectronic Scanners: Resolution Constraints and Possible Ways of Improvement. In Optoelectronic Devices and Properties. InTech.

Sergiyenko, O. (2010). Optoelectronic System for Mobile Robot Navigation. *Optoelectronics, Instrumentation and Data Processing*, *46*(5), 414–428. doi:10.3103/S8756699011050037

Shan, J., & Toth, C. (2008). *Topographic Laser Ranging And Scanning*. Boca Raton, FL: CRC Press. doi:10.1201/9781420051438

Sharma, S. (2013). Template matching approach for face recognition system. *International Journal of Signal Processing Systems*, *1*(2), 284–289. doi:10.12720/ijsps.1.2.284-289

Shih, H.-C. P. (2015). High-resolution gravity and geoid models in Tahiti obtained from new airborne and land gravity observations: Data fusion by spectral combination. *Earth, Planets, and Space*, *1*(67), 1–16.

Singla, N., & Sharma, S. (2014). Advanced survey on face detection techniques in image processing. *International Journal of Advanced Research in Computer Science & Technology*, *2*(1), 22–24.

Sixta, Z., Linhart, J., & Nosek, J. (2013). Experimental Investigation of Electromechanical Properties of Amplified Piezoelectric Actuator. *Proceedings of the 2013 IEEE 11th International Workshop of Electronics, Control, Measurement, Signals and their application to Mechatronics (ECMSM)* (pp. 1-5). Toulouse: IEEE.

Slovak, R., Goschl, C., Su, X. X., Shimotani, K., Shiina, T., & Busch, W. (2014). A scalable open-source pipeline for large-scale root phenotyping of *Arabidopsis. The Plant Cell*, *26*(6), 2390–2403. doi:10.1105/tpc.114.124032 PMID:24920330

Snyder, W. E. (2010). *Machine vision.* Cambridge University Press. doi:10.1017/CBO9781139168229

Solchenbach, T., & Plapper, P. (2013, December30). Mechanical characteristics of laser braze-welded aluminium–copper connections. *Optics & Laser Technology*, *54*, 249–256. doi:10.1016/j.optlastec.2013.06.003

Song, B. (2000). Nyquist-Rate ADC and DAC. In E. W. Chen (Ed.), *VLSI Handbook*. VLSI.

Song, D., & Chang, J. (2012, February29). Super wide field-of-regard conformal optical imaging system using liquid crystal spatial light modulator. *Optik (Stuttgart)*, 2455–2458.

Sonka, M. A. (2014). *Image processing, analysis, and machine vision.* Cengage Learning.

Starostenko, O., Rodríguez-Asomoza, J., Sánchez-López, S., & Chávez-Aragón, J. (2008). Shape indexing and retrieval: a hybrid approach using ontological descriptions, In K. Elleithy (Ed.), Innovations and advanced techniques in systems, computing sciences and software engineering (pp. 381-386). Springer Science+Business Media B.V.

Starostenko, O., Cortés, X., Sánchez, J. A., & Alarcon-Aquino, V. (2015a). Unobtrusive emotion sensing and interpretation in smart environment. *Journal of Ambient Intelligence and Smart Environments, 7*(1), 59–83.

Starostenko, O., Cruz-Perez, C., Uceda-Ponga, F., & Alarcon-Aquino, V. (2015b). Breaking textual-based CAPTCHAs with variable word and character orientation. *Journal Pattern Recognition, 48*(4), 1101–1112. doi:10.1016/j.patcog.2014.09.006

Stemmer, M. R., Costa, C. P., Vargas, J., & Roloff, M. L. (2014). Artificial Intelligent Systems for Quality Assurance in Small Series Production. *Key Engineering Materials, 613*, 279–287. doi:10.4028/www.scientific.net/KEM.613.279

Strelkov, A. I., Zhilin, Y. I., Lytyuga, A. P., Kalmykov, A. S., & Lisovenko, S. A. (2007). Signal Detection in Technical Vision Systems. *Telecommunications and Radio Engineering, 66*(4).

Strelkova T. A. (1999). The Potentialities of Optical and Electronic Devices for Biological Objects Investigations. *and Radio Engineering, 53*, 190-194.Telecommunications

Strelkova, T. A. (2014). Studies on the optical fluxes attenuation process in optoelectronic systems. *Semiconductor Physics Quantum Electronics & Optoelectronics, 17*(4), 421-424.

Strelkov, A. I., Stadnik, A. M., Lytyuga, A. P., & Strelkova, T. A. (1998). Comparative Analysis of Probabilistic and Determinate Methods for Attenuating Light Flux. *Telecommunications and Radio Engineering, 52*(8), 54–57. doi:10.1615/TelecomRadEng.v52.i8.110

Strelkova, T. A. (2014). Influence of video stream compression on image microstructure in medical systems. *Biomedical Engineering, 47*(6), 307–311. doi:10.1007/s10527-014-9398-1

Strelkova, T. A. (2014). Statistical properties of output signals in optical-television systems with limited dynamic range. *Eastern-European Journal of Enterprise Technologies, 2*(9 (68)), 38–44. doi:10.15587/1729-4061.2014.23361

Sukegawa, S., Umebayashi, T., Nakajima, T., Kawanobe, H., Koseki, K., Hirota, I., & Fukushima, N. (2013). A 1/4-inch 8Mpixel Back-Illuminated Stacked CMOS Image Sensor. Proceedings of ISSCC 2013 (pp. 484–486). IEEE.

Surmann, H., Nuechter, A., & Hertzberg, J. (2003, December). An autonomous mobile robot with a 3D laser range finder for 3D exploration and digitalization of indoor environments. *Robotics and Autonomous Systems, 45*(3-4), 181–198. doi:10.1016/j.robot.2003.09.004

Suzuki, A., Shimamura, N., Kainuma, T., Kawazu, N., Okada, C., Oka, T., & Wakabayashi, H. (2015). A 1/1.7 inch 20Mpixel Back illuminated stacked CMOS Image sensor for new Imaging application. Proceedings of ISSCC '15 (pp. 110–112). IEEE.

Suzuki, S., & Be, K. (1985). Topological structural analysis of digitized binary images by border following. *Computer Vision Graphics and Image Processing*, *30*(1), 32–46. doi:10.1016/0734-189X(85)90016-7

Syms, R. R. (1992). *Optical guided waves and devices*. McGraw-Hill.

Szymanski, C., & Stemmer, M. R. (2015, June6). Automated PCB inspection in small series production based on SIFT algorithm.*International Symposium on Industrial Electronics*. doi:10.1109/ISIE.2015.7281535

Takayanagi, I., Yoshimura, N., Sato, T., Matsuo, S., Kawaguchi, T., Mori, K., & Nakamura, J. (2013). A 1-inch Optical Format, 80fps, 10. 8Mpixel CMOS Image Sensor Operating in a Pixel-to-ADC Pipelined Sequence Mode.*Proc. Int'l Image Sensor Workshop* (pp. 325–328).

Tame, B., & Stutzke, N. (2010). Steerable Risley Prism antennas with low side lobes in the Ka band. *Proceedings of the 2010 IEEE International Conference on Wireless Information Technology and Systems (ICWITS)* (pp. 1-4). IEEE.

Tang, F., Cao, Y., & Bermak, A. (2010). *An ultra-low power current-mode CMOS image sensor with energy harvesting capability*. ESSCIRC. doi:10.1109/ESSCIRC.2010.5619822

Tavernier, F. A. (2011). *High-speed optical receivers with integrated photodiode in nanoscale CMOS*. Springer Science and Business Media. doi:10.1007/978-1-4419-9925-2

Tilman, D., Balzer, C., Hill, J., & Befort, B. L. (2011). Global food demand and the sustainable intensification of agriculture. *Proceedings of the National Academy of Sciences of the United States of America*, *108*(50), 20260–20264. doi:10.1073/pnas.1116437108 PMID:22106295

Timmermann, K. E., & Nowak, R. D. (1999). Multiscale modeling and estimation of Poisson processes with application to photon-limited imaging. *Information Theory. IEEE Transactions on*, *45*(3), 846–862.

Topp, C. N., Iyer-Pascuzzi, A. S., Anderson, J. T., Lee, C.-R., Zurek, P. R., Symonova, O., & Benfey, P. N. et al. (2013). 3D phenotyping and quantitative trait locus mapping identify core regions of the rice genome controlling root architecture. *Proceedings of the National Academy of Sciences of the United States of America*, *110*(18), E1695–E1704. doi:10.1073/pnas.1304354110 PMID:23580618

Trépanier, J., Sawan, M., Audet, Y., & Coulombe, J. (2002). A Wide Dynamic Range CMOS Digital Pixel Sensor. *Proceedings of the 2002 45th Midwest Symposium on Circuits and Systems MWSCAS '02*. IEEE. doi:10.1109/MWSCAS.2002.1186892

Tribolet, P. A. (2008). Advanced HgCdTe technologies and dual-band developments.*SPIE Defense and Security Symposium*. International Society for Optics and Photonics.

Unar, J. A., Seng, W. Ch., & Abbasi, A. (2014). A review of biometric technology along with trends and prospects. *Journal Pattern Recognition*, *47*(8), 2673–2688. doi:10.1016/j.patcog.2014.01.016

Vahelal, A. (2010). *Digitization.* Gujarat, India: HasmukhGoswami College of Engineering.

Valluru, B. R., & Hayagriva, R. (1995). *C++ Neural Networks and Fuzzy Logic* (2nd ed.). M & T Books.

Van Grinsven, H. J., Ten Berge, H. F., Dalgaard, T., Fraters, B., Durand, P., Hart, A., & Willems, W. J. et al. (2012). Management, regulation and environmental impacts of nitrogen fertilization in northwestern Europe under the Nitrates Directive; a benchmark study. *Biogeoscience*, *9*(12), 5143–5160. doi:10.5194/bg-9-5143-2012

Vauhkonen, J. a. (2014). Tree species recognition based on airborne laser scanning and complementary data sources. In Forestry applications of airborne laser scanning (pp. 135-156). Springer. doi:10.1007/978-94-017-8663-8_7

Vezzetti, E., Marcolin, F., & Fracastoro, G. (2014). 3D face recognition: An automatic strategy based on geometrical descriptors and landmarks. *Robotics and Autonomous Systems*, *62*(12), 1768–1776. doi:10.1016/j.robot.2014.07.009

Villa, J. L., Sallán, J., Llombart, A., & Sanz, J. (2009). Design of a high frequency Inductively Coupled Power Transfer system for electric vehicle battery charge. *Applied Energy*, *86*(3), 355–363.

Viola, P., & Jones, M. (2001). Rapid object detection using a boosted cascade of simple features, *Proceedings of IEEE Conference On Computer Vision And Pattern Recognition* (Vol. 1, pp. I-511 - I-518). Retrieved from https://www.cs.cmu.edu/~efros/courses/LBMV07/Papers/viola-cvpr-01.pdf

Viola, P., & Jones, M. (2004). Robust real-time face detection. *International Journal of Computer Vision*, *57*(2), 137–154. doi:10.1023/B:VISI.0000013087.49260.fb

VisTex. (2002). *Vision texture database.* Maintained by the Vision and Modeling group at the MIT Media Lab. Retrieved from http://vismod.media.mit.edu/vismod/imagery/VisionTexture/

Vogel, E. M., Voelkl, E., Colombo, L., & Ruoff, R. S. (2010). The optocouplers functions. *Journal of Electronics Systems*, *12*(6), 11–34.

Voisin, A. S., Salon, C., Jeudy, C., & Warembourg, F. R. (2003). Root and nodule growth in Pisum sativum L. in relation to photosynthesis. Analysis using 13C labelling. *Annals of Botany*, *92*(4), 557–563. doi:10.1093/aob/mcg174 PMID:14507741

Wang, L., Liu, L., Sun, J., Zhou, Y., Luan, Z., & Liu, D. (2010, September). *Large-aperture double-focus laser collimator for PAT performance testing of inter-satellite laser communication terminal.* Academic Press.

Wang, Y., Sun, Y., & Zhang, W. (2009). Automatic Inspection of Small Component on Loaded PCB Based on Mean-Shift and Support Vector Machine. *Fifth International Conference on Natural Computation.* doi:10.1109/ICNC.2009.407

Waske, B. a. (2007). Fusion of support vector machines for classification of multisensor data. *Geoscience and Remote Sensing. IEEE Transactions on, 122*(45), 3858–3866.

Wasson, A. P., Richards, R. A., Chatrath, R., Misra, S. C., Prasad, S. V. S., Rebetzke, G. J., & Watt, M. et al. (2012). Traits and selection strategies to improve root systems and water uptake in water-limited wheat crops. *Journal of Experimental Botany, 63*(9), 3485–3498. doi:10.1093/jxb/ers111 PMID:22553286

Weckenmann, A. D.-R., Jiang, X., Sommer, K.-D., Neuschaefer-Rube, U., Seewig, J., Shaw, L., & Estler, T. (2009). Multisensor data fusion in dimensional metrology. *CIRP Annals-Manufacturing Technology, 2*(58), 701–721. doi:10.1016/j.cirp.2009.09.008

Wendt, K., Franke, M., & Härtig, F. (2012, December). Measuring large 3D structures using four portable tracking laser interferometers. *Measurement, 45*(10), 2339–2345. doi:10.1016/j.measurement.2011.09.020

Willett, R. M., & Nowak, R. D. (2007). Multiscale Poisson intensity and density estimation. *Information Theory. IEEE Transactions on, 53*(9), 3171–3187.

Wong, H. P. (1997). *CMOS Image sensors - Recent Advances and Device Scaling Considerations.* IEDM.

Wong, H.-S. (1996). Technology and device scaling considerations for CMOS imagers. *IEEE Transactions on* Electron Devices, *43*(12), 2131–2142. doi:10.1109/16.544384

Wooldridge, M. (1997). Agent-based Software Engineering. *EE Proceedings on Software Engineering* (pp. 26-37). London: IEEE.

Wu, H.-H., Zhang, X.-M., & Hong, S.-L. (2009). A visual inspection system for surface mounted components based on color features.*2009 International Conference on Information and Automation* (pp. 571-576). Zhuhai: IEEE. doi:10.1109/ICINFA.2009.5204988

Xiang, S., Chen, S., Wu, X., Xiao, D., & Zheng, X. (2010, February). Study on fast linear scanning for a new laser scanner. *Optics & Laser Technology, 42*(1), 42–46. doi:10.1016/j.optlastec.2009.04.019

Xie, F., Dau, A. H., Uitdenbogerd, A. L., & Song, A. (2013). Evolving PCB visual inspection programs using genetic programming.*28th International Conference on Image and Vision Computing New Zealand.* doi:10.1109/IVCNZ.2013.6727049

Xu, Q. (2013). Impact Detection and Location for a Plate Structure using Least Squares Support Vector Machines. *Structural Health Monitoring.*

Xu, R., Ng, W. C., Yuan, J., Yin, S., & Wei, S. (2014). A 1 / 2. 5 inch VGA 400 fps CMOS Image Sensor with High Sensitivity for Machine Vision. *IEEE Journal of Solid-State Circuits, 49*(10), 2342–2351. doi:10.1109/JSSC.2014.2345018

Xu, W., & Liu, G. (2010). Multi-Layer Image Segmentation Based on Fuzzy Algorithm in PCB Inspection.*International Conference on Multimedia Technology*. doi:10.1109/IC-MULT.2010.5630842

Yale Extended Face Database B. (2015). Retrieved from http://vision.ucsd.edu/content/extended-yale-face-database-b-b

Yallup, K. A. (2014). *Technologies for smart sensors and sensor fusion*. CRC Press.

Yang, H., Heo, J., Park, S., Song, H. J., & Seo, D. H. (2010). Detection of microbial organisms in soils. In Use of optoelectronics in the agriculture (10th ed.). Elsevier Books.

Yang, F., Lu, Y. M., Sbaiz, L., & Vetterli, M. (2012). Bits from photons: Oversampled image acquisition using binary poisson statistics. *Image Processing. IEEE Transactions on*, *21*(4), 1421–1436.

Yang, S., Lee, K., Xu, Z., Zhang, X., & Xu, X. (2001, October). An accurate method to calculate the negative dispersion generated by prism pairs. *Optics and Lasers in Engineering*, *36*(4), 381–387. doi:10.1016/S0143-8166(01)00055-0

Yang, Y. (2008, November21). Analytic Solution of Free Space Optical Beam Steering Using Risley Prisms. *Journal of Lightwave Technology*, *26*(21), 3576–3583. doi:10.1109/JLT.2008.917323

Yao, Q. (2013). *The design of a 16*16 pixels CMOS image sensor with 0.5 e- RMS noise*. TUDelft.

Yoshida, K. (2005). *Japan Patent No. 3706385*. Japan.

Yoshida, K., Kamruzzaman, M., Jewel, F. A., & Sajal, R. F. (2007). Design and implementation of a machine vision based but low cost standalone system for real time counterfeit Bangladeshi bank notes detection. *Proceedings of the 10th international conference on Computer and information technology ICCIT '07*.

Ytting, N. K., Andersen, S. B., & Thorup-Kristensen, K. (2014). Using tube rhizotrons to measure variation in depth penetration rate among modern North-European winter wheat (*Triticum aestivum* L.) cultivars. *Euphytica*, *199*(1-2), 233–245. doi:10.1007/s10681-014-1163-8

Zappala, S., Helliwell, J., Tracy, S., Mairhofer, S., Sturrock, C. J., Pridmore, T., & Mooney, S. et al. (2013). Effects of X-ray dose on rhizosphere studies using X-ray computed tomography. *PLoS ONE*, *8*(6), e67250. doi:10.1371/journal.pone.0067250 PMID:23840640

Zhang, C., & Zhang, Z. (2010). *A survey of recent advances in face detection*. Technical Report MSR-TR-2010-66. Retrieved from http://research.microsoft.com/pubs/132077/facedetsurvey.pdf

Zhang, F.-M. A.-H.-H. (2008). Multiple sensor fusion in large scale measurement. *Optics and Precision Engineering*, (7), 18.

Zhang, Y., Hornfeck, K., & Lee, K. (2013). Adaptive face recognition for low-cost, embedded human-robot interaction. In *Intelligent Autonomous Systems* (Vol. 193, pp. 863-872). Springer. Retrieved from http://www.case.edu/mae/robotics/pdf/ZhangIAS2012.pdf

Zhang, D., & Lu, G. (2004). Review of shape representation and description techniques. *Pattern Recognition*, *37*(1), 1–19. doi:10.1016/j.patcog.2003.07.008

Zhang, M., & Bermak, A. (2010). Compressive Acquisition CMOS Image Sensor: From the Algorithm to Hardware Implementation. *IEEE Transactions on Very Large Scale Integration (VLSI) Systems*, *18*(3), 490–500. doi:10.1109/TVLSI.2008.2011489

Zhang, Q., Su, X., Xiang, L., & Sun, X. (2012). 3-D shape measurement based on complementary Gray-code light. *Optics and Lasers in Engineering*, *4*(50), 574–579. doi:10.1016/j.optlaseng.2011.06.024

Zhang, X. J., Fan, X. Z., Liao, J., Wang, H. T., Ming, N. B., Qiu, L., & Shen, Y. Q. (2002). Propagation properties of a light wave in a film quasiwaveguide structure. *Journal of Applied Physics*, *92*(10), 5647–5657. doi:10.1063/1.1517731

Zhao, W., Wang, T., Pham, H., Hu-Guo, C., Dorokhov, A., & Hu, Y. (2014). Development of CMOS Pixel Sensors with digital pixel dedicated to future particle physics experiments. *Journal of Instrumentation*, *9*(02), C02004–C02004. doi:10.1088/1748-0221/9/02/C02004

ЛевинБ.Р. (1968). *Теоретические основы статистической радиотехники*. М.: Сов. радио.

About the Contributors

Oleg Sergiyenko (M'09-09) was born in February, 9, 1969. He received the B.S., and M.S., degrees in Kharkiv National University of Automobiles and Highways, Kharkiv, Ukraine, in 1991, 1993, respectively. He received the Ph.D. degree in Kharkiv National Polytechnic University on specialty "Tools and methods of non-destructive control" in 1997. He has editor of 1 book, written 8 book chapters, 87 papers and holds 1 patent of Ukraine. Since 1994 till the present time he was represented by his research works in several International Congresses of IEEE, ICROS, SICE, IMEKO in USA, England, Japan, Italy, Austria, Ukraine, and Mexico. Dr. Sergiyenko in December 2004 was invited by Engineering Institute of Baja California Autonomous University for researcher position. He is currently Head of Applied Physics Department of Engineering Institute of Baja California Autonomous University, Mexico, director of several Master's and Doctorate thesis. He was a member of Program Committees of various international and local conferences. He is member of Scientific Council on Electric specialties in Engineering Faculty of Autonomous University of Baja California and Academy of Engineering. Included in the 2010-2015 Edition of Marquis' Who's Who in the World.

Julio C. Rodríguez-Quiñonez received the Ph.D. degree from Baja California Autonomous University, México, in 2013. He is currently Professor of Electronic Topics with the Engineering Faculty, Autonomous University of Baja California. He is involved in the development of optical scanning prototype in the Applied Physics Department and research head in the development of a new stereo vision system prototype. He has written 27 papers and is currently reviewer for the IEEE Sensors Journal and participates as a reviewer and Section Chair of IEEE conferences in 2014 and 2015. His current research interests include automated metrology, stereo vision systems, control systems, robot navigation and 3D laser scanners.

* * *

Vicente Alarcon-Aquino received the Ph.D. and D.I.C. degrees in Electrical and Electronic Engineering from Imperial College London, London, U.K. in 2003. Before pursuing his PhD studies at Imperial College, he was an Assistant Professor in the Department of electronics at the Universidad de las Americas Puebla, Mexico, where he is now full professor and Chair in the Department of computing, electronics and mechatronics. During his stay at Imperial College, he was Lab Demonstrator and Research Assistant in the Communications and Signal Processing Group. Dr. Alarcon-Aquino has supervised several theses and taught courses at undergraduate, master and doctoral level in areas related to communication networks and signal processing. He has authored over 150 research articles in several refereed journals and conference proceedings, has written a research monograph on MPLS networks, and has over 500 citations to his research articles. His current research interests include signal processing for network security, biomedical signal processing, wavelet-based signal & image processing, and communication networks. Dr. Alarcon-Aquino has been active in program committees for international conferences and has also served as manuscript reviewer for numerous refereed journals. He is a Senior member of IEEE, belongs to the Mexican National System of Researchers (SNI Level I) and has been elected to membership of the Mexican Academy of Sciences (AMC).

Gustavo Lopez Badilla was born in Santa Ana, Sonora, México and made studies in Electronics Engineering from 1989 to 1994 in the Faculty of Engineering of the Universidad Autonoma de Baja California, a Masters in Systems Engineering from 1998 to 2002 and a Ph.D. from 2005 to 2008 at the Institute Engineering of the Universidad Autonoma de Baja California and Postodoctorado Center of Nanoscience and Nanotechnology at the Autonomous University of Mexico in Ensenada from 2009 to 2010. The areas of expertise are in Electronics Engineering, Computing, Biomedical, Industrial and Environmental Chemistry and Nutrition. With five years in industrial engineering in the area of programming automatic insertion machine in the company LG Electronics from 1994 to 1998 and Skyworks in the area of maintenance of machines and equipment for quality testing of Electronic Devices Cellular Phone 1998 to 1999. Member of the National System of Researchers (SNI) level Candidate, Research and Academic with twelve years experience in areas of environment, corrosion, materials and surface analysis and over 30 scientific and popular articles, conference participation since 2006 in engineering, materials, nutrition and health, with participation in three books with chapters and two complete books. Research has been on people and companies with engineering approaches, technology and health in organic and inorganic chemistry and physical chemistry of materials with specialized equipment and nanoscopy and microscopy. He has participated in four research projects supported with funding from the Autonomous University of Baja California to be responsible in two investigations have involved students as human resource training support in experimental field and laboratory. It has links with industry since 2005 with the Coca Cola and mainly BIMBO. Now is working as a Researcher in the Engineering and Competitiveness Department, in the Manufacturing Engineering School of the Universidad Politecnica de Baja California in Mexicali, Baja California, Mexico.

Luis Carlos Básaca-Preciado was born October 25, 1985. He received a B.S. degree in Cybernetics and Electronics Engineering from CETYS University, Mexicali, Baja California, Mexico in 2007. On 2013 he obtained his Ph.D. in Optoelectronics at the Engineering Institute of the Autonomous University of Baja California. Currently he teaches courses in the area of Robotics, Sensors & Actuators, Electronics and Mechatronics at the Engineering School of CETYS University, where he coordinates the Cybernetics Electronics Engineering Major. He participates in applied research projects related to industry and does research in areas such as 3D Vision Systems and Robotics. He is also mentor of a FIRST FRC Robotics team since 2013, to promote science and engineering among the young. He has written two book chapters, five journal papers, ten conference papers including international conferences such as IEEE ISIE, IECON, ROC&C, PAHCE, and IPC; in Italy, USA, Brazil and Mexico where he holds two patents.

Bastien Billiot is a Computer Vision and Statistics researcher at R&D department of Roullier group.

Javier Rivera Castillo received the M.S. degrees in Baja California Autonomous University, México in 2006. He has written 2 book chapter and 9 Journal and Proceedings Conference papers. M.S. Rivera was in charge of Calibration Laboratory at Engineering Institute of Baja California Autonomous University and then he was Head of Mechatronic Program at Baja California Polytechnic University Since 2006 to 2011. He is now full researcher in Applied Physics Department of Engineering Institute of Baja California Autonomous University, Mexico.

Frédéric Cointault gets his Ph-D in Instrumentation and Image Computing in the University of Burgundy in 2001 and his accreditation to supervise research in 2010. He is currently an associate professor at Agrosup Dijon, the Higher Education Institute of Agronomy and Agrofood of Dijon, and leads research in Image acquisition and processing in the UMR 1347 Agroecology Agrosup/INRA/University of Burgundy. His research are turned on the development of image processing methods for the determination of morphometric parameters for crops in a context of plant phenotyping, and on the development of imaging systems for plant disease detection in precision agriculture and viticulture.

Claudia Cruz is PhD student of the Computer Science program in the Department of Computing, Electronics and Mechatronics at University de las Americas Puebla, Mexico. She received her Master degree in 2009 from National Institute of Astrophysics, Optics and Electronics, Mexico. Her main research interest is focusing in computer vision and pattern recognition.

Wendy Flores-Fuentes was born in Baja California, Mexico in 1978. She obtained the PhD degree at Autonomous University of Baja California (UABC), Mexicali, B.C., México, as Doctor of Science in Applied Physics, with emphasis in Optoelectronic Scanning Systems for SHM, in June 2014. She obtained the Master in Engineering in May 2011 at Technological Institute (ITM) and graduated from College of Electronic Engineering in October 2001, at UABC. She has teaching experience in Engineering Faculty of Autonomous University of Baja California, Mexico. She performs as Basic Sciences Professor, focused on Physics and Mathematics learning units of Engineering. She is the institutional representative of the thematic network of physics at the ANFEI. She has also taught at Polytechnic University of Baja California, Mexico from 2012 to 2014, collaborating at Industrial and Renewable Energy Engineering Programs as Professor. Currently she is a Research Professor at Autonomous University of Baja California and Authorized member of SNI-CONACYT *(Mexican National Research System), contributor on multiple research programs on electro mechanic systems for machine vision applications, working toward completion of an optical scanning system for SHM, focused in the system loop feedback control by artificial intelligent methods.

Juan Manuel Terrazas Gaynor graduated from CETYS University in Electronic Cybernetics Engineering in 2002, obtained his master's and Ph.D degree in Semiconductor Engineering in 2007 and 2009 respectively at the Institute of Engineering of Baja California Autonomous University. Since 2002 till the present time he has worked on material characterization, electrical analysis, reliabililty and failure analysis of semiconductor electronic devices. Currently works as a full time professor at the Engieering School of CETYS University and as a researcher at the Center of Excellence in Innovation & Design – CETYS University. He is member of the National Research System from the National Science and Technology Council (CONACYT) from 2012 till the present time. His main areas of interest are: Industrial Electronics, Semiconductors, Material Characterization, and Solid State Devices Reliability and Failure Analysis.

Daniel Hernández-Balbuena was born in July, 25, 1971. He received the B.S. degree from Puebla Autonomous University, Puebla, México, in 1996 and the M.S degree from Ensenada Center for Scientific Research and Higher Education, Baja California, México, in 1999. He received the Ph.D. degree in Baja California Autonomous University in 2010. Dr. Hernández has been with the Engineering Faculty of Baja California Autonomous University, Mexicali since 1999, where he currently is a Professor. His research interests are in the areas of time and frequency metrology and RF measurements.

Sylvain Jay received his M.Sc. degree in Image and Signal Processing from Paul Cézanne University, Marseille, France, in 2009, and his M.Eng. degree from Ecole Centrale Marseille, France in 2009. He received his Ph.D. degree in hyperspectral remote sensing and signal processing from Ecole Centrale Marseille in 2012. He is currently a post-doctoral research fellow within Irstea, Montpellier, France. His research interests include hyperspectral remote sensing of vegetation and coastal environments, as well as canopy 3D modeling.

Volodymyr M. Kartashov, Prof., Doctor in Radio i Telev Systems, M. Sc. in Radio Engineering. Citizenship: Ukraine. Date and place of birth: Jule 3, 1958, Russia. Research interests in optimal signal processing; radar investigations of the atmosphere, particularly by the radar-acoustic technique. The total number of publication is 150, among them 10 manual for student and 20 patents.

Lars Lindner was born on July 20th 1981 in Dresden, Germany. He received his M.S. degree in mechatronics engineering from the TU Dresden University in January 2009. He was working as graduate assistant during his studies at the Fraunhofer Institute for Integrated Circuits EAS in Dresden and also made his master thesis there. After finishing his career, he moved to Mexico and started teaching engineering classes at different universities in Mexicali. Since August 2013 he began his PhD at the Engineering Institute of Autonomous University of Baja California in Mexicali and deals since with the development of an optoelectronic prototype for measuring 3D coordinates using dynamic triangulation.

Alexander P. Lytyuga, Senior Researcher, PhD in Radio and Television Systems, Citizenship: Ukraine. Research interests in statistical properties of signals in optoelectronic systems, signals processing. The total number of publications is 87.

Arjuna Marzuki obtained his B. Eng (Honours) from the Department of Electronic & Electrical Engineering at the University of Sheffield in United Kingdom, MSc from Universiti Sains Malaysia and PhD from Universiti Malaysia Perlis. After finishing his study he joined Hewlett-Packard as R&D engineer in Wireless Semiconductor Division. His primary tasks are to develop radio frequency (RF) and radio frequency integrated circuit (RFIC) products. Such products are high frequency transistors, RF gain blocks, I/Q demodulator and etc. He then later joined IC Microsystems Sdn. Bhd. in Cyberjaya, Selangor, Malaysia as IC design staff engineer. In the company he involved in designing 12/10/8 bits digital-to-analog converter ICs and family of RFIC devices. He later joined Agilent Technologies as IC design engineer in Optical Product Division. His main jobs is to lead the analog design which cover from operational amplifiers, band-gap circuits, data-converters, I/Os, Power on reset and etc. His latest small contribution in the industry is "Industry's 1st Digital Colour Sensor IC with I/O via 2-wire Serial Interface" released in February 2006 by Avago Technologies. Arjuna has been granted 1 US patent and authored more than 20 technical papers. He is presently a lecturer with School of Electrical and Electronic Engineering, Universiti Sains Malaysia. Arjuna is also a member of society/editorial board, such as "The Far East Journal of Electronics and Communication".

Alexandre Reeberg de Mello graduated from the Pontifical Catholic University of Paraná in Control and Automation Engineering in 2013, obtained his master's degree at the Graduate Program in Automation and Systems at UFSC in the area of Control, Automation and Systems in the period from 2013 to 2015. Currently is a PhD student at the Graduate Program in Automation and Systems at UFSC, also in the area of Control, Automation and Systems. His main areas of interest are: Computer Vision and Artificial Intelligence.

Viktor Ivanovich Melnik was born January 8, 1958. In 1980 was graduated from Poltava Agricultural Institute (Poltava, Ukraine) and was qualified as a mechanical engineer in agriculture. PhD degree obtained at the Kharkov State Technical University of Radio Electronics (Kharkiv, Ukraine) on specialty "Technology, equipment and production of electronic devices" in 2000. He get the Degree of Doctor of Sciences (in Technics) at the Kharkov National Technical University of Agriculture named after Peter Vasilenko (Kharkiv, Ukraine) on specialty "Machinery and mechanization of agricultural production" in 2011. He is the author of 4 books, 151 articles, 39 patents of the Russian Federation and 6 patents of Ukraine. Since 1986 participated in the international scientific conferences in the United States, Belarus, Russia and Ukraine. He is currently a professor in the Department of technologic systems optimization, the Head of the research laboratory Engineering of nature management, as well as the deputy director for scientific activities in Institute of Mechatronics and management systems at the Kharkov National Technical University of Agriculture. He was tutor of several masters' and two PhD thesis. He is deputy chairman of the specialized scientific council for doctoral dissertations on specialty "Machinery and mechanization of agricultural production" at the Kharkov National Technical University of Agriculture. Since 2014 he is the Editor-in-Chief of the scientific journal Engineering of nature management.

Gilles Rabatel is Research Director in Computer Sciences at National Research Institute of Science and Technology for Environment and Agriculture (IRSTEA), specialized in multispectral and hyperspectral imagery for in-field vegetation characterization.

Moisés Rivas-López was born in June, 1, 1960. He received the B.S. and M.S. degrees in Baja California Autonomous University, México, in 1985, 1991, respectively. He received PhD. degree same University, on specialty "Optical Scanning for Structural Health Monitoring", in 2010. He has written 5 book chapters and 35 Journal and Proceedings Conference papers. Since 1992 till the present time he has presented different works in several International Congresses of IEEE, ICROS, SICE, AMMAC in USA, England, Japan, turkey and Mexico. Dr. Rivas was Head of Engineering Institute of Baja California Autonomous University Since 1997 to 2005; was Rector of Baja California Polytechnic University Since 2006 to 2010 and now is full researcher in Applied Physics Department of Engineering Institute of Baja California Autonomous University, Mexico.

Javier Rivera-Castillo Received the M.S. degrees in Baja California Autonomous University, México in 2006. He has written 2 book chapter and 9 Journal and Proceedings Conference papers. M.S. Rivera was in charge of Calibration Laboratory at Engineering Institute of Baja California Autonomous University and then he was Head of Mechatronic Program at Baja California Polytechnic University Since 2006 to 2011. He is now full researcher in Applied Physics Department of Engineering Institute of Baja California Autonomous University, Mexico.

David Rousseau is full professor in signal and image processing with Université Lyon 1 since 2011. His research focusses on image processing applied to life sciences including the biomedical and plant sciences domains.

Christophe Salon is Deputy Manager of the UMR AgroEcology INRA Dijon, Research Director in Genetics and Ecophysiology of leguminous plants. Scientific director of the 4PMI Phenotyping platform of INRA Dijon.

Jean-Claude Simon is assistant professor in Computer Science at Agrosup Dijon.

Marcelo Ricardo Stemmer graduated from the Federal University of Santa Catarina in Electrical Engineering in 1982, obtained his master's degree at the Graduate Program in Electrical Engineering at UFSC in the area of Control, Automation and Industrial Informatics in the period from 1983 to 1985 and his doctorate in the period 1986-1991 at the Institute WZL from the RWTH-Aachen, Germany, in the area of computer networks for industrial automation. He held a post-doctoral internship at the Institute LIP6 University Paris VI, France, in 2004. Currently works as a full professor at the Department of Automation and Systems (DAS) at UFSC. His main areas of interest are: Industrial Networks, Computer Vision, Robotics and Artificial Intelligence.

Alexander I. Strelkov is Prof, Doctor in Radio and Television Systems, Citizenship: Ukraine. Research interests in optical-electronic systems, statistical properties of signals, signals processing, images processing. The total number of publications is more then 450, 150 patents among them.

Tatyana Strelkova, Associate Professor, PhD in Radio and Television Systems, Citizenship: Ukraine. Research interests in stochastic-deterministic methods of signal processing. The total number of publications is 128.

Vira V. Tyrsa was born on July, 26, 1971. She received the B.S., and M.S., degrees in Kharkiv National University of Automobiles and Highways, Kharkiv, Ukraine, in 1991, 1993, respectively. She received the Ph.D. degree in Kharkiv National Polytechnic University on specialty "Electric machines, systems and networks, elements and devices of computer technics" in 1996. In April 1996, she joined the Kharkiv National University of Automobiles and Highways, where she holds position of associated-professor of Electrical Engineering Department (1998-2006). She has written 3 book chapters, more than 20 papers and holds 1 patent of Ukraine. From 1994 till the present time she was represented by her research works in several International Congresses in USA, England, Italy, Japan, Ukraine, and Mexico. In November 2006 was invited by Polytechnic University of Baja California, Mexico for professor and researcher position. She is currently associate professor in the Department of technologic systems optimization in Institute of Mechatronics and management systems at the Kharkov National Technical University of Agriculture.

Index

E

F

G

H

I

L

M

N

O

P

R

S

T

V

Printed in the United States
By Bookmasters